BEASTLY

Library of Congress Control Number: 2022947984

ISBN: 978-1-4197-6703-6
eISBN: 978-1-64700-961-8

Printed and bound in the United States
10 9 8 7 6 5 4 3 2 1

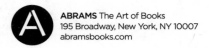

ABRAMS The Art of Books
195 Broadway, New York, NY 10007
abramsbooks.com

BEASTLY

THE 40,000-YEAR STORY
OF ANIMALS AND US

KEGGIE CAREW

ABRAMS PRESS, NEW YORK

To Patrick Walsh, for the extra miles

To Jonathan, for everything

CONTENTS

I set out to assemble the fragments that should spell to us the lives of our Wild Things as they really are – knowing that we had so long been told that they were dumb, brute beasts; suspecting that we ourselves were cowards and liars in so placing them. I have had some shocks and joys in my search for the truth . . .

      Ernest Thompson Seton, *Lives of Game Animals*, Part II, 1925

INTRODUCTION

This damn book. It began as a low-grade hum. Something was bothering me. It had been gathering over the years, getting louder, and it wasn't going away. An ever-present feeling of loss and anxiety, and I am not an anxious person by nature. But I wasn't alone. There was a whole movement of Worry. We were in trouble and we were being told . . . Yet at the same time, we weren't listening. For me, all roads kept coming back to one thing, the same thing, over and over: humanity's big error, our interactions with the planet's *other* inhabitants. Our fellow beasts. I wanted to understand how on earth this had happened. Then I was sent a photograph of a large boar, a girl, a loaf of bread, a candelabra and a clock with both hands on 12, and it was like a spell and I couldn't turn back. So began a long journey.

I wanted to track the gargantuan story of our paradoxical relationship with the animal world. Our clock starts around 40,000 years ago when a human hand carved five finger holes into the slender hollow radius bone of a griffon vulture's wing for its breathy notes to mingle with the rushing river and the rustling leaves . . . to now. A lot happens in between. We follow our wilder hearts along the road to 'civilisation', into the reins of religion, the windows of science, and the murder of God (spoiler alert: like any actor in a play, he doesn't die). Hurtling along now with our god-like technology, all life in the biosphere has fallen into our hands. We, the cleverest creature that ever breathed, a large primate from Africa. Plumbed, wired and upholstered the same as so many creatures, the result of 3.8 billion years of evolution by natural selection, adapted to a world we are recklessly taking apart. Rivet by rivet.

I went in search of our most revealing encounters with the animal world, from the smallest creatures to the loftiest minds, that could teach us about the world we live in – the home we *could* have – and ourselves. Only today I watched footage of the world's largest known fish-breeding colony, discovered in 2021: 60 million icefish nests across nearly 150 square miles, 1,000 feet below the frigid surface of Antarctica's Weddell Sea. *Holy cow!* said one of the biologists. Scientists were in disbelief. Jonah's icefish are a ghostly blue-grey about 2 feet long with spiny dorsal fins, fan-shaped pelvic fins, and blood that is gin clear – no need for haemoglobin when you absorb oxygen through your skin. The nests they excavate are circular depressions about a metre across in which they lay, on average, 1,735 yolky eggs. Picked up on remote camera by a research vessel during a survey of the seafloor, nest after nest after nest, there was no end in sight for miles and miles. Before this discovery, the largest known breeding colony contained 60 nests. Which shows just how much there is yet to discover, how much we don't know, and why we should pause before we mine another ecosystem.

Animals have shaped our minds, our lives, our land, our civilisation, and they will shape our future too. We would not have got far without them. Their labour, their protection; their filtration, pollination and transportation 'services'; their dung, their flesh, their skins, their bones and their babies. Animals matter to us (while I am sure we don't matter to them). But the boxes we put animals in cannot contain them. These days, the whole of nature has become a last-minute add-on in the disembodied, faceless, colourless, lifeless catch-all: biodiversity. It sounds like a soap powder. BIODIVERSITY! Easy to rinse away. Why would we care about it? A desiccated name devoid of all the head-whirling, heart-thumping, mind-glowing, skin-tingling, glittering psycho-pageant of life.

Yet we are the animal who so eagerly forms deep attachments across the species divide. Many of us *do* care about a wren's song (64 different notes in 8 seconds), a whale's song (range up to 10,000 miles), a wolf's howl, a bower bird's love-shack, a beetle's iridescent carapace, a swift asleep on the wing at 5,000 feet, a seahorse with a pouch full of fry.

'The world shall perish not for lack of wonders, but for lack of wonder,' said the British geneticist Jack Haldane.

What *more* do we want? It took nearly 4 billion years to build a rhino, a river dolphin, a pink-toed mussel, a passenger pigeon, but only decades to disappear them. For ever. It seems to me there is *nothing* more important than to rethink our relationship with the creaturely world. Animals are the key. Variety and abundance are the strengths. Our saviours are all around us.

This book is a love letter. You will detect strong feelings. They are biologically wired and proportionate to the problem; passive objectivity, here, is not a virtue. I have skin in the game, we all do. We are the Story Animal. Stories of our relationships are what we are made of. They are how we read the world, and stories will be how, if ever, we move on.

The whole nation of Croatia awaits the return of a single stork.

An Austrian zoologist watches a tropical freshwater fish make a decision.

A man who cannot speak finds his voice in an encounter with a jaguar.

An American hurls starfish out to sea and discovers how nature functions.

A Japanese yakuza gang has a shoot-out on the high seas over a cargo of sea cucumbers.

An Indian hunter rows a wild leopard and her cub across a surging river.

A gorilla cracks a joke.

A beluga whale mimics the voice of a human.

A schoolboy in Africa receives his most important lesson from the crocodile who ate his best friend.

A photograph of a girl, a boar and a candelabra make me write this book.

Author's note
I use 'animal' to mean beings other than humans, except when it doesn't. Let common sense prevail.

PREFACE

Close encounters

I am waiting for Jonathan. He should have been home by now. I crane my neck over the hedge to see if he is coming down the path, but no sign. He was late taking the dogs for a walk, so he decided to do 'the round' – along the meadow, over the bridge, up the hill and then down the track behind the house. But now it's getting dark. Where is he?

He is standing halfway up the hill with a barn owl on his head.

He was watching her noiseless reconnaissance along the hedgerow, over the blackthorn, hovering for a moment, wings quivering above her like a warrior headdress, her searchlight eyes and ears sweeping the sward. She was out early, so must have had chicks to feed. He stood still as the dogs went on ahead. Then she landed on a post 30 metres away. She was watching the dogs. Jonathan remained motionless. The dogs, no doubt wondering what was keeping him, turned around to come back up the path in the direction of the owl. As they approached, she stood up on the post, then lifted off to fly away from them. Towards Jonathan. The owl was flying at almost head height, straight for him. He held his breath, riveted. He could see her glowing face, almost human, her round black eyes in the heart-shaped ghostly whiteness, her short sharp beak, the tiny diamonds trapped in her caramel wing feathers.

Just inches away, she rose above him and dropped her feet. To land. Softly, weightlessly. On his head.

So, there he is, with an owl on his head, standing in the middle of the track, silhouetted against the darkening amethyst sky. The owl turns round to face the direction of the dogs. Her claws are sharp as pins. He feels the warmth of her body through her plumage as she sits down. The dogs stop to look up at Jonathan with an owl on his head. Disconcerted, their ears twitching backwards and forwards. And there he stays. Completely still. Thinking, *I have an owl on my head! I have an owl on my head!*

And the owl watches the dogs, and the dogs watch the owl, and everyone stays like that for what is probably minutes but seems so much longer. Which is when I am wondering what on earth he is doing. When he is doing nothing but feeling everything. Human and wild thing joined in a fleeting silvered moment in the twilit hour. Nothing to her, but branded sharp as her pin-claws into his memory. Jonathan remains, the violet sky deepening, the dogs in the middle ground looking up at the man and the owl in this four-way silent conversation, until finally . . . finally, he tips his head slowly to the side, and the owl flies off. Thinking whatever an owl thinks when her post suddenly walks away.

I: WILD THING

Are you or were you, once part of a part
of us, something for us or from us, you
the most precious of all prepositions
for with us, undoubtedly, you are.

Grace Ingoldby, 'Blackbirds from the point of no return'

GOOD TO THINK WITH

For as long as I can remember, encounters with the creaturely world are characterised by a rich slowing down of time. Watching a spider wrap a fly; a community of ants bearing their eggs to safer territory; sweeping my hand through the halloween-menagerie of a phosphorescent sea. These moments transport me yet root me to the here and now. I lose myself and, simultaneously, join the sun-sharing, star-blooded, global throng.

When epiphanies come to us in our youth, they imprint primal pathways that can rescue us later. In *The Seabird's Cry*, the author Adam Nicolson recalls the time his father first took him to the seabird colonies of the remote Shiant Isles in the Hebrides when he was a boy. He was overwhelmed, 'the air and the sea around us filled with 300,000 birds, a pumping, raucous polymorphous multiversity in which everything was alive and nothing refined'. The chaos, the stink, the vulgarity, the vitality, the unedited full-throttled spectacle was a rite of passage which Adam experienced as a drilling-down through Creation, as if the maw of life opened in front of his eight-year-old eyes. 'It became a baseline and touchstone for me of what the world might be.'

So that's it. Something lights up inside us. And this touchpaper between humans and other animals has been lighting up, and indeed chronicled, for more than 40,000 years.

When we were wild

The first modern eyes to fall on prehistoric cave paintings belonged to an eight–year–old girl, María Sanz de Sautuola. In 1879, as María's amateur archaeologist father, Don Marcelino, searched for bones and carved flints on the floor of a cave in the Altamira hills in northern Spain, she looked up at the ceiling in the flickering light and cried out, '*Toros, Toros!* Papa, look at the painted bulls!' Sealed off for millennia by a rockfall, the caves had been found by a hunter chasing his dog in 1868, but little investigated. Don Marcelino's belief that the paintings were Palaeolithic was dismissed by both the Church and archaeologists. The animals were too fresh, too astonishing, too vivid, too real. How could such primitive people produce art of this calibre? They had to be modern forgeries. Sautuola was publicly humiliated. Fourteen years after Sautuola's death in 1902, the leading French archaeologist Émile Cartailhac admitted his mistake in the journal *L'Anthropologie*: 'Mea Culpa d'un sceptique.'

In 1965, my family visited Altamira. I was eight years old, the same age as María. The network of twisting passages, low ceilings and huge caverns seemed infernal and dangerous yet exhilarating. In those days you were allowed to climb down narrow ladders over vertiginous drops to the chilled depths of the decorated caverns below. I was scared and mesmerised. The underworldly breath of dank limestone was alien and cold. There was a petrified waterfall, gigantic stalactites and all around us the shivery presence of unknowable lives. Bison stampeded above our heads; boar, stags and ancient horses in ochres of burnt blood, their muscle-power pushing against the charcoal lines through the abyss of millennia. Creatures galloping, charging, once real and once alive, like me. The guide pointed out the claw marks of bears. Why would people want to paint this wild chase in the pitch-black bowels of the Earth? Yet even to a child it was evident that the animals and the people who drew them were intimates.

Prehistoric artists were far better at depicting the way animals move than their nineteenth-century counterparts. What the photographer Eadweard Muybridge demonstrated about equine footfall sequence with his 1878 photographic series *The Horse in Motion*, the cave painters knew all along. They had to. Life and death depended on acute observation. These paintings give us a glimpse into the prehistoric mind. For as the anthropologist Claude Lévi-Strauss put it, animals were 'good to think with'. What is striking in these pantheons is that human beings are *not* the centre of this world, we rarely feature; a few stick men with spears, a flock of stencilled hands, but nothing close to the reverence given to beast after beast. I see these subterranean galleries as flame-lit theatres for the dramatic plays of life and death. The performers wield spears in front of the tableaus reliving a great hunt, a choreography of shadows in flickering movement, silhouettes of man-beasts in antlers, voices from behind rocks, the whisperings of storytellers, the mimicry of owls, the bellows of bison, the flurried piping of flutes whittled from the hollow bones of birds, the rattle of seedpods, drumming on skins, beating on wood. A readymade set decorated with crystals and stalagmites; a roaming participatory audience in punch-drunk fear and screaming frenzy. Gateway to another world. From sensory overload to a darkened stage closed for 20,000 years while the backdrops charge on.

We lived in small family groups, 20, 30 of us, hunting, foraging, protecting our young, bringing firewood to our dens in caves or under rock ledges. What we lacked in strength and speed, we made up for with sharp spears, stone tools, tricks, teamwork, butchery skills, cooking, cooperation and language. An aurochs or bison could last us days. We knew every plant, nut, fungus, root, berry, seed, every drinking place, every termite mound, every nest. We knew every bird song, every frog croak. Our tracking instincts were honed, our senses alert, our hearing supple, our feet nimble. Our omnivorous diets were organic; our air and water clean; our tasks were various; no forms to fill in or taxes to pay, but we had to stay as fit, if not quite as fleet, as a deer. Our ancestors' prehistoric bones tell us that if we made it through the perils of childhood, didn't break a leg, catch tuberculosis or get eaten by a lion, we might live for 60 to 80 years.

Fire was our power. Roasted meat took less time to digest, which helped build and supply bigger brains. Aside from cooking, we could heat *and* torch the place. Which helped to flush out game. Because we pooled knowledge, hunted together and shared our gains, we had Time. Time to carve mammoth ivory, stitch cloaks, share stories around the fire and paint on cave walls. Fire set us apart and with its light we chose to summon animals. We were entwined. Our life-taking and their death-giving held the mysteries of the spirit world. At a 30,000-year-old grave site in Sungir, 200 kilometres northeast of Moscow, a man and two boys were buried with antlers and decorated ivory spears. They were wearing ivory brace-lets and fox-tooth headdresses. Just one headdress required the jaws of 60 foxes. What remained of their burial clothing was the decoration, over 13,000 mammoth ivory beads.

For a million years we had been middle management, preyed on and preying upon. But now, our technology raced ahead: traps and snare lines, nets, spear-throwers. Those clever bows and arrows, multiplying speed with power and distance; soon nothing was too big or fierce for us. Sometimes we just chased a herd over a cliff then scrambled down for the *yabba-dabba-doo* meat harvest. Dried it, salted it, saved it for later.

We went where the meat went. In the last Ice Age, around 30,000 years ago, humans followed herd migrations to the vast frigid grasslands of the mammoth steppe that stretched across the northern hemisphere from central France to Alaska. The supply of woolly mammoth, bison, elk and horse must have seemed inexhaustible. We constructed temporary dwellings with mammoth tusks and dug ice cellars to preserve meat. We hunted and hunted; as herds dwindled our skills only improved. As mammoths moved further north, they survived by snowploughing in the tundra with their curved tusks, trampling, fertilising and opening feeding areas for horses and bison. Until 10,000 years ago when, apart from in the most far-flung inaccessible places, they were gone. Large mammals with long pregnancies and a big investment in few offspring could not reproduce in number, or in time to replace themselves. One population on Wrangel Island, cut off by rising seas 200 kilometres north of the Siberian coast, held on for another 6,000 years. Then, just 3,700 years ago, 1,000 years after the Giza pyramids had been built, humans arrived on Wrangel and mammoths disappeared.

As the permafrost gives up its Ice Age bestiary we read individual biographies in the contents of stomachs, and the fine ivory growth rings of tusks like a tree. Good seasons, bad seasons. Dima, not a year old, falls into a clay pool. There is nothing in his stomach to sustain him, just soil and his own hair; he is weak from parasites and unable to climb out. His mother circling, trumpeting, prevents attacks from predators, until Dima drowns and sinks in the mire. Younger still, Lyuba, well nourished, 30 days old – suffocates by inhaling a thick batter of mud which clogs her trachea, perhaps while crossing a river with her herd, or falling through ice on a frozen lake. Her mother's milk has been preserved in her stomach for 41,800 years.

A male steppe mammoth with 4-metre tusks would have weighed over 6 tons. The hump, familiar from cave paintings, was a fat store like a camel's; the matted greasy guard-hairs over their undercoat could be a metre long. They had small ears to minimise heat loss, and a flap like an apron over their anus for the same purpose. In the 10-kilometre cave network of Rouffignac there are 158 depictions of mammoths, some showing anal flaps and the delicate two-fingertip trunks they used to forage. However, the first complete skeleton of a mammoth, found in Siberia in 1799, was assembled with its tusks placed in the wrong sockets, to curve outwards instead of in. Their noise we can only imagine; the body rumblings, the trunk orchestras carried on the glacial winds, the crunchings of ice-splitting fissures across the trembling ground.

We could never have imagined their loss, let alone that it would put paid to the whole ecosystem. For 100,000 years the largest terrestrial biome on the planet had thrived, the soils were fertile, and the foraging rich in a cold, dry climate supporting a biomass of mega-herbivores on a par with the African savannah. One adult mammoth could consume 200 kilos of herbage a day and scatter a vast tonnage of fertiliser in urine and droppings, ensuring a continuous cycle of nutrients. Without the mammoth converter, grassland became dominated by tundra vegetation which, uneaten, became waterlogged and frozen, slowly turning to acid peat where grasses struggled to regrow. Held in ice cores, the climate record for that period shows a drop of CO_2 in the atmosphere. Could, as mammologist and palaeontologist Tim Flannery suggests, the loss of the mammoth engine of productivity in this vast area account for the missing carbon, trapped in uneaten plant matter submerged in Siberian bogs?[1] Were we capable, by proxy, of influencing climate even then? And thus this cold snap at the end of the Ice Age, around 13,000 years ago, allowed sea levels to stabilise and remain so for 8,000 years, assisting the inexorable spread and growth of human populations.

Wherever we showed up, extinctions followed. Sabre-tooth cats who had lived on the planet for 30 million years, gone in 2,000 years. In Europe, the mammoth, the rhinoceros, the cave bear. In the Americas, the horse, the mastodon and a 4-ton giant sloth measuring 20 feet from head to tail. In Australia, 45,000 years ago, the giant marsupials were easy pickings. Adieu, Demon Duck of Doom, a 10-foot-tall flightless bird weighing half a ton; goodbye, Diprotodon, a 3-ton wombat; farewell, 70 kilos of rat-kangaroo. As numbers diminished, ungrazed vegetation grew coarse and forests became susceptible to fire, helping humans clear land and catch more prey. Terrestrial megafauna, unused to being predated, neither knew to be wary nor had time to evolve. Those who survived were those who'd had the time to learn to avoid us, who had co-evolved with us in Africa, where they cling on.

Back to the future
In France's Chauvet caves, the animals painted 36,000 years ago stampeded unseen until a draught gave them away in 1994. Nineteen species, with lions, woolly rhinos and mammoth joining the throng. The film director Werner Herzog, given special access to make a 3D documentary, *Cave of Forgotten Dreams*, supplies an unsettling epilogue. Picking our way past the drawings and bone heaps of prehistoric beasts (190 skulls of cave bears), we are plunged into his offering of a post-human future. Herzog tells us that just 20 miles from Chauvet is one of France's largest nuclear power plants, where nuclear-warmed water, sloughed from the cooling towers, is fed along subterranean arteries to heat a nearby tropical biosphere. The camera attends to a white reptilian foot, clawed toes stretching out

in black water. Herzog's commentary explains these are the mutant albino offspring of the biosphere's crocodile inhabitants. The uncanny creature floats, luminous as the moon. A close-up of a pink crocodilian eye like a polished gemstone in the hooded entrance of its own fathomless cave. A zoologist would identify the captive as an albino *alligator*, one of fewer than 100 such albinos in the world – a film extra, it turns out, shipped in from Louisiana. Details like this don't concern Herzog. His imagination spools out. From the quiver of a spear to our silenced apocalyptic powers. From our cradle to our grave. Here we *aren't*, the postscript appears to suggest. Here crocodilians, around for 200 million years, are the consummate survivors. Yet what can our freak-show inmate do in his glass confinement but slowly twirl between water and air? While down the road at Chauvet the prehistoric hunt gallops on.

How to think of all this? The credits roll until the final still. The spattered spray of ochre, the colour of dried blood, outlining the human hand held against the rock 32,000 years ago. One handprint of the many heavy handprints of the 100 billion humans who have ever lived.

TAKING LIBERTIES

But should you tame me, we will each need the other.
Antoine de Saint-Exupéry, *The Little Prince*

The Greek slave Aesop was both clever and fabulously ugly. The story goes he earned his freedom 2,500 years ago by telling animal fables. One night a lean, hungry wolf falls in with a well-fed house dog and trots along beside him. He envies the dog's full belly and warm shelter and is tempted by his life of ease, until he notices a bald mark on the dog's neck. The wolf asks the dog what it is. The dog dismisses it as nothing, just where his collar is fastened when he is chained up. The wolf high-tails it as fast as he can back to the forest, shunning any chance of food or luxury for the price of his freedom.

The wolf at the door

Wild things flee from us. I have been feeding generations of songbirds in our garden, yet when I replenish their bird table, ungratefully they fly away. Yet many animals are comfortable with each other, the mouse and the robin, the deer and the woodpigeon, even the fox and the rabbit unless the predator is in hunting mode. But not us. Only when a species evolves in surroundings devoid of humans can we experience the close curiosity of the wild. Lindsay McCrae, a wildlife cameraman, described 4-foot-tall emperor penguins shuffling over to where he was sitting and plonking themselves down beside him; he said it felt both fantastical and ordinary, and profoundly humbling. Orphaned young are more trusting; newly hatched goslings will famously imprint on a human parent ready to foster them; otter kits, lion cubs and chimpanzee babies can all be tamed easily enough. But only one animal has been part of our society for tens of thousands of years. The paradox is that the canine creature we most love is descended from one we have relentlessly persecuted. Our best friend and our worst enemy.

One theory goes: wolves began hanging around hunter-gatherer caves for bones and leftovers with some quid pro quo in the relationship: food in exchange for guarding the cave from bears or chasing quarry. I doubt it. Ever tried taming an adult wolf? No, we knew where the dens were and we stole their babies. In tribal societies hunters often bring wild baby animals home as gifts for the children. Some, like toys, are expendable and short-lived, and others grow until they become too edible to ignore, but the survivors can become part of the household. Wolves have particular qualities in their own society that suited us: strong hierarchies, keenly aware of status; highly social, loyal, affectionate, with firm family bonds, both defenders and attackers; they don't mess in their den; and they learn quickly. The domestic dog evolved from their wild ancestor around 33,000 years ago, according to DNA analysis of a canid skull discovered in the Altai Mountains of Siberia.★

Although in one sense *Canis lupus familiaris* connects us closer to the animal world, we are about to take an enormous leap that will irrevocably distance us. We are going to become farmers.

The compatible and combatable

Twelve thousand years ago, as the glacial period ended and the climate warmed, our populations grew, we settled by rivers and lakes, and tied ourselves to tilling the land. No need to chase after food any more: grow it. Or corral it into warm living larders.

We set our sights on herd animals who stick together in social hierarchies, who would yield to our control. We needed specific traits and surprisingly few creatures ticked all the boxes. Some animals could be tamed but not domesticated, which is a different thing. (A domesticated animal has been selectively bred in captivity and remodelled to be genetically distinct from their wild ancestors to better suit human use.) So, not too nervous or flighty, nor trying to escape all the time. Gazelles, for instance, the most hunted creature in the Fertile Crescent, running at 50 miles per hour, leaping 30 feet, smashing themselves against the palisade, won't do at all. We can't farm fussy eaters, like panda bears. Or the big brutes who would like to kill us, so that's hippos out. Or herbivores

---

★ The skull was anatomically doglike, and DNA analysis corroborated that it was more closely related to a domestic dog than a wolf, which makes the origin of *Canis lupus familiaris* about 15,000 years older than originally thought.

who bite and kick us, which is why zebras were never hitched to the plough. Zebras are notoriously filthy tempered. You can't lasso them either, they duck away every time. Zebras injure more zookeepers than tigers do. Lord Walter Rothschild was one of the few people to tame one. He famously drove a carriage pulled by five zebras (and a disguised horse) to Buckingham Palace and was nervous in case Princess Alexandra wanted to pat them, for they had already killed one of his grooms 'who was careless'. Watch Mama Zebra in action defending her foal. She will take a cheetah on, teeth bared, raging, both front feet boxing out. That's another prerequisite: young we can take away from their mothers. And they need to grow up fast, which disqualifies elephants. They need to be robust enough to survive, easy to subdue, tolerant and amenable enough to crowd together. They must not fight over territory. They must breed easily in captivity. Which only left five main candidates, give or take the odd yak: sheep and goats,* who we domesticated from the Asiatic mouflon and the Bezoar ibex about 11,000 years ago; cows, from aurochs around 10,000 years ago; pigs, from Eurasian wild boar 9,000 years ago; horses about 6,000 years ago; chickens came later, in Asia around 5,000 years ago. It's hard to imagine the slow task of restraining wild aurochs, 6 feet high at the shoulder with bloody great horns, let alone the guile required to persuade them to be milked. These were gradual processes with a lot of bother, with herds breaking out, trampling and eating crops, fouling the water supply. And domesticated livestock present easy pickings for other predators. Here are the seeds of our 12,000-year war against nature.

Then we started playing God. We selected characteristics useful to us: meatiness, hardiness, woolliness, adaptability, docility, and we dispatched the individuals who didn't suit. Quite sinister. A bit like *The Handmaid's Tale* for farm animals. With no need for speed or agility, animals became slower and fatter. If a sheep falls over these days, she can hardly get up. No use for their wits any more. In a few generations these animals became dependent on us, more bulk, less brain. What we couldn't change was their natural behaviour and their need to socialise, to play, to protect and suckle their young. As far as we were concerned, by giving food, shelter and protection, *their* concerns were taken care of. While we got the meat, milk, fur, leather, wool, fertiliser and pulling power.

---

*   Same sub-family, Caprinae.

As our numbers multiplied we created frameworks of governance with hierarchies. There were leaders and there were workers. Gone were the freedoms of the hunting-gathering life; no time for conjuring poetry, gone the Tygers Tygers burning bright; gone also the rich varied diet, the rich varied day; gone the call for quick thinking, skilled prowess and multitasking. Things are different in settled societies. Our bones tell the story. Cultivation is backbreaking and harvests are vulnerable to bad weather and predation. Domesticated animals have to be fed and are susceptible to disease. Our life expectancy decreases. We shrink. Our brains shrink! We domesticated ourselves.

Of course, we didn't turn back. We were tied to our investment with too many mouths to feed. Now crops and livestock needed safekeeping. That's when we began demonising the smarter brother of our best friend.

The Big Bad Wolf

Can there be a creature more embedded into the Eurasian psyche? He won't just take our sheep, he will eat our grandmothers. Aesop, Ovid, Luther, Grimm, Bram Stoker, Ma Zhongxi and Milton have all led us through lupine corridors of malevolence and the uncanny. In Gustave Doré's seductive etching, the wolf stands shoulder to shoulder with Red Riding Hood, his back to us, head turned slightly towards her. The coy, knowing face of the chubby-armed girl, the polite, towering wolf with oh, that tiny sliver of white moon in his eye. You know what he's up to. But you yearn to put your arms around him. The appeal, apparently, is anxiety and excitement. Which is where the eerie power is, and why perhaps this love-hate tug-of-war has endured for millennia. For are we more vicious, more unforgiving, or more punishing towards any creature than the wolf? The rancher's retribution for predating on livestock was a mercenary violence exacted as if the culprit had swallowed his daughter. The wolf is a crepuscular creature, stealthy, potent and smart, whose sins are countless: greed, lust, deception; they are liars, seducers and thieves. The wolf is the brooding presence that thumped our medieval heart. If we didn't take care we might become one. The wolf has had it rough. As an enemy of the flock, so, therefore, of Jesus. Wolves crawl around sacred texts, spring off our churches, suckle our children. The wolf must die, in literature, myth, reality. We banged our saucepans to the rhythmic primal heartbeat; chased the villains into the pits; expunged their night howls. And now we miss them. And if we don't miss them, the landscape misses them . . . as we shall see.

In 1998 Jonathan and I set off on foot to follow a rumour of wolves along the remote mountainous Spanish–Portuguese border. How badly I wanted to see one. We walked for a month, schist crunching underfoot, sleeping under apple trees outside tiny hamlets. The hills were an oven, my pack a torment, my plea to buy a donkey thwarted (by guess who). At around 1,500 metres, above the village of Montesinho where boulders littered the landscape like giant eggs, we saw in the distance a shimmering white carpet across the escarpment about 3 miles away. We stared, baffled, thinking the heat had fried our brains, until our binoculars revealed a gargantuan flock of sheep. As we approached we could hear the tinkling of their bells on the wind. We could make out two shepherds and their wolf-like dogs. That night, drawn by our fire, one of the shepherds and a small dog visited our camp looking for wine. The dog pirouetted around his feet, ballet-begging, as the man roared in delight through his crooked dolmens, *Bravo! Bravo!* An untethered laugh, feral and filled with childlike joy. I was reading John Berger's *Pig Earth.* As the shepherd picked up the book and laughed, holding it upside down, I wanted to say, *It's about you.* We shared soft cheese and dried apricots. He was 63 years old, 100 kilometres from his home in Spain, and would be away for three months. I asked how many sheep in the flock. He said 6,000. 'Y los lobos?' I asked. He exploded with laughter. *Nunca, nunca!* Never see them, he said. But when he was young he would shoot them, *Bang, bang!* The next day we spied him through the binoculars, the 6,000 sheep in his care bounding over rocks and aromatic scrub, their long tails swinging, their bells jingling. An unearthly undulating whiteness against a searing blue sky.

Some days later, in the nearby village of Fafião, we saw an eighteenth-century wolf trap. Two stone walls, 2 metres high, converged over a distance of 64 metres to funnel into the circular corral of a deep pit. The end of the beat. From above it would have looked like the eye of a giant keyhole. You could hear echoes in this poignant place.

On our last day in the high sierras the humidity was debilitating. We were walking in thick cloud, unable to see more than a foot in front. I sat on a rock and leant back on my heavy pack like a bug upside down. I was staring into the dim monochrome, filthy and tired, when a long loud howl came out of the gloom. I sat bolt upright. Then another. Unmistakable.

The spine-shiver trace of a primal memory. We stared wide-eyed at each other. We waited, spellbound. We howled back. But we did not hear it again.★

★

Then, 6,000 years ago we laid our covetous eye on the creature who would carry us into the future. More than a beast of burden, more valuable than dinner, the horse gave us speed and distance. The knit of his thoracic and lumbar vertebrae, the springy suspension of his tendons, the toughness of his hoof and those powerful rump muscles; we wanted his strength, his stamina, his spine (a bit uncomfortable, but we'll come up with something). The key was his breakable nature. We broke it.

★   In 2021, Jonathan and I returned to northern Portugal where they now boast a popu-
    lation of 250–300 Iberian wolves. Conversely, the Sierra Morena population of two
    healthy wolf packs became extinct in 2005.

Now we could go to war on a horse. We could steal, invade and colonise. What a distance we'd travelled from the animism and animation of the wild stampede of prehistoric cave paintings.

Cities grew and spawned rulers whose displays of power acted as a check to political rivalry. Why should they not be entertaining too? What could be a more magnificent symbolic display than to vanquish the fiercest animal of the wild? Just in case anyone has any doubt who the ruler is, the *spectacle* of the hunt is born.

THE BEASTLY BUSINESS

Witness this. The swipe of a lion's claw. An arrow shaft jams between his eyes; another tunnels under buff hide; flesh flows into muscle. A lioness drags her dead legs behind her, eyes blazing, ears flat back, flared nostrils rucked into tight waves.

I am staring at stone. The air in my lungs has snap-frozen. Room 10a, the British Museum: The Lion Hunt of Ashurbanipal, 645 BCE. Panel after gypsum panel, these carved reliefs from the North Palace of Nineveh, ancient Assyria, chart a king's prowess. Ashurbanipal, the last Assyrian king, enters a palisaded ring in a horse-drawn chariot. For this hunt is not in the Assyrian wilderness, but an enclosed arena built for the purpose. Thirty lions or more are let loose, but not loose, circumscribed by soldiers wielding swords, spears and three-tailed whips, guards with muscly mastiffs straining at the leash. At my feet a felled lion rests his heavy head, eyes closing, to lick his great paw. I stare with horror-wonder at his moment of tenderness, his self-kiss. A lioness bristling with arrows flipped upside down. Nothing has prepared me for this. Only the ghastly roars have been silenced. The pain is exact, palpable, unequivocal. This happened. Two and a half thousand years ago these real lions poured their warm blood into the earth and were raised out of rock. The skill is breathtaking. A lion rears up between the chariot's wheels as an attendant thrusts his spear into his ribs and Ashurbanipal plunges his sword into his throat. A king's sport with a king's odds; a king against kings. Thus the king protects his people and his power, for there were a multitude of lions, royal records report, terrorising the neighbourhood, killing cattle and men, their roars ricocheting from hill to hill. A lion squats on his haunches, every sinew straining to stay upright, veins pump under cold alabaster, his claws clench at the ground. An arrow pierces a lion's forehead, his tongue shoots out; his mane becomes his wreath. What is striking is how each lion, *not* the king, is the focus here. The king is a cut-out decoration of a man, hardly a breath in his stone chest. He is remote to me. Unreal. The human figures are uniform, unmoving, unfeeling, but the creatures flare into life (and death). Wild, willed, sensate, intense. The lions are what matter. This artist knows them better than his own kind. His fingers plait their blood rivers through the stone. I crouch down, fill my gaze, then flee.

Today, Ashurbanipal's enclosed hunting technique is no longer the sport of kings, but of dentists, oligarchs and wealthy businessmen. Canned hunting, it is called. It costs £4,000 to shoot a lioness in a South African enclosure, which is half the price of hunting a wild lion in Tanzania. Crossbows are popular. Battery lion farms keep up the supply and offer further attractions: you can pay to cuddle the cubs; you can even pay for the privilege (rather a lot at £1,500 for two weeks) to work on these farms, raising the cubs to 'rewild'. Heads are destined for walls, bones for sorcery. The lion-bone trade in Vietnam, Laos and Thailand is flourishing. After the ban on farming tigers in 2007, the lion became a tiger in death. In 1970, 400,000 lions roamed the African continent, now there are fewer than 20,000. In South Africa's 200 lion farms there are about 8,000 captive lions padding up and down. What a business. Which is, when you think about it, down to the sorry state of human penises: traditional Chinese medicine buyers of lion bone and rhino horn have frightened or exhausted ones; trophy hunters, as we know, need bigger ones.

Blood thirst feeds on blood thirst. Like anything, to get the same hit you need more and more. Ancient Rome plumbed depths that implode the imagination. Step into the labyrinthine bowels of the Colosseum, where the enslaved spent their last crazed moments, and you can taste the residual aura that sickens the air. Eighty vertical shafts winched the prisoners, gladiators and demented animals into the arena through a trapdoor. From cramped cage into the blinding light, a confused moment of freedom, the smell of blood, fear and excrement, then the thunderous cheering. From every corner of the empire: elephants, giraffes, aurochs, bison, Barbary lions, panthers, leopards, Caspian tigers, crocodiles, rhinoceroses, hippopotamuses, bears. The Emperor Trajan held games for 120 days to spill the blood of 11,000 animals. Emperor Commodus invented crescent-shaped arrows which could decapitate ostriches, affording the crowds the diversion of watching them run without heads. Prisoners *damnatio ad bestias* were executed by beasts, and to spice it up might have a part to play in a re-enactment of a Greek myth: Prometheus, with a bit of creative license, instead of having his liver consumed by an eagle was tied to a post and torn apart by a bear. Because it was unnatural for lions to attack humans, they had to be starved or trained to do so. The Colosseum had 80 entrances and could seat 50,000 spectators. From the inaugural games in CE 80, the fun went on for nearly 500 years.

★

It's mind-boggling to imagine how they caught, kept alive and transported the sheer numbers of dangerous animals brought to Rome 2,000 years ago. Carl Hagenbeck, a German animal dealer in the 1880s, gives us an inkling in his memoir, *Beasts and Men*. His father, a Hamburg fishmonger, ran a menagerie of lions, cheetahs and monkeys as a sideline, so when 14-year-old Carl was asked if he wanted to be a fishmonger or an animal dealer, there was no contest. So began a lifelong career in the trade that coupled his love for big wild animals with their capture.

Here's how: to catch a giraffe, antelope or ostrich, chase them on horseback until they are out of gas. To catch a zebra, hire up to 2,000 men to surround the herd, drive them somewhere they cannot escape, like a dry river gorge with cliff sides, and then (sorry) whip them with long lashes until they are so exhausted they can be fettered and tied. Elephants and young hippos were caught in pitfalls. Baboons were trapped at waterholes, pinned to the ground with forked sticks, muzzled, bound, wrapped up in cloth and carried suspended from a pole by two men, so that each captive looked like 'a great smoked sausage!' On one momentous occasion in Abyssinia (now Ethiopia) a group of captured baboons brought down a herd of 3,000 silver-grey hamadryas baboons from the hills by the din of their screaming. The raiders made short work of the cage, released their clan, and beat off their human captors by the sheer force of their numbers. Hurray! Their manes erect, they fought baring their teeth, beating the ground with their hands as they came. In this epic battle an injured infant was seen being swept up 'by a great male from the very midst of the enemy'.[2]

Alas, most captives were less fortunate. Long caravanserais crept across desert sands in the cool of moonlight, the shadows of giraffes or elephants walking in harness with drivers at their side. Hagenbeck reports wild baboons running beside the cages of their captive brethren, screaming to each other in an 'ear-splitting chorus'. Lions and panthers were pulled by camels; other creatures were strapped to palisades and carried aloft. The six-week journey, from Atbara (Sudan) to 'the port of embarkation on the Red Sea,'★ required huge entourages of water carriers and shepherds driving sheep and goats to provide fresh milk for the young animals, or fresh meat for the carnivores. From the Red Sea to Suez, then by train to Alexandria, to catch a ship to Trieste, Genoa or Marseille, for the train to Hamburg. An unimaginable three-month journey. Those who didn't make it fed those who did.

The losses were enormous but nothing deterred Hagenbeck, whose hunters scoured the globe for exotic animals to supply the world's menageries.★★ For all the hapless cruelty, the 'Wild Animal King' cared for the animals in his charge, and animals isolated from their own kind were permitted to make unusual friends. Thus a male kangaroo became inseparable from a female elephant who stroked him with her trunk. Hagenbeck's lasting impact was the zoo makeover – to recreate an impression of the animal's natural home. Out go the iron bars and in come moats and nifty invisible devices of internment. Scatter-rocks, painted palms and exotic panoramas bestowed fake freedoms where distance was illusory and unlikely appreciated by the animals themselves. What troubled his conscience were his 'anthropoid apes'. The psychological toll on his gorillas suffering 'homesickness', hunched in misery and refusing food, ended only one way. 'Perhaps in the course of time I shall succeed in discovering the proper method of treating these great apes,' he wrote, wistfully. Maybe not the solution Lieutenant Heinicke came up with in 1908 for the young gorilla he brought back to Europe with 'two negro boys' as 'constant associates' . . .[3]

Mum took me to London Zoo when I was a child. It wasn't a success. I remember thinking, what is the giraffe *doing*? Of course, the giraffe was doing absolutely nothing at all. The elephant was standing in a saggy kind of a way. Where did she go after we'd gone? Nowhere. I was embarrassed. Ahead of us there was a magpie strutting sprightly along the path, doing something at least *he* seemed to know about. The other animals were not sprightly. To my mother's frustration I wanted

---

★   Port Sudan, about 300 miles from Atbara.
★★ Animals were not his only 'exotic' exhibits; Laplanders wandered around with their reindeer, Egyptians lolloped past papier-mâché pyramids, with similar casualties; eight Inuit died of smallpox; a group of Sioux Indians died of measles and pneumonia.

to watch the magpie. The magpie flew away. My heart crunched into a walnut. Something, like Peter Pan's shadow, had been taken away. I dragged my feet and gulped my freedom down. I wanted to go home, like the animals couldn't.

Now, as then, there is little chance of reprieve or a dash to freedom for zoo inmates . . . unless, that is, you happen to be an octopus. The only successful Houdinis of the captive world. Being boneless, octopuses are able to pour themselves through a gap little bigger than the size of one eye. In 2016 Inky escaped from his aquarium in New Zealand, then found his way to a 50-metre drainpipe that led back into the sea. An octopus in Otago won his freedom by spraying jets of water at the lights; the cost of having to fix the short-circuited power supply began to outweigh the fascination of his behaviour. Another learnt to plug the outflow valve with a tentacle to get more water, and flooded the whole lab, not an uncommon event if you put octopuses in tanks, it turns out. In Seattle Aquarium an octopus named Lucretia McEvil dismantled everything in her tank. Because SHE WAS SO BORED. In 1959, Harvard scientist Peter Dews put the errant behaviour of octopus Charles (who spent his time with his eyes above the surface of the water, aiming water jets at anyone who came close, instead of pulling a lever for a reward like the amiable Bertram and Albert) down to something that must have *happened* to him. It was unthinkable that an animal without a backbone had an individual personality. Another sceptical scientist was ruffled by an octopus she had just fed with thawed squid (a third-rate food for an octopus) who waited until she returned to make a conspicuous show of putting it down the drain. Back at the Seattle Aquarium Roland Anderson devised the cunning plan of dressing two keepers identically, but allocating one the task of feeding the eight resident octopuses, and the other the task of prodding them with a bristly stick. Surprise, surprise, the octopuses steered clear of the bristly-stick guy and swam towards the keeper who fed them.

Today we are far closer to understanding how clever they are. Being prey themselves, we shouldn't be so surprised by their skills as expert and subtle observers. We also might deduce that their proclivity for escape demonstrates their awareness of their own captivity, and that they do not take kindly to it. Why otherwise would a marine animal climb out of her natural medium into the air where she cannot exist for very long? Because she is being chased by a shark, or because she really, *really* doesn't want to be there. We know octopus escapes are premeditated, because they do it when no one is looking, usually at night. Which is also when they raid neighbouring tanks for a crab snack. I think we can guarantee that once back in the sea they will not try to escape out of that. Life for an octopus, it seems, is not just about being fed, but about hunting, hiding, swimming and following their octopus wiles. In the Oscar-winning documentary *My Octopus Teacher*, cameraman Craig Foster dives near his home in the kelp forests off Cape Town every day for a year to learn more about an octopus he met hiding on the sea floor in a camouflage cloak of collected shells. Among the many moments of wonder is one when he is confused by what she is doing with a shoal of silver fish. She floats beneath them, her long tentacles swirling like chiffon among their glittering sparks. Then he sees it. Not hunting, or hiding, or fleeing. She is playing.

★

Sometimes good things come from bad things. In 1966 the blockbuster film *Born Free* told the story of Joy and George Adamson, who rescued and raised a lioness cub called Elsa, and successfully returned her to the wild. Virginia McKenna and Bill Travers, the actors who starred in the film, went on to make *An Elephant Called Slowly*, with a two-year-old elephant called Pole Pole (pronounced Polypoly) who had been taken from the wild by the Kenyan government as a gift for London Zoo. When the actors met Pole Pole with the elephant conservationists David and Daphne Sheldrick, she was smashing herself against the side of a pen in Nairobi, and so distraught it seemed impossible they could film with her. The Sheldricks took the traumatised elephant into their care and she soon became, as Virginia described, 'the most gentle, adorable, sweet, trusting elephant that I have ever met'. However, the Kenyan government would not change its mind. If Pole Pole didn't go to London after filming, another baby elephant would be taken from the wild in her place. What an irony. As the soundtrack of *Born Free* blasted from every radio in Kenya with Matt Monro's golden voice purring *As free as the wind blows*, the government was woefully ignoring the sentiment behind what had brought so many admiring eyes to their country. Pole Pole was shipped to Regent's Park where she lived on her own in the infamous concrete Elephant House cut off from the world by the water-less moat. An island in an island.

In 1982, Daphne Sheldrick wrote from her home in Kenya to Bill and Virginia in the UK about Pole Pole, whom she'd heard had become 'difficult to manage'. Bill and Virginia steeled themselves to visit her. There is a heart-crushing black and white photograph of the moment Pole Pole recognises them, stretching her trunk as far as she can across the moat to reach them. And here is the paradox: this magnificent animal, trapped in her elephant skin, is both utterly able and yet utterly unable to communicate to the people who needed to hear her. Virginia and Bill launched a campaign to give Pole Pole a better life, and in 1983 the zoo agreed to send her to Whipsnade Wildlife Park where she could be among other elephants. Her transportation was a fiasco: she collapsed, damaged a foot, and a week later it was concluded she had 'lost the will to live' and was euthanised. She was just a teenager.

So that was the very bad thing. So bad and so devastating for Bill and Virginia that they set up the watchdog organisation Zoo Check in 1984 with their eldest son, Will. Zoo Check became the Born Free Foundation, which is the very good thing that rose from the ashes of Pole Pole's short life; it campaigns relentlessly on behalf of captive wild animals around the world. In a telephone conversation, Will Travers told me he had identified a place with the kind of topography, climate, rainfall and temperature range suitable to establish the first elephant sanctuary in Europe that could be home to 50 or 60 elephants. He needs £22 million to make it happen, he said. Spare change for the rich list these days. Imagine the somersaulting joy one would have to be able to do that.

There are arguments for places of safety, like zoos, for threatened species; and many do support conservation. But overall, my child's-eye view of zoos has not changed. We still 'host' gorillas in 'enclosures' so we can stare at them. A new enclosure for three gorillas cost London Zoo £5.3 million. It's business. Meanwhile there is a scarcity of resources for vehicles and trained men for anti-poaching patrols to conserve gorillas in the wild. Captive breeding programmes can give a false impression that a species is being saved from decline, but animals bred in zoos are rarely released into the wild. Zoos are miserable places for large animals. Can't we empathise? Lockdown was claustrophobic enough. Multiply that by a lifetime. Polar bears, tigers and lions range across vast territories. Tigers and lions have 18,000 times less space in a zoo than in the wild.* Polar

---

* London Zoo's £3.6 million tiger enclosure is .6 of an acre, 2,500 square metres.

bears have one million times less space. Their biology shouts at them: Move. Set off. Travel! Lions spend 48% of their time pacing; 54% of elephants show behavioural problems. African elephants live three times longer in the wild. Even working Asian elephants live longer than zoo elephants. In Gaza Zoo, lion cubs' claws are removed so children can play with them. In the UK, 75% of elephants are overweight, and only 16% can walk normally. Yes.

Before lockdown zoos were hired out as party venues, with music, fireworks and drunkards throwing bananas. It's time to phase out life sentences for these wild creatures in zoos. What can we learn from trapped animals devoid of their natural context, unable to do their thing? Waiting from one meal to the next, padding the same trail round and round. Waiting. And waiting.

In their dreams, Aesop's wolf, bounding fast away.

BACK TO THE WILD

It is Good Friday, 21 March 1913. The writer Edward Thomas has set off on his bicycle from South London towards the Quantock Hills in Somerset *in pursuit of spring*.★ In a heavy downpour of rain he shelters under the canopy outside a bird shop beneath the meagre chirps of chaffinches in the row of tiny cages overhead. The 'battered ones' cost a shilling; the sprucer ones, 1s 6d. Through the window he can see linnets (2s 6d) dash themselves against the bars of their 6-inch cages; a goldfish swims his 6-inch circle of life round and round and round. As a knowledgeable birdwatcher, Thomas would have been aware that March was an unnatural time for a goldfinch to lose feathers – which normally happens at the end of the summer after breeding – nevertheless, 'house-moulted' gold-finches are 5s 6d. By now, three other shelterers have joined him under the canopy looking out at the rain. One enters the shop. The proprietor comes out to fetch down a chaffinch cage. A few minutes later the customer comes out with 'something fluttering in a paper bag', and with the rain now stopped, departs on his bicycle. Thomas also leaves, heading in the same direction. The man ahead stops halfway up the street by a garden. Thomas watches as he opens the paper bag to let the cock chaf-finch fly away into a lilac bush.

The liberator of the chaffinch shadows Thomas on his spring odyssey in both body and mind. Half-admiring, half-despising, Thomas crosses paths with this 'Other Man' again and again. Thomas watches him being jeered at by some young bicyclists for standing in the middle of the road sketching a weathervane. The man who freed the chaffinch. As Thomas had not. Or had he? Was the man ahead of him the stranger-self Thomas was pursuing? There is an unease in these encounters which suggests the 'Other Man' might be a trope for us all. A doppelgänger or literary device, the self we want to be. The caged self we would like to free. I will hold

---

★   The title of Thomas's book that came out of this journey.

that thought as I make my way between freedom and captivity in this story of animals and us.

★

May 2019. I am standing on the wooden verandah of an old forester's lodge, in a sunlit glade deep in the heart of the ancient Białowieża Forest in the far eastern corner of Poland. The nearest road is more than 8 kilometres away down a forest track. The grass has been cut around a bed of uncoiling young ferns and dotted clumps of blue iris, although no one has lived here for a decade. The smooth belly of a grey-pink granite boulder balances a perfectly round egg stone, just as they left it. This is a pilgrimage. This place and the two people who lived here are the spirits that gave breath to this book. I trail my hand along the wooden balustrade. I touch the root-bones of an ancient oak mobile strung from the rafters, stopping for a moment the hover of a giant prehistoric bird. Here is the well, and the woodshed. Here is the path into the forest. Jonathan has wandered off somewhere; he knows. Dziedzinka. Time present and time past. Where the wild and tame blurred, where chance conspired to bring together two unlikely strangers. Poles apart. It is here that for 36 years a magnetic force of creatures slunk from forest into human hearts. First a wild boar piglet, then everything followed. The rules of nature flipped. A preposterous beauty.

Two years ago, my agent, Patrick Walsh, sent me a black and white photograph. He is in the habit of emailing me snippets like: 'Man looking like a mummy is rescued after surviving a month in a bear den in Russia', or 'Snowball, the dancing parakeet'. But this floored me. Centre stage, in a dining room crowded with nineteenth-century clocks and ornate brass lamps, was a giant boar standing on her hind legs with both front feet and hairy snout on a thick oak-ring table. The enormous boar – and she *is* enormous – snuffles at breadcrumbs scattered by a girl in long plaits sitting at the table. Behind the boar's head a candelabra is precariously alight. The time is 12 o'clock. Like a parable or strange fairy story. For all the strangeness, there is something else going on. The dramatic quality is outlandish, yet somehow also entirely natural. It is as if one is spying through a private keyhole yet beholding a theatrical stage set. Surreal, yet real. Intimate yet inscrutable. I stared and I stared.

I clicked on the link. Three more photographs: the same girl curls up on a rug beside a bed in which the same boar and a large dog are fast asleep (note: boar head on pillow); to their right, an older woman sews lace with an owl perched asleep on the arm of her chair.

I scrolled down: same girl crouches in a forest clearing with a raven on her knee, the raven leans towards her, his beak slightly open, to whisper or kiss, as sunlight catches the white shine of his black feathers. Same girl walking through a snowy forest with a small band of young deer. With more looking comes more wonder.

She is not 'a girl'. She is Simona Kossak and she is 27 years old. The year is 1970. The country is Poland. The place is the lodge, Dziedzinka, in the Białowieża Forest. As I feed Polish text into Google Translate, out come strange English sentences that give me the first disjointed fragments of her tale. Born into a famous old Polish family of painters and poets, Simona Kossak knew the moment she saw Dziedzinka in the snowy

moonlight that it would become her home. The search for her future
was over, 'It's here or nowhere,' she said. It had no electricity, just a well
in the garden, and was only accessible by a forest track. Simona is at a
point in her life when she wants to live simply, as far from humans as
she possibly can. The forester's lodge, divided into two apartments, has
one irritation – the man from Warsaw who has rented the other half. He
won't last long, she thinks. He also wants to live simply and close to
nature, as far from humans as he possibly can. He is not exercised about
this tenant. It's obvious that she won't survive the winter. Kossak, a zoologist, is employed by the Forestry Research Institute to study the wild
forest mammals and their ecological interrelations with the environment.
Her position is controversial and gains her more enemies than friends.
You will already have guessed that the irritating man becomes her life
partner. He is Lech Wilczek, the wildlife photographer who brings home
a one-day-old wild boar piglet.

Wilczek's photographs are compulsive. I download as many as I can
find. Meet Korasek. Ruler of the roost, possibly the biggest character of
all. A smarter, more scheming, more guileful, treacherous, magnificent,
mercenary raven one could not imagine. Here he is with his glittering
eye, his coal mirror wings slicking the light, his beak ready to steal. Look
at him upside down on his back, feet akimbo, rolling in a nest of grass

with a flower in his beak, eyeing you. All wickedness and anticipation. Rough him up a bit, try to steal the flower, and he will fly away, *Cwvwarrrk! Cwvwvwwrk!* Then back for more. Korasek tugging Simona's plaits. Korasek riding on the back of Lech's motorbike. Korasek pulling the dog's tail. Korasek pulling the boar's tail. Korasek untying Lech's shoelace. Korasek dive-bombing a stork. Korasek's beak latched onto the rear end of a white chicken. Korasek peering over Simona's shoulder as she fixes her moped. Korasek flying straight at the camera, eyes blazing, beak zeroing in. Hitchcock would have sold his cine-arm for Korasek. I peer into his staggering eye. He has the gait of a comedian. In courting mood he ruffs his dashing black coat into a feathered boa. Cute as Satan.

The boar is called Zabka. Simona lies in the long summer grass, her head resting on Zabka's shoulder, her palms cupped beneath her snout. Zabka wears a smile on her bristly face. The sun warming their faces. Sweet dreams. The eternal lie of Time. This was once. This is no more. The mingling musky humus smell of boar, girl's hair and leaf litter. Sprite, goblin, sylph. A girl asleep in the forest with a boar. These damn photographs.

For two years Simona has been on my mind. Now here we are. In the forest garden where the badger cubs played, where Korasek chased the donkey, where Zabka snoozed, where the fox curled in the sun, where elk came, where deer wandered as if humans did not live here. Simona died in 2007. Lech died in December, 2018. I baffle my face to the window; I can see the dusty wallpaper, the nails where coats once hung, the door connecting the apartments which Lech put in after years of living together.

Tomorrow we will meet Simona's niece, Joanna. I feel a nervous anticipation. I sit crosslegged on the grass by the boulder and cup my hand over the egg stone.

The evolutionary biologist E.O. Wilson used the word biophilia, love of life, to describe what he believed is our deep connection with the natural world and its other inhabitants. An innate and fundamental affinity rooted in our distant evolutionary past, when our existence alongside other creatures was closer and more intimate. It manifests in our desire, or need, to watch and be close to other forms of life, and is rewarded by uncomplicated feelings of peace, joy and well-being. It is the motor and the heartbeat of our story.

★

Let us go to one life-changing day in 1993, outside the small village of Brodski Varos in Croatia, when a retired janitor, Stjepan Vokic, found a female stork with a wounded wing. She had been shot by a hunter and survived but was unable to fly. Stjepan took her home and named her Malena. He caught fish to feed her. He built a summer nest around his chimney for her, and a winter nest in his garage. Stjepan and Malena went for walks and drives together, and down to the river on fishing trips. Stjepan's days began to revolve around Malena. They became attached. 'The stork is my whole life,' Stjepan said. He worried for her stork loneliness. For 10 years they lived together, Malena sitting in her nest on Stjepan's roof each summer, migrating into his garage each winter. Then in 2003, out of the spring sky, a male stork flying over Brodski Varos came down and landed beside Malena's nest. Stjepan watched Malena and her visitor bend their necks back in courtship to each other. Stjepan named the male stork Klepetan. Malena was soon sitting on a clutch of eggs with Klepetan flying off and returning with food. Malena and Klepetan reared three chicks. At the end of the summer Klepetan flew 13,000 kilometres back to South Africa. The following spring Klepetan returned, and they raised another family. Every year Klepetan migrated to South Africa for winter, and every spring Stjepan and Malena waited for him to return.

When Klepetan left, Stjepan said Malena mourned for 10 days. After a few years of the couple's growing fame, the whole of Croatia began to wait for Klepetan's return, watching a livestream video of Malena's nest. In 2018 Klepetan was three weeks late. When he arrived, millions of followers rejoiced. Klepetan had to negotiate a route on which 2 million birds are killed every year by hunters' guns. The thought that Klepetan might not return scared Stjepan, he feared he would hardly be able to take the emotional strain.

Despite rumours of his death, Klepetan returned to Malena in April 2020. Malena died at a grand unknown age in 2021.* In her long life she and Klepetan raised 66 chicks between them. In their different ways the three of them, Stjepan, Malena and Klepetan, have raised empathy and a nationwide awareness for the protection and possible plight of migratory birds.

★

Dragonflies are among the fastest and oldest insects in the world. Helicopters eat your hearts out. A dragonfly's agile aeronautic prowess has not been matched, fast forward, backward, up, down, cruise, hover, thrust, quick turns, combat, capture, patrol. They have been clocking in for 300 million years, but like everything else, they were being squeezed out in the endless shrinkage of places for wild things to be. In 2014, Jonathan and I went to a talk by the dragonfly enthusiaser Ruary Mackenzie Dodds, about how his life changed the day a dragonfly landed on him. He didn't particularly like insects, but then he looked at it. What he saw utterly astonished him. The eyes, the abdomen, the markings, the colour, the wings, the legs. The emissary from the order Odonata had perspicaciously chosen the man who would devote his life to their service. In 1989, Ruary began work on a lake in Northamptonshire to create the first dragonfly sanctuary in Europe. In a few years the number

---

* Storks can live for 35 years. The oldest recorded wild stork was 39 years old.

of dragonfly species at the sanctuary had increased from 5 to 16, while his dragonfly museum in the nearby mill attracted 22,000 visitors. One sanctuary followed another. He'd cadge any pond. Clear out the omnivorous, water-muddying carp. Introduce native water plants. Ruary's enthusiasm was infectious. We wanted to *do* something to help wild things too. We could buy a bit of land and dig a lake and plant native species and have dragonflies. So we did. We found 22 acres of undulating unproductive clay on the Wiltshire/Dorset border. There was a small stream, an area of marshy sedge, some big oaks and some recent plantings of ash, lime and field maple – with the ubiquitous plastic tree guards – from a forestry grant scheme. We took the forestry saplings out of the sedge marsh, fixed up the barn for talks, put up a barn owl box, and bat boxes, and a log hive for wild bees, and so began our own back-to-the-wild, life-changing venture at Underhill. Soon to see, when you restore habitat for one species, a host of others will come.

All-consuming

Bill Hamilton, a visionary British biologist, credited his capacity to formulate theories about the world to an obsession with natural history which began in childhood. 'Incurably fascinated' by insects, his early specialism had been colony-forming insects such as ants and bees, and how natural selection acts on social behaviour. Later, he turned his attention to the interconnection of living things and their influence on the global self-regulating organism that James Lovelock called Gaia. Intrigued by the curious phenomenon of extreme weather events that lifted organisms high into the atmosphere – that caused herrings to rain in Scotland, frogs to be found in American hailstones and fungal spores to be discovered 50 kilometres above the Earth's surface – he began to investigate how the Earth's atmospheric convection might actually be aided by the plankton, bacteria and microorganisms from the ocean surface that are carried high into cloud formation – to then fall across the planet's surface and infiltrate every available habitat.*[4] Hamilton's work was cut short when he died in London in 2000, aged 63, from complications after catching malaria during a field trip to the Congo. Among his papers was

---

* Hamilton speculated that various species might help to produce winds, others ice nuclei, which drop the microorganisms back to the surface. Thus side effects could arise from activities that are adaptive for dispersion.

one entitled 'My Intended Burial and Why'. In the Amazon rainforest he
had watched the enormous *Coprophanaeus* scarab beetle – size of a golf
ball with iridescent violet wings – dispose of a carcass:

> I will leave a sum in my last will for my body to be carried to Brazil
> and to the forests. It will be laid out in a manner secure against the
> possums and the vultures, just as we make our chickens secure; and
> this great Coprophanaeus beetle will bury me. They will enter, will
> bury, will live on my flesh; and in the shape of their children and
> mine, I will escape death. No worm for me nor sordid fly, I will
> buzz in the dusk like a huge bumble bee. I will be many, buzz even
> as a swarm of motorbikes, be borne, body by flying body out into
> the Brazilian wilderness beneath the stars, lofted under those beautiful
> and un-fused elytra which we will hold over our backs. So finally I
> too will shine like a violet ground beetle under a stone.[5]

It was never going to happen. Think of the paperwork, the logistics.
However. In the churchyard at Wytham village near Oxford, where he
lived, his Italian partner Luisa has had a bench inscribed with the words
she spoke beside his grave:

BILL. NOW YOUR BODY IS LYING IN THE WYTHAM WOODS,
BUT FROM HERE YOU WILL REACH AGAIN YOUR BELOVED
FORESTS. YOU WILL LIVE NOT ONLY IN A BEETLE, BUT IN BILLIONS
OF SPORES OF FUNGI AND ALGAE BROUGHT BY THE WIND HIGHER
UP INTO THE TROPOSPHERE, ALL OF YOU WILL FORM THE CLOUDS
AND WANDERING ACROSS THE OCEANS, WILL FALL DOWN AND FLY
UP AGAIN AND AGAIN, TILL EVENTUALLY A DROP OF RAIN WILL JOIN
YOU TO THE WATER OF THE FLOODED FOREST OF THE AMAZON.

In other words: back to the wild.

II: GOD DAMN

What a piece of work is a man! How noble in reason!
How infinite in faculty! In form, in moving, how express and admirable!
In action how like an angel! In apprehension how like a god!
The beauty of the world! The paragon of animals!

William Shakespeare, *Hamlet*

DESPOT OR SHEPHERD

Welcome to a moment of my six-year-old consternation. The first colour illustration in my King James school bible. Two men each tend their burning pyre. The pyre in the foreground is a mess. The stone altar is roughly built, and the straw kindling ragged and damp – responsible for the billowing smoke. This pyre is tended by a dark, shaggy-haired man in a shaggy, black fur loincloth; he twists away, his bearded face in anguish. Behind him, a clean-shaven Adonis with shining blond hair, wearing a smooth, pale fur loincloth, raises his arms to the golden dawn, the smoke plume of his fire plaiting its neat way straight up to heaven. Look closer for the shocking revelation: a baby lamb, his little legs sticking out among the straw of the tidy fire.

God, inexplicably, is pleased with Abel's burning lamb, and makes Cain feel bad about the measly fruit he is offering. It seemed deeply unfair. For the lamb, for the mother of the lamb, and not very fatherly of God to favour Abel, or reward killing. I was Cain, of course, getting smoke in my eyes, the dark-haired child in a blond family, a bit scruffy, in trouble at school, having done the wrong thing, having not pleased my elders, and incensed at the injustice of it. I did not like the blond man at all. That the first blood to be spilled on Earth was to *please* God? Was God going to eat the lamb? I didn't think so. If the Garden of Eden had been an idyll of peaceful coexistence, why now was he encouraging this? What did God really want? I understood that these stories were parables supposed to help us make sense of the world but I didn't get the message. Cain was a 'tiller of the ground'. What was he supposed to do, steal a sheep?

There was a lot of trickery in Eden. This paradise, where boughs were heavy with fruit, where birds sang, where animals did not run away and everything was fecund, naked, unashamed — and vegan — had one rule: Don't eat from the tree of knowledge of Good and Evil. Just don't. But God had already created the naughty serpent. We know what happens. (And to be nit-picky, if Adam and Eve didn't *already* know the difference between good and evil, then how could they have been beguiled?) Adam and Eve are banished from the Garden in garments of skins. Where did they come from? The astonishing thing is, screwing up in Eden has been the dark tale of our wretched condition for more than 2,000 years. Bad for women, bad for men and very bad for snakes. It has survived even our belief in it.

From the fruitful garden with friendly beasts, we were banished to a stony land of thistles (good for goldfinches) to till the earth. And then we died. Nevertheless, we had been created in God's image and had dominion over the fish, the birds and every living thing 'that creepeth upon the earth'. Who were now wild and unruly and lived in the wilderness.★ Our orders were to subdue them and multiply ourselves. For ancient scribes in the fifth and sixth centuries BCE, the Holy Land might well have seemed in need of subduing for it was not the parched landscape it is today, but a verdant 'wilderness' teeming with lions, wolves, bears, leopards, hyenas, oryx, eagles, vultures, storks, ostriches, crocodiles and hippopotamuses.

Here's a vision of desolation from the enraged Old Testament God:

no one shall pass through it forever and ever. But the pelican and the porcupine shall possess it, also the owl and the raven shall dwell in it. And he shall stretch out over it the line of confusion and the stones of emptiness . . . And thorns shall come up in its palaces, nettles and brambles in its fortresses; it shall be a habitation of jackals, a courtyard for ostriches. The wild beasts of the desert shall also meet with the jackals, and the wild goat shall bleat to its companion; also the night creature shall rest there . . . There the arrow snake shall make her nest and lay eggs and hatch and gather them under her shadow; There also shall the hawks be gathered. Every one with her mate.[6]

★ The word wilderness comes from the Norse *will*, meaning that which is unruly, wilful, and the old English *déor*, meaning beast. Wilderness was the place of beasts beyond human control.

Wilderness was a curse. (Now we must bring it back for all its free ecosystem services.)

After much begetting our numbers multiply, but then God sees 'the wickedness in man's heart' and regrets creating us (QED my point about rewarding Abel's offering). His solution is to send the rains to wipe us off the face of the Earth, except Noah and his family (big mistake, God), and a seed bank of creatures. When the flood subsides, Noah thanks God with burnt offerings of beast and fowl. God smells the 'sweet savour' and promises never to flood the Earth again. So what does Noah do? He plants a vineyard and gets drunk. A family punch-up ensues, followed, predictably, by lots of begetting. And off we go again. Was this another failed experiment for God? And with a climate crisis upon us, one he might be about to break his promise on? It's hard to know, with so many mixed messages, what this God wanted. These days it might be called an abusive relationship – trying to please the powerful controller and never knowing how things might go.

Then God says to Noah, 'the dread of you shall be upon every beast of the earth' and 'every moving thing that liveth shall be meat for you'. Thus the fate of the creatures is sealed. Old Testament scribes might not confer much, but as the smell of roasting flesh becomes increasingly pleasing to God, it is clear they all like a good BBQ. Indeed, meat swoons so permeate the Old Testament that passages are quoted today by the Texan meat industry: 'When the LORD your God has enlarged your territory as he promised you, and you crave meat and say, "I would like some meat", then you may eat as much of it as you want. (Deuteronomy 12:20).'

Farming and God became a cooperative that justified the gory sacrifice of calves, bullocks, goats, rams, turtle doves and a never-ending supply of lambs as prepayment for abundant harvests.

Piece of work

Most origin stories allow humankind authority over animals, and most peoples have degraded the land they have settled, and killed the creatures who lived there. Nevertheless, few stories have been so puzzled over or been so resilient as the Creation story in the Book of Genesis. Whether man is created in God's image or God in man's, a certain arrogance can be expected. Nonetheless, it's a breathtaking leap from 'subdue' to annihilate in nano-geological time. And by God's own admission, he was fully aware he left the fox in charge of the hen coop.

Animals fared better around Buddhist, Hindu and Jain communities, where the key virtue was *Ahimsa*, the principle of non-violence and respect for all life. From Sanskrit it translates as 'absence of injury' and dates back 4,000 years to the ancient texts of the Vedas (meaning 'knowledge'). Animals were kin, and in some cases, sacred. Lucky to be a sacred cow in India. The giver of the milk that nurtured the population. Mahatma Gandhi saw the cow as a 'poem of pity' who spoke through her eyes: 'you are not appointed over us to kill us and eat our flesh or otherwise ill-treat us, but to be our friend and guardian'. Which rather sidesteps the snag of male calves produced to keep the sacred cow in milk.* The Dao, or Way, that influenced life in China for more than 2,000 years considered that humans could learn from observing animals, who were closer to their innate being. Daoist precepts warn against taking all life, 'be it flying or merely wriggling'; or perverting an animal's natural state (like keeping them in cages). 'Do not in winter dig up insects hibernating in the earth', 'Do not startle birds or beasts . . .', exhort the tender precepts of Lord Lao.[7]

Jerks

There is a Gary Larson cartoon of God creating the world. God is at the kitchen bench with his chef's hat on, the corner of a clock is just visible. The plum pudding in the saucepan is not a pudding, but planet Earth with its clearly demarcated land and sea. On the shelves behind God are packets and jars of ingredients: 'Birds'; 'Insects'; 'Krill'; 'Med. Skinned People'; 'Light-Skinned People'; 'Dark-Skinned People'; 'Reptiles'; 'Amphibians'; on the bench beside the saucepan is a green packet labelled 'Trees'. The thought-cloud beside God's head reads: 'And just to make it interesting . . .' The salt cellar in his hand, with which he is just about to season his Creation, is labelled 'Jerks'.

---

*   Gandhi eventually abandoned drinking milk because of the cruel practices of the dairy industry.

SUGARCANDY MOUNTAIN

Four legs good, two legs *better*!

George Orwell, *Animal Farm*

In George Orwell's *Animal Farm*, the parable he wrote about farm animals who want to be free, the wise old boar, Major, has had a dream. Sensing the end of his life, he calls the animals to the big barn. There, in a rousing speech, Major lays down the stark reality of their lives of slavery that will end in horror before their natural span; he spells out the injustice of the contract: the theft of their young, the theft of their labour, their bodies, their lives. Then he prophesies that justice *will* come, and with those magic words, the dream of a better world enters the animals' imagination. I am not suggesting Martin Luther King was inspired by Orwell's boar, but the visionary dream of a future of freedom for the oppressed inspires revolutions. By casting farm animals (to play out the Russian Revolution), Orwell draws with uncomfortable clarity the tyranny of our pecking orders. Our own nature is reflected back to us and we are forced to gulp. In the real world, and indeed the world of *Animal Farm*, Good doesn't vanquish Evil and God doesn't sort it all out. Hope is preyed upon. Major dies three nights later, and the clever younger pigs craft his speech into an ideology, Animalism, which boils down to seven commandments.* A wild excitement sweeps the farm as the animals, desperate to escape their brute lives, become faithful disciples singing 'Of the golden future time'. However, the pigs' golden future has competition from 'lies put about' by Moses, a raven, about a place called Sugarcandy Mountain, a little distance beyond the clouds, just out of view. Sugarcandy Mountain is where all animals go when they die, where they can live freely with lump sugar, linseed cake and clover all year round. That it sounds familiar is its brilliance. But Major's dream is

---

* Those who have two legs are enemies; those with four legs or wings are friends; no animal can wear clothes, sleep in a bed, drink alcohol, kill another animal; and 'All animals are equal'.

perverted by the pig administrators for their vested interests. The power-hungry pig, Napoleon, takes the helm with his attack dogs and propaganda, and the pattern of power follows the same pattern as it ever did. Indeed, 'the golden future time' for Boxer, the horse in *Animal Farm*, will be the glue factory, while the raven's sugar mountain remains unburdened by proof, hidden behind the clouds. In *Animal Farm* it was not the original ideology that was a problem, but its perversion. We'll return to this farm.

It is hard to square the idea of an omnipotent, beneficent god with the suffering in the world. As the worldly paradise gets a pasting. Eden is such a seductive thing, and our punishment so harsh, the treadmill so eternal. The doctrine of redemption looks forward to a time of peace. A compensation later for the wretchedness now. That is the hardest sell for me. The lure .of a peaceable time in the future feels too close to having a fight now in order to make up afterwards. The pull of paradise, of course, is where the power lies. For animals, Christianity cannot make up its mind. I would *like* to think of the flogged horse, the poisoned wolf, the dancing bear, the poached rhino in a kinder place with meaning. If only I could. Just one more load, Donkey, God will look after you . . . later. There is something else that bothers me. The glimpse of a time when the wolf lies down with the lamb. How would the wolfness of the wolf be fulfilled? Her sharp-toothed earthy is-ness. No zizz, no heart-mouth, no life-spirit, for what *is* life without death but the tedium of angels? A Christian redemption somehow doesn't grab me. Love for the natural world with all its fascinations and struggles has diminished the lure of Eden. *This* is the world I want saved. For many Christians it remains that animals – unlike themselves – only have this mortal existence. Which is, paradoxically, where I am with it. For us *and* them. It is all I have. It is all the sparrow has. It is all the mouse has. It is all the wolf has. Let her, for *God's* sake, have it.

God Damn

A short drone-shot video on Twitter shows a calf housed outside in a veal crate. His body is under cover, but his head is sticking out through the bars. It is snowing. The calf stretches his head up to the sky, opens his mouth as his great lovely tongue winds out to lasso the snowflakes.

His moment, maybe his only moment, of earthly cowly pleasure. He takes it with his tongue.

Pecking orders

On the Greek island of Lesbos, wading in the warm rock pools of the Pyrrha lagoon, a man is transfixed by the different crabs, sea anemones, tiny fish, starfish and vast assortment of life. His mind is not constrained by one god rustling up creatures in a matter of days. It's 350 BCE. There are still plenty of gods and they are usually busy fighting among themselves. Aristotle is free, robe slung over shoulder, to ponder and observe. He collects specimens, cuts them open, inspects their anatomy. He watches creatures running, swimming, flying, chasing, hiding, eating, nesting, dying and giving birth. He begins to sort them into groups: with blood, without blood; two legs, four legs; fur or feathers; teeth or beaks; bones or cartilage, or neither; reproduction with live birth or egg (wet and dry). In the bloodless group (insects, spiders, scorpions, centipedes, crustaceans and cephalopods) he includes a curious class of plant-animal, baffled as he is by the sedentary nature of anemones and sponges. Aristotle orders the animal world into a hierarchy, his *Scala Naturae*, a narrowing ladder that climbs, getting warmer (and better), all the way up to Man.* He wants to ascertain facts in a scientific way about the animals he observes and so becomes the first zoologist in the Western world. His *Historia Animalium* (nine volumes) mentions a total of 560 species, with many eyewitness observations. The drawings we know he made are lost, but his description of a sea urchin's mouth is so accurate – he likens it to a five-sided lantern, without panes – that the organ is still known as 'Aristotle's lantern'. He recorded how an octopus can change colour, squirt ink and, most astonishingly, described the hectocotylus arm, with which the male literally hands over the sperm to the female, an organ ignored as fanciful until scientists in the nineteenth century discovered he was right.

Aristotle believed everything in nature had a purpose. That while the perfect structure of each species lent the greatest advantage to itself, nature had made all things to benefit the next rung up the ladder, to lead inexorably (and purposefully) to us.[8] To see the natural world as an ascending

---

* 'Man' indeed, ascribing them as hotter than women, and so the most perfect of all animals.

progression from the lower to higher orders became known as the Great Chain of Being. It is an outlook that persists deep in the human psyche and directs our attitudes (and language – spineless, bloodless) towards animals.

Aristotle lived in an age without microscopes, without knowledge of the circulation of blood, or the movements of the planets, or that the Earth was round. Of the few mistakes he made, the most curious was to maintain that men had more teeth than women. A lapse, perhaps, from not personally fact-checking the specimens.* Nevertheless, no one did better, scientifically, for *two thousand* years.

In the centuries that follow, medieval Christianity saw the Great Chain of Being as being decreed by God, ascending through the lower forms to man, then to angels, and finally to God himself, whose Word must be spread. The fate of animals would depend on interpretation. Enter Augustine of Hippo (CE 354–430).

A naked madman possessed by demons is running around, screaming. Jesus comes along and banishes his demons into a herd of 2,000 pigs on a hillside. The herd rushes over a cliff into the sea and drowns. The madman becomes miraculously 'clothed and in his right mind'. This gospel story clinches it for Augustine. For if Christ doesn't worry about animals, why should we? To care about drowned swine would be ridiculous. Aristotle had paved the philosophical way – nature was to serve us. Animals were irrational, therefore conveniently outside the moral realm. We have Augustine's teachings to thank for separating our (inferior) temporal body away from our (superior) eternal soul, which can transport our rational egos to an exclusive (animal-free) afterlife. But there can be no salvation for the ox or the sacrificed lamb because they have no soul.** Our way was up to heaven, theirs was down to earth.

Fast forward on the anthropocentric highway to St Thomas Aquinas (1225–74), another disaster for the animal world, fond of quoting:

---

*   Or, as it has been quipped, one that could have been avoided if only he'd allowed his wife to open her mouth.
** In early versions of Genesis 2, the Hebrew word *nephesh*, translated as soul, was used for both humans and animals. This lack of distinction between us seems to have been intolerable to subsequent translators, who transcribed the word – in regard to animals – more simply as *life*. A subtle difference that conveniently would be understood as a life without an immortal soul.

'Everything that moves and lives shall be meat to you.' Even kindness shown to animals was for man's benefit: treat them well and you get more out of them; treat them badly and sadistic ideas might warp your mind.★

Advocacy for kindness to fellow creatures under the Abrahamic god never really caught on. Judaism has a stab at trying to be thoughtful, by giving cows a day off on Sunday; not killing a cow and her calf on the same day (Leviticus 22:28); or the dark consolation of not cooking goat kids in their mother's milk (Deuteronomy 14:21). Which we can safely assume the ungrateful beasts did not appreciate. Not even the charity of St Francis of Assisi, patron saint of animals, was a free lunch.★★ Except at Christmas, when he recommended that the Nativity animals, the ox and the ass, receive extra rations of fodder of the best quality. And he did have a fondness for larks: 'If I were to speak to the Emperor, I would, supplicating and persuading him tell him for the love of God, and me to make a special law that no man should take or kill sister Larks . . .' They were to be given a hailstorm of grain at Christmas too.

The paragon of animals

Put me out in the woods on my own and I'd struggle to make a fig leaf costume. Humans were never the fastest, strongest or best suited to their environment. We can't fly, breathe in water or swim very well, our teeth are not very sharp, we can't echolocate or see in the dark, we have good-for-nothing fur, no feathers to shout about, and let's face it, we're not the most attractive creature on this Earth. No furry bambinos bounding around at 10 days old, either; irascible, bald, old man lookalikes who can't walk, can't talk, can't feed themselves. For years! It's a wonder we got here. Yet what sets humans apart obsesses us. How we are so damn smart. One could diagnose a superiority-inferiority complex.

The hierarchy of how we rank animals reflects how we rank human society. While Orwell used what we do to farm animals to expose what we do to ourselves (in this case, Stalin's tyranny in Russia), the extra kick in *Animal Farm* is that his chosen metaphor shines its light back just as

---

★ Those very reasons would become the basis for animal protection centuries later.

★★ When St Francis saved the wolf who was terrorising the town of Gubbio, the deal was he not only had to quit eating sheep and killing shepherds, but supply protection to the town. Francis got his sermon too, reminding the townsfolk how much they should dread the jaws of hell, if the tiny jaws of a wolf so terrorised them.

strongly on its synonymous self. For the farm animals could just as easily
be representing themselves. The fable is all too recognisable as being not
as if, but as how it is. Where we are baldly reminded that we take not
only mother's milk, but mother's child. The extreme end of a continuum
of the same apparatus of supremacy in the history of the subjugated. In
Animal Farm, the corrupt Napoleon removes rivals, makes false promises
and sells 'the farm' down the river. Now who *else* could that be? It is
remarkably fresh and apposite. As Napoleon's behaviour becomes ever
more human, so does his upright posture.

We are in the Transylvanian town of Sibui, standing in the Lutheran
Cathedral of St Mary, where I'm looking at a stained-glass window of St
George and the dragon. Just a normal dragon minding his own business,
miles from the town, not bothering anyone, in a modest rocky cave, doing
what dragons do, a bit of fire-breathing at the entrance of *his* cave, that
kind of thing. When along comes a white knight (they're always white)
on a white horse (ditto), both dressed, we should point out, not in his
own scales but *blacksmith*-forged armour, with a *blacksmith*-forged lance,
ditto shield, so hardly a fair contest. The knight, George, raises his lance
and smites the dragon right in the mouth and down his throat! I appreciate there is the quest of saving a maiden, but she might well have been
in the process of talking to the dragon and quietly persuading him to let
her go. In the better-known representation of Paolo Uccello's painting,
she certainly looks unharmed. Not a hair out of place. Isn't that just a
typical first response? Extirpate the dragon. And probably the last dragon
in the land. That's men for you. No thinking this whole thing through
to see if there is a smarter solution, or employing the precautionary
principle in case the dragon plays an important ecological role that we
don't yet know about, or considering that the dragon might have some
other value and killing it might be killing the golden goose, like a tourist
opportunity for the town. The charge that it has 'envenomed the country'
sounds like the same old tiresome rumour-mongering that we hear about
the wolf, or the fox, just being wolves or foxes. Okay, in some depictions
I do remember a few human skulls lying around, but the dragon really
shouldn't have been fed slaves in the first place; start supplementing a
dragon's diet and you could end up with all sorts of consequences you
hadn't bargained for. We mess with the food web at our peril. Take out
our apex predator and who is to keep the deer on the move? Which is

probably why the landscape is so bare. Keep a healthy population of dragons and the hills remain healthy, not too overgrazed, and the aspen groves thicken. Uccello's Italian landscape could do with a dragon. But George had to kill it. Furthermore, not content to slay the dragon there and then, he has to humiliate the beast by putting the maiden's girdle around his neck to lead him through the town. Which with some weird logic instigates the town's people to be baptised as Christians. Of course, George takes the maiden as his prize. Which sort of puts me in mind of religions which promise their martyrs virgins. But it didn't do George much good, because God later decided it was his will for George to die for his faith, so he had him tortured and then beheaded. Oh, and there was fire which came down from heaven. Okay for God, but not the dragon.

BACKWARDS INTO BESTIARIES

What an astonishing feat, to keep our knowledge of the natural world in total stasis century after century, while being surrounded by *real* live animals. And of course, dead ones. Apothecaries filled their jars with dried and powdered animal parts: paws, claws, ears, eyes, tails, teeth, skin, liver, gonads, entrails and excrement to cure our ills and keep the Devil away (who, of course, had animal horns and an animal tail). Natural history was imprisoned in the dark ages (and in some places, sadly for the rhino and pangolin, still is). Any interest in creatures that stirred the medieval mind was directed towards the moral order of God, strictly overseen by the Church as *the* institute of learning. Folklore and curious beasts were drafted in to provide stand-ins for Good and Evil and allegories of Redemption and Resurrection. Fabulous creatures prowled the pages of illuminated bestiaries – though some no more fabulous than the wondrous beasts wildlife documentaries bring into our sitting rooms in the twenty-first century. One bestiary influenced ideas about the natural world for a staggering millennium: *The Physiologus*, written in the second century CE by an unknown scribe from Alexandria, translated from Greek into Latin, and then into the many vernacular languages of Europe. The *Physiologus* (Greek for 'The Naturalist') dispensed the meaning of nature by preaching Christian morals in sanctimonious and superstitious stories against animals. Of all the slanders, our old friend the wolf came off worst. Hardly a coincidence that the Latin words for whore, *lupa*, and female wolf, *lupa*, are the same. In English law the Latin term *caput lupinum*, meaning 'wolf's head', was used in medieval England to refer to an outlaw whose legal rights had been removed and who could be slain without penalty like a wolf.* The net of hatred had transgressed the human–animal boundary to ensnare them both.

---

* First recorded in a law at the time of Edward the Confessor, 1042–66.

It's a small step from ascribing to a person the qualities of a beast to *becoming* the beast. As the medieval Christian Church scoured the land for heretics, the werewolf of ancient folklore was commissioned into reality. The Devil who hid in 'the flock', who shifted shape from human by day to wolf at night. In these harsh, bitter times of hunger, pestilence and fear, accusations spread fast. Werewolves could be identified by long, slanting eyebrows that met over the nose, by a long third finger, curled fingernails, low-set ears or a swinging stride. I know a few myself. Where once they could be exorcised with a potion (½ oz of sulphur, ½ oz of devil's dung, ¼ oz of castor★ in spring water), the Church began to burn them at the stake along with the witches.[9] In 1487, what the writer Barry Lopez described as possibly 'the most odious document in all human history' blew fresh foul air into ancient superstitions to whip up hatred and justify the persecution. The *Malleus Maleficarum* summoned Leviticus: 'I will send the beasts of the field against you, who shall consume you and your flocks.'[10] One might call it Medieval Popularism. The Inquisition did much to further the campaign to extirpate the wolf from the forests of Europe. And in the bargain purge undesirable humans, the simple-minded, the epileptic and disturbed.

The wolf, savage and lustful, would stand in for our darker depths, to conjure a violence and sexual depravity that had no reflection in any form of animal behaviour, aside from our own.

Going astray

Never was the battle of the Christian will against pagan wild nature more brutal, more visceral, more exhilarating, than in The Madness of Sweeney, *Buile Suibhne*, the ancient Irish tale of the Celtic king turned into a bird by a vengeful Catholic missionary.[11] Admittedly, Sweeney *did* throw the cleric's book of psalms into a lake and kill one of his monks who had sprinkled him with holy water.

I was in my twenties when I read Seamus Heaney's translation from the Gaelic, *Sweeney Astray*, and in my own exile in Ireland living on a cliff in a place as wild as I'd ever been. What fevered my young head, and still does, was the actual *becoming*. Sweeney's mind rends apart, his fingers stiffen and his arms begin to flap dementedly. In a wild flurry he

---

★ Fluid from a beaver's castor gland.

coils into the air. From human to wild creature. The hollowing of bones,
the toughening of skin, the contractions of harsh necessity. A raving owl-
eyed, sharp-beaked man-bird. A curse. We are taken to whistling branches
at the tops of trees where we must consider what it is to be wild and
are imagined into creatureliness. We fly from crag to brook to glen to
estuary to island. We perch with Sweeney in the howling gaps of tangled
branches. We tremble with cold. Hunger howls within us. We watch
suspiciously. We flee in fear at the slightest thing. Our world is no longer
the human realm, but the brute realm of bitter winds, of combing the
forest floor, of groping among bog-berries and roosting in ivy. That is
what got me. A king is emptied into a bird.

But here's the thing: while Sweeney pines for his lost kingdom and
suffers great hardship, he is nevertheless beguiled. The curse intended to
diminish him to a birdbrain instead awakens a raw elation in the power
of wild places, consolingly far away from 'the sheepish voice of the cleric
bleating out plainsong'. Sweeney's rebuke is the view of the gods. Flying
over an incantation of Irish mountains: Slemish, Cooley, Ben Bulben, the
Mournes and over the wave-smash, Sweeney tastes a pristine world of
cold rivers and moonlit nights. (All the more astonishing, for a medieval
bard could never have experienced a bird's-eye view.) Once a great hunter,
Sweeney now identifies with wild creatures, preferring the squeal of
badgers 'to the hullabaloo of the morning hunt' and 'the belling of a stag
among the peaks to that terrible horn'. He knows the old matriarch of
one deer herd and claims his roost 'among her mazy antlers'. This is not
medieval nature from a moralising Christian bestiary. This is nature penned
by an intimate, exposed in a land of deer, fox, hawk, dove – and wolf,
with their 'low-slung speed' and the howling of their 'vapoury tongues'.
You cannot describe that without knowledge of them. This was an Ireland
with wolves.★

Yet the punishments inflicted on Sweeney reflect the types of penance
required by the Church. He is 'wind-scourged', stripped, 'clad in black
frost', flailed by thorns. He crouches under thunderstorms, the weather
withers him, blizzards bury him, the wind winnows him down, he is
'crucified in the fork of a tree'. Though his feathers represent the freedom

★ Ancient Celtic versions of *Buile Suibhne* go back to the ninth century. The last wolf
in Ireland was reported in 1786, 300 years after their extermination in England. In
the seventeenth century one of Ireland's nicknames was 'wolf-land'. Cromwell thought
wolves such a problem he banned the export of Irish wolfhounds.

of flight, they are also the curse of the hairshirt. Our imagination is strung between rapture and misery. There he is again, that punishing benevolent god. Being human in a bird world stirs our admiration of how hard it is to be wild. The icy wind picking through our feathers. We are too weak to be wild. Sweeney laments a woman gathering cress (his only available sustenance) without considering his plight. If she left the cress, she would not have been poorer, yet he will starve the night in the sharp wind, 'Woman, you cannot start to know,' he says. A cleric, with the common resentment towards wild creatures helping themselves to free board and lodging, scolds him for stealing cress and drinking from the church well. Sweeney flies back into the tree. Terrified. Panicky. He dare not blink. Sweeney is ultimately hobbled by his suffering. In a churchyard he laps the milk left for him in the scoop of a cowpat made by the heel of the cook's shoe. Her jealous husband, the swineherd, smites him (as fate has decreed) with a spear. From life to death the Church gets their man. No clifftop demise, no hollow bones turning to flutes in the wind, Sweeney's mortified body (bird? man?) gets a Christian burial and his soul flies to heaven. Of course it does, for this is medieval literature. In counterpoint to Sweeney's embodiment of a bird, I wonder, where is the real estrangement? Our pious refusal of our own animalness? Or the strangement of setting oneself apart? Sweeney roosts in the tree of my imagination, but in life, as a bird. He takes me to the owl on a cold night, to the red kite's hollow-boned frame in the cold dawn, to the kestrel's sharp gaze on the thermal. Ever since we met, Sweeney reminds me, I could hide among the twisted branches of a tree and 'never come out'.[12]

★

The nature of belief is that it is not about facts. That we will die for a belief, and not a scientific conclusion, has depended on the promise of heaven and fear of hell. Beliefs involve huge emotional investments, life-long commitments, and are hard to change. Nevertheless, the plight of animals has been helped by Jorge Mario Bergoglio, the Argentinian who upon becoming pope in 2013 took Francis, patron saint of animals, as his papal name. (More helpful than Pope John Paul II, who, although declaring St Francis the Patron Saint of Ecology in 1979, suggested St Francis set Christians an example by inviting the animals of Creation to praise the Lord.) In 2015 Pope Francis wrote an 80-page letter, *Laudato Si': On Care for Our Common Home,*★ in which he informed the world's 1.2 billion Catholics that animals *will* join humans in the kingdom of heaven. In other words, he allowed them a soul. 'We must forcefully reject the notion that our being created in God's image and given dominion over the earth justifies absolute domination over other creatures.' Then, in 2019, Pope Francis went further and declared ecocide a sin, calling for it to become a 'fifth category of crimes against peace' at the international level.[13] Amen to that.

Walking on water

The Central American basilisk 'Jesus Christ' lizard is around 2 feet long, mossy-green, with a yellow eye and dragon frills along his head, body and long, striped tail. He lives in the rainforests beside rivers. If threatened, he dives out of the trees, hits the water running and tears across the water at 5 feet a second. His long toes have skin flaps that unfold in the water and smack the surface, creating air pockets as his legs whizz round in a pinwheel motion whisking up air bubbles to keep him from sinking.

Water striders are insects who have been walking on water for millennia. They sweep their middle legs in a circle like oars to propel themselves forward in bursts. Their hydrophobic (water-repelling) legs have tiny hairs that capture air and help increase the surface tension on the water, creating a delicate membrane on which they walk.

The raft spider's hairy legs also repel water. They dash across the surface of a pond to grab small prey such as fish. Sometimes they put a couple of legs in the air to act as sails. They can stay under water for half an hour surrounded in trapped air bubbles in their hydrophobic hairs.

Male and female Clark's grebes run across water together in an exuberant mating ritual at 22 steps a second.

★   The title, meaning 'Praised Be You', comes from St Francis's 'Canticle of the Sun', also
    known as *Laudes Creaturarum* – 'Praise of the Creatures'.

DIVIDE & RULE

In a long-forgotten Chinese encyclopaedia entitled *Celestial Emporium of Benevolent Knowledge*, the estimable Jorge Luis Borges was struck by a classification of the animal world divided into 14 categories. Outlandish as the taxonomy seems: the 'embalmed'; the tame; the crazed; those who 'look like flies' at a distance; those 'belonging to the Emperor' – and that it might just as easily have included: the injured; not mice; those with runny noses – we might muse under the Borgesian lamplight how absurd or arbitrary is *any* system that attempts to pigeonhole the world.

Keggie Carew, *Beastly*

In 1607, almost 2,000 years after Aristotle described the octopus's sperm-presenting arm, the good clergyman Edward Topsell, in his *Historie of Foure-Footed Beasts*, divided animals into three categories: edible and inedible; wild and tame; useful and useless. Weasels give birth through their ears; dragons are partial to lettuce; wolves eat their prey in cold water; and true toads have a toadstone in their heads which protects humans from poison. Not great for the toads, and a universe away from Aristotle accurately recording a female wolf's gestation period at 59–63 days, and that her pups are born blind. The road from science to religion to science was, for animals, a miserable ride. Bestial nature was brute in behaviour and loose in morals. Animals could be cowardly, treacherous or noble. They could smell foul and sound hoarse, but can't we all?

And all the while we had rigged ships; sailed the seas; built pyramids; made gunpowder; fired muskets; we had invented the compass, the mechanical clock and spectacles to see with . . . indeed, in 1610 Galileo will point his telescope at the night sky. Men of science began to wrest ideas away from the Church but provided no better fortune for animals. Perhaps it was not so surprising, in 1637, an age beginning to tick-tock with the new mechanical marvels, that the renowned French philosopher and scientist René Descartes dismissed animal behaviour as no different from the workings of a clock.[14] Animals were complex automata, so if they yelped when you kicked them it was because of a conditioned reflex. Animals had neither language, nor intelligence, nor feelings, nor reason. All mental activity was located in the mind, which was located in the incorporeal soul — which animals did not have.* Animals acted purely out of instinct; they could not fully experience pain or pleasure or know anything.** No mind, no soul. What a convenience. It was Descartes' followers who nailed live dogs to the dissection table and heard their howls as the screeching of gears. In 1674, ardent disciple Nicolas Malebranche (whose name incidentally means Evil Claw) wrote ecstatically: 'They eat without pleasure, cry without pain, grow without knowing it; they desire nothing, fear nothing, know nothing.'[15]

The consequence to animals was a head-on collision into the widening wall that separated the human–non-human boundary. What gave it traction was our egos. How agog we were to hear about the great divide between special us and inferior them. But the appeal was far from universal. Voltaire, in 1733, thought it pitiful, and enough to observe a dog's joy at the return of his master as empirical proof against Descartes. Why, he wanted to know, would nature arrange the very same organs of feeling, revealed beneath the vivisectionist's knife, in order *not* to have feelings? Thankfully, we did not really take to the beast-machine, and it never settled into our consciousness.

*

I post you, dust of Descartes, the video of a young, 12-metre humpback whale being freed from the stranglehold of a nylon gill net.*** It is Valentine's Day, 2011, off the coast of Mexico in the Sea of Cortez, when

---

* Animals, innocent of original sin, did not need redemption so had no use for a soul. Nor would God give them one, because then they would be subject to pain and suffering, which is reserved for us, sinners with souls . . . so crunched the cogs of seventeenth-century God-fearing minds.

** The French verb *sentir*, to feel, can mean to feel passions, fear, anger, etc; or to feel sensations. Descartes denied animals passions, but not entirely sensations, which he saw as a mechanical process that produced reflex responses that excited the nerves, but without conscious awareness (letter to Henry More, 1649). So we can assume he did not walk his little dog, Monsieur Grat, of whom he was fond, to the vivisectionist's parlour.

*** A gill net is a mesh net designed to let the fish get his head through the netting but not his body, so catching him by the gills.

Michael Fishbach's small cruising boat approaches the whale, who appears to be dead. Neither he nor his crewmates see any signs of life until she rises in the water and exhales. Michael goes overboard in his snorkel and can see that the whale is severely entangled in fishing net. As he floats beside the animal their eyes meet. Michael is scared because he knows the whale is frightened and tired, and he fears the force of her sudden panic. He wants her to understand they are going to try and help. The whale is tangled so badly and unable to move that she is being pulled down 15 feet below the surface. Both of her pectoral fins are pinned to the side of her body. You can make out the floats of the gill net against her shiny skin. We hear her bursts of breath. Michael manages to get the whale's dorsal fin free, but the rest of the knotted nylon is impossible to undo, and the whale is becoming increasingly distressed, so Michael swims back to the boat to radio for help. Help turns out to be an hour away, and the whale is ailing. Meanwhile, two of Michael's crewmates have managed to drag some of the net over the side of the boat and are cutting away at it with a small knife. They free a pectoral fin. You can feel the tension. The whale raises her blow hole out of the water, *phhwwooooh*. The crew believe she knows they are trying to help. Sensing freedom the whale tries to swim away, dragging the gill net and the boat behind her for quarter of a mile. She soon tires and they resume cutting. 'Haul and cut, haul and cut,' someone is saying. After half an hour the other pectoral fin is almost free, and they begin cutting the net around the tail fluke. She tries to sound. Even when you know the outcome it is tense. After an hour they think they have enough net on board to make the final cut. 'Cut, cut!' The whale expels air, her back rolls beneath the surface, you can feel everyone holding their breath. She's gone. The camera pans out across the sea. Like a great torpedo exploding out of the water, she breaches. We hear whooping and hollering from the boat, we see the crew's mouths open with disbelief. She flips over backwards. They swoon. She breaches again, this great magnificent animal, splashing over on her back. For the next hour she gives the boat an astonishing display, more than forty breaches, tail lobs and pectoral fin slaps.

I can watch this film again and again; it never loses its ecstatic wonder. It is *not* sentimental anthropomorphic imagination. It is sensitive creaturely imagination. The real emotional connection of concern for another being. The sea is white with the broiling foam of a humpback whale's freedom. She is slapping her tail fluke again and again and again. 'We all believed it was a show of pure joy, if not 'thanks,' said Michael. At the time of writing, the film has had more than 34 million views.[16]

Beasts of the breast

One momentous day in 1674, the startled eye of a Dutch draper hovered over the lens of his homemade microscope beholding 'very little animalcules' in a drop of water. It was the day that would, years later, reward Anton van Leeuwenhoek's fascination for optical lens-making with the title Father of Microbiology. He was to discover blood corpuscles, bacteria, protozoa and human spermatozoa.★ But it was a Swedish botanist, zoologist and physician who roused natural history from its slumber. In 1735, Carl Linnaeus published his first edition of *Systema Naturae,* in which he sorted the animal kingdom into different classes based on their physical characteristics. These groups (taxa) divided into classes, then orders, then genera and species. Shockingly for the time, he placed humans alongside apes, sloths (who do have remarkably human-like faces) and a dog-headed man (cynocephalus) in the order Anthropomorpha. The first edition was a bit wayward: satyrs crept into the class Paradoxa, along with deathwatch beetles; hippos and shrews were classed as beasts of burden (Jumenta); and in the ferocious order (Ferae), lions and tigers were joined by hedgehogs, moles, armadillos and bats. But Linnaeus persevered. By 1748 Paradoxa was out, and by the 10th edition (1758) the new term, Mammalia, to replace Quadrupedia, was in. We were now animals 'of the breast'.

Linnaeus might have chosen a different name by selecting a different defining feature of the class.★★ The hairy ones, maybe. Yet, he chose *mamma,* Latin for breast, when strictly speaking it is the gland, not the organ (which the platypus does not have), that is the defining feature. By electing this prominent feature, he most explicitly included us, our qualifications impossible to refute. That was a big deal. But where

---

★   Ironically, the microscope did not scotch the idea of 'spontaneous generation' but instead revealed new populations of tiny organisms that seemingly appeared spontaneously. All you had to do to make some 'animalcules' was add a little hay to water, then wait a few days to examine your creation under a microscope.

★★ The six mammalian features are: the mammary gland; an articulated jaw; three ear bones; an aortic arch that bends to the left; cheek teeth with divided roots; and either hair or fur.

Mammalia spotlights the female, the name he gave our species, *Homo sapiens*, could barely sound more male. Female breasts would join us with the other animals, while the 'wise man' would stand alone.* Linnaeus had put us in the animal kingdom and managed to separate us at the same time.

Linnaeus's great legacy was his binomial naming system, which gave each species a universal scientific label – so we know exactly who we are talking about.** Each species was assigned a two-part name in Latin, the first part (capitalised) indicated genus, as in *Homo*; the second part indicated the species name, as in *sapiens*. In this way we can tell that *Mustela erminea* (stoat) is related to, yet distinct from, *Mustela nivalis* (weasel). Ditto *Panthera lio* (lion) and *Panthera onca* (jaguar). The code helps us understand family relationships, a bit like our surnames, and since Linnaeus's day has become refined and expanded, though the method remains the same. By 1766 his 12th edition of 2,400 pages identified almost 4,400 animal species. By grouping animals with similar features together, Linnaeus demonstrated (unintentionally) the distant relationships that would reveal the clues to their evolutionary past.

Naming new species is given serious thought and can confer great honour. In 2017 a new moth species was discovered in southern California with a remarkably prominent 'hairpiece' of feathery blond scales on its head; it was named *Neopalpa donaldtrumpi*. Donald's friend, Greta Thunberg, was chosen for a new species of beetle less than a millimetre long, *Nelloptodes gretae*. Michael Darby, the Scientific Associate at the Natural History Museum in London who honoured her in naming the beetle said, 'The family that I work on are some of the smallest known free-living creatures. They are not parasitic and are not living inside other creatures. Few of them measure more than a millimetre long.'

*

---

* Linnaeus had been impressed on his travels to Lapland by the rude health of the breast-fed babies compared to the progeny of upper-class women who employed wet nurses – which he believed violated the laws of nature. His advice to human mothers was to look at nature, where no wild animal denies newborns her breast milk.

** And to avoid confusion where common names for the same creature can vary from place to place, in the way a mountain lion can variously be called a cougar, a puma or a panther.

In the quiet Hampshire parish of Selborne, a clergyman (another one, they seem to have all the time) is finding delight in the meticulous study of the natural world around him. His name is Gilbert White and he is, without knowing it, England's first ecologist. For White is an 'outdoor naturalist' who prefers to observe animals alive, unlike his contemporaries who opt to shoot and dissect dead specimens, or as the writer Richard Mabey has pointed out, sometimes not bothering to shoot them first. White observes how animals behave and writes about how they relate to each other. His letters and journal jottings that form *The Natural History of Selborne*, published in 1789, have never been out of print.

Aside from famously wondering if swallows hibernate underwater, White discovers that worms are hermaphrodite; that male and female chaffinches join separate flocks in winter; that owls hoot in B flat and cuckoos in D; that swifts mate on the wing. He notes how small insects rising from the dust of galloping horses attract swallows. White gives us glimpses of a Britain unblemished by the Industrial Revolution that was about to let off steam.

> White owls seem not (but in this I am not positive) to hoot at all: all that clamorous hooting appears to me to come from the wood kinds. The white owl does indeed snore and hiss in a tremendous manner; and these menaces well answer the intention of intimidating: for I have known a whole village up in arms on such an occasion, imagining the church-yard to be full of goblins and spectres. White owls also often scream horribly as they fly along; from this screaming probably arouse the common people's imaginary species of screech-owl, which they superstitiously think attends the windows of dying persons. The plumage of the remiges of the wings of every species of owl that I have yet examined is remarkably soft and pliant. Perhaps it may be necessary that the wings of these birds should not make much resistance or rushing, that they may be enabled to steal through the air unheard upon a nimble and watchful quarry.[17]

There has been a shift. Nature is being seen as benign and endlessly fascinating, instead of something to be subdued or tamed, or worse,

as our adversary. For Gilbert White's clear and curious eye has seen a
community of creatures with interdependencies, that included us.

Though still ruled and very much divided, the next person who will
drag us into the future is a German by the name of Alexander von
Humboldt.

THE MAKER & THE MURDERER

Certain germs, falling upon water, become duckweed. When they reach the junction of the land and the water, they become lichen. Spreading up the bank, they become the dog-tooth violet. Reaching rich soil, they become wu-tsu, the root of which becomes grubs, while the leaves come from butterflies, or hsü. These are changed into insects, born in the chimney corner, which look like skeletons. Their name is ch'ü-to. After a thousand days, the ch'ü-to becomes a bird, called Kan-yü-ku, the spittle of which becomes the ssǔ-mi. The ssǔ-mi becomes a wine fly, and that comes from an i-lu. The huang-k'uang produces the chiu-yu and the mou-jui produces the glow-worm. The yang-ch'i grafted to an old bamboo which has for a long time put forth no shoots, produces the ch'ing-ning, which produces the leopard, which produces the horse, which produces man. Then man goes back into the great Scheme, from which all things come and to which all things return.

The Book of Chuang Tzu[18]

Soul to the Devil

Doctor Faust, astrologer, physician, philosopher, magician, necromancer, has been the subject of legend since his death in 1541 when his body was found 'grievously mutilated' after an alchemical experiment exploded. His clerical enemies put it about that the Devil had come to collect him in person. The motif of bargaining one's soul to the Devil for infinite understanding had Christian ballads damning Faust for valuing human knowledge of the world over the divine. When the poet, polymath and playwright Goethe gave Faust the words, 'That I may detect the inmost force / Which binds the world, and guides its course' it might, as Andrea Wulf points out in *The Invention of Nature*, have been Alexander von Humboldt speaking, Goethe's young friend. Until recently Humboldt (1769–1859) had been 'the great lost scientist'. Now, thanks to Andrea, we have him back. The revolutionary idea that shaped his understanding of the natural world was the realisation he reached on his expeditions that everything seemed somehow connected. And what journeys. In 1800 he and fellow explorer Aimé Bonpland arrived at the vast South American tropical grassland plains of the Llanos, south of Caracas. Humboldt was struck by how much life gathered around the tall, solitary Mauritia palms – birds fed on their fruit and their fanned fronds shaded the wind-blown soil that collected around their trunks, keeping in moisture to provide perfect conditions for insects and worms. Each tree created a community of life. Humboldt began to see nature as a dynamic living organism, and with that he was more than a hundred years before his time.

As they were paddled down the 'Oroonoko' hundreds *and hundreds* of large crocodiles lined their route, so numerous they were never out of view. Grazing along the banks were huge herds of capybaras, enormous guineapig-like rodents, at 50 kilos or more – twice as heavy as a Labrador dog. The capybaras, escaping the river crocodiles, were as likely to run headlong into a jaguar's jaws. Humboldt wrote of river dolphins, enormous tapirs, thousands of flamingos and spoonbills, boa constrictors so big they could 'swallow a horse'. This was a thriving world of creatures, independent of man. At night wild pandemoniums broke out, sparked by jaguars hunting tapirs in the dark, scaring the monkeys sleeping in the trees and waking a commotion of birds. The abundance was, he noted, limited only by the pressure 'of themselves'. Humboldt rejected the man–centred conception of Nature of his predecessors. What he also observed was that rather than improving nature, man's interference most usually upset the natural balance. The indigenous tribes showed him how Spanish monks took *all* the turtle eggs they could find from the riverbanks for oil to light their makeshift churches in their remote missions, and how turtle numbers had fallen in consequence. On the Venezuelan coast he saw oyster stocks decimated by unrestrained pearl fishing.

How vivid is the impression produced by the calm of nature, at noon, in these burning climates! The beasts of the forest retire to the thickets; the birds hide themselves beneath the foliage of the trees, or in the crevices of the rocks. Yet, amid this apparent silence, when we lend an attentive ear to the most feeble sounds transmitted by the air, we hear a dull vibration, a continual murmur, a hum of insects, that fill, if we may use the expression, all the lower strata of the air. Nothing is better fitted to make man feel

the extent and power of organic life. Myriads of insects creep upon
the soil, and flutter round the plants parched by the ardour of the
Sun. A confused noise issues from every bush, from the decayed
trunks of trees, from the clefts of the rock, and from the ground
undermined by the lizards, millepedes, and cecilias. There are so
many voices proclaiming to us, that all nature breathes; and that,
under a thousand different forms, life is diffused throughout the
cracked and dusty soil, as well as in the bosom of the waters, and
in the air that circulates around us.[19]

What Humboldt saw in South America fortified his holistic appreciation of animals functioning together in a web of life. 'Nothing appears
isolated; the chemical principles, that were believed to be peculiar to
animals, are found in plants; a common chain links together all organic
nature.'[20] Humboldt's writing blazed with vivid descriptions that combined
his clear-eyed scientific instinct with an ability to transport readers deep
into these pulsating landscapes, where 'man is nothing'. It was Humboldt's
travelogue of this tropical South American journey, *Personal Narrative*, with
its otherworldly images of insects that glowed red light, dragon trees and
cave-dwelling nocturnal oilbirds, that fired the imagination of a young
Englishman who really did *not* want to study theology as his father wished.
His name was Charles Darwin.

As Humboldt peers over the side of the canoe into the Oroonoko
River with his great big thoughts, 4,700 miles away Reverend William
Paley's pen hovers over a sheet of linen paper. The English clergyman
and philosopher is pleased with himself. He has just come up with an
analogy to demonstrate the existence of a Divine Creator. He calls it,
'The Watch on the Heath' and it will be the ubiquitous lesson drummed
into Victorian English minds. It goes like this: You are crossing a heath
when your foot glances a stone. How did that stone come to be there?
It was always there, you surmise. But suppose you found a watch on
the ground. Why would you not think the watch had always been there?
Because it had been clearly made for a purpose – telling the time – and
so must have had a maker. So it followed, apparently, that the complex
structures that made up each creature or plant must also have been
designed for their purpose, by *the* Maker of course. To study God's
Creation is to get closer to God and becomes a devout practice. Paley's
book, *Natural Theology*, published in 1802, unleashes thousands of amateur

Victorian naturalists; ladies prod rockpools with fishing nets; boys trail meadows with butterfly nets; vicars crawl after beetles. William Bingley, cleric, ranks British songbirds in a hit parade of 'comparative merits' under the headings: 'Mellowness of tone', 'Sprightliness', 'Plaintiveness', 'Compassion' and 'Execution', awarding marks out of 20. At the top of the leader board the nightingale scores 19 in almost all categories; the goldfinch receives 19 for Sprightliness; while the blackbird's highest score is 4, with 0 for Plaintiveness, and only 2 for the Execution of his own song.[21]

This was the early nineteenth century young Charles Darwin grew up in. That science was the underdog to religion did not mean questions were not raised, or new theories about life's origins put forward. In 1794, in his book *Zoonomia*, Charles's grandfather, Erasmus Darwin, had written: 'Would it be too bold to imagine, that in the great length of time since the Earth began to exist, perhaps millions of years . . . that all warm-blooded animals have arisen from one living filament . . . possessing the faculty of continuing to improve by its own inherent activity, and of delivering down those improvements by generation to its posterity, world without end!' The concept of evolution was not new, just taboo in the Christian doctrine.

Thanks to Archbishop James Ussher, who in 1650 meticulously added up the lifespans of the Old Testament patriarchs, Christians knew that the Earth had been created at 6 p.m. on Saturday, 23 October, 4004 BCE. But this did not of course address the snag of geology, nor all those irksome fossils, metre-long thigh bones sticking out of bogs and the scimitar canines of dragons and giant monsters. It was hard for geologists to accept the Earth was only 6,000 years old. But what would have been the point of this pre-human menagerie if humans were not around to admire or make use of them? Extinctions implied God's Creation was imperfect if a creature was allowed to disappear. Maybe God had been honing his skills in a dress rehearsal for the opening night of the gala performance. French scientist George Cuvier's explanation for the fossil record promoted extinction from a catastrophe, like Noah's flood, neatly reconciling Genesis with geology.*

Then along came Charles Lyell's first volume of *Principles of Geology*,

---

* To account for so many fossils, Cuvier suggested there were four or five catastrophes following previous 'Creations', but as more and more fossils were found, more catastrophes had to account for them.

published in 1830, just in time for Darwin to pack in his bag with Humboldt's *Personal Narrative* before he boarded the HMS *Beagle* as the ship's naturalist at the age of 22.

The Murder

> Of bodies chang'd to various forms, I sing . . .
>   Ovid, *Metamorphoses*

Ardent beetle collector and worm worrier Charles Darwin had always been fascinated by the animal world. Determined to follow in Humboldt's footsteps, the opportunity came in 1831 when his father agreed to pay his passage as a naturalist on board the *Beagle*'s expedition to survey the coastlines of South America. The ship set sail from Plymouth on 27 December 1831; when his head wasn't buried in his pillow with seasickness, Darwin was immersed in the gradual geological processes of continuous change and slow action of natural forces (wind, rain, ice, glaciers, rivers and sea) that shaped the Earth's surface. If Lyell's principles of geology spanned such deep time, why not organic life? At Punta Alta in Argentina he chipped out a trove of fossil bones from a cliff, enormous versions of existing animals particular to the continent, including giant sloths as big as elephants – some of the largest mammals ever to have lived.

After almost four years at sea, on 15 September 1835 came Darwin's first sight of a remote sun-baked archipelago 600 miles west of Ecuador. The rugged volcanic Galapagos Islands, not long thrown out of the sea, were home to a unique community of reptiles. Giant land tortoises stared at him, then 'slowly stalked away'; the tameness of the birds and multitudes of 'hideous-looking' marine iguanas along the shores gave him the impression of an ecology at the beginning of time. Here was Darwin's key to the origin of species. Not that he realised it. The significance of the 13 different finches he collected, each adapted to a different ecological niche and diet with varying appropriate beak shapes, did not sink in until the ornithologist John Gould identified them in England as separate species, implying a common ancestor. And it was a throw-away parting remark by the vice governor of the islands, who said he could tell which island each tortoise came from by the difference in their shells, that Darwin later realised demonstrated the mutability of species. But it also, as he well understood, undermined the tenet of God's perfect design that held all creatures had been created in their present form. After 36 days of collecting specimens and surveying the coast, the *Beagle* sailed for Tahiti. Darwin, unaware he bore the evidence to unravel 'the mystery of mysteries'.

When the *Beagle* sailed into Falmouth on 2 October 1836, Darwin had been away almost five years. His father remarked on the increased size of his forehead. Ahead lay five years of careful reflection as bits of the puzzle gradually fell into place, and more than 20 years of painstaking work to provide the evidence and the mechanism to bear out his heretical theory of 'descent with modification'. From Charles Lyell came the necessary expanse of immense evolutionary time. Then Thomas Malthus's *Essay on the Principle of Population* (1798) supplied the crucial jigsaw piece to show how natural selection works across species: the competition for resources favour the fittest, who survive to reproduce. Darwin knew the implications. His devout wife, Emma, demonstrated the steadfast nature of religious belief. In a letter to the botanist Joseph Hooker in 1844, Darwin described his growing ungodly conviction: 'that species are not (it is like confessing a murder) immutable'.[22] His challenge was to demonstrate it.

A book published anonymously that year would jog him along. *Vestiges of the Natural History of Creation* rocked Victorian Britain by suggesting man was descended from the lower orders. The theory went that the Divine Maker had designed nature in his 'terraqueous theatre' to progress, so that primitive life arose out of a vague electro-chemical process to develop gradually, from fish to reptile to mammal, upwards to man, as 'the true and unmistakable head of animated nature upon this earth'. Couched in pious tone, *Vestiges* remained respectful to God, if not to Genesis. The mystery author was rumoured to be Prince Albert, while men of the establishment suspected a female hand behind it for the guile and 'hasty jumping to conclusions'.* It became an international best seller. The buzz made its way into Benjamin Disraeli's 1847 novel, *Tancred*, when Lady Constance explains how everything is proved by geology: 'First, there was nothing, then there was something; then – I forget the next

---

* In 1884 the author was revealed to be Robert Chambers, of the *Chambers's Encyclopaedia* family, twelve years after his death.

– I think there were shells, then fishes; then we came – let me see – did we come next? Never mind that; we came at last . . . Ah! that's it: we were fishes, and I believe we shall be crows.'[23]

★

In ninth-century Baghdad, the prolific Muslim polymath known as Al-Jahiz (CE 776–868) wrote books on many subjects: *The Book of Misers* about greed; *The Art of Keeping One's Mouth Shut*; *Against Civil Servants* . . . His *Book of Animals* (*Kitāb al-Hʾayawān*) ran to seven volumes. His name, meaning 'The Bug-eyed', referred to his protruding corneas. Maybe his humorous writing style was compensation for his affliction. For fear of boring his readers, wit, poetry and satire weave through the *Book of Animals*, alongside the more serious influence of Aristotle's zoological studies, sailors' anecdotes and his own observations. His death at his home in Basra at the grand age of 93 was reported as the consequence of being crushed by a toppling pile of books in his library. But this is what he wrote:

> Animals engage in a struggle for existence; for resources, to avoid being eaten and to breed. Environmental factors influence organisms to develop new characteristics to ensure survival, thus transforming into new species. Animals that survive to breed can pass on their successful characteristics to offspring.[24]

Pigeon club

In ramshackle sheds in back gardens across the land, Darwin discovered an unexpected community of devoted pioneers whose skill revealed more knowledge than most scientists about some of the finer vagaries of evolution. Darwin took up with the fraternity of pigeon fanciers. In 1855 he got his first domestic pigeons and within months he was hooked. Whereas natural selection took thousands of years, by manipulating evolution with *artificial* selection he could observe mutations and traits in just three generations. Every day he visited his pigeon loft, breeding ever fancier birds. He was fascinated how one species could produce so many dramatically different-looking varieties: Fantails, Nuns, Jacobins, Trumpeters, Crested Helmets, Giant Runts; Frillbacks with their fancy feather-dos of extravagant curls across their back; Pouters, who can inflate their crop into a ball; Ice pigeons, who look like they have wings on their feet. By pairing certain mutations he could maintain the type; in producing fertile offspring by crossbreeding he was able to show that all known pigeon breeds originated from *Columba livia*, the rock dove. Never slow to exploit a resource, 5,000 years ago humans had provided nesting houses and regular food to wild rock doves in the Middle East. In India, Persia and the Mediterranean their homing skills were deployed as a sport – our city pigeons are descended from racing pigeons who never made it home. Darwin noted how each pigeon had their own unique personality: bossy, quiet, gentle, aggressive or shy. Tumbling pigeons, who flipped over backwards in flight, enthralled him:

> We may believe that some one pigeon showed a slight tendency to this strange habit, and that the long-continued selection of the best individuals in successive generations made tumblers what they now are; and near Glasgow there are house-tumblers, as I hear from Mr Brent, which cannot fly eighteen inches high without going head over heels.[25]

From rock doves came tumbling pigeons came Parlor Rollers who cannot fly but do involuntary backward somersaults on the ground – a breed selected for a defect in the balance centre of the brain. Which is where pigeon fancying tips *head over heels* into somewhere dark. The purpose of these pigeons is for competitions where breeders throw them like bowling balls to see how far they continue to backflip. The videos are disturbing. As I write, the world record is 201.85 metres.

Darwin's pigeon examples went to the heart of *On the Origin of Species*, and were easily recognised by lay readers, but what he had no explanation for was how characteristics were passed on. He originally favoured the blending idea, like mixing blood, but aside from it being observably wrong (why then did a blue-black pigeon crossed with a white pigeon not produce a melange misty-grey?), it would average out beneficial characteristics before selection could materialise. Humans had been selectively breeding domesticated animals since the birth of agriculture, but the mechanics of inheriting traits – ironically, rather fundamental to Darwin's carefully argued theory – would remain unresolved until the turn of the twentieth century. Even though, at the same time Darwin was feeding his pigeons, a monk in a monastery in Austria was busy watering his peas . . .*

Year after year, Darwin worked on his theory, his every proposition backed up by example, his every statement tight with evidence and fact. He was plagued by migraines, stomach pains, boils, ulcers, flatulence, nausea, trembling, vomiting and insomnia. Could his incendiary preoccupations be the cause of his ill health? Progress was painfully slow. Until a disturbing letter arrived in December 1857. Perhaps the most famous letter in natural history. Alfred Russel Wallace, a young naturalist collecting specimens in the Malayan Archipelago, was thinking along exactly the same lines. The following spring, in 1858, another letter from Wallace arrived. In the throes of malarial fever he had had an epiphany: he'd remembered Malthus's *Essay on Population* . . . As Darwin read the explanation of his *own* theory of natural selection, his life's work began to dissolve before his eyes. It was too cruel. The timing, the ethical conundrum . . . to be pipped at the post like this by an unknown. What to do? He couldn't suppress Wallace's paper without appearing in a bad light, yet he could hardly publish his own. Darwin forwarded Wallace's manuscript to Lyell, miserably commenting that it couldn't be a better résumé of his own work. Lyell's honourable solution was to submit Wallace's paper to the Linnaean Society, together with Darwin's 1844 outline, making clear where precedent and priority lay. Wallace accepted gracefully, and with the theory out, Darwin rushed to complete his great work. *On the Origin of Species by Means of Natural Selection* was published on 24 November 1859.

Here was the explanation of the mechanism by which organisms changed over deep time. Competition in the natural world ensures that it is the individuals with advantageous traits who survive to reproduce and pass on those traits to their offspring. As Darwin put it, 'I have called this principle, by which each slight variation, if useful, is preserved, by the term of Natural Selection . . .'[26] All 1,250 copies of *Origin* sold that very day.

---

* Gregor Mendel demonstrated the basic laws of genetic inheritance of traits in his experiments growing peas. Although he published his results, 'Experiments in Plant Hybridisation', in 1866, his paper did not surface until it was rediscovered in 1900. So, although the mechanism for the transference of traits from parent to offspring had been discovered, Darwin would go to his grave without this piece of the jigsaw.

Going ape

> But Faith, like a jackal, feeds among the tombs, and even from these dead
> doubts she gathers her most vital hope.
>
> Herman Melville

At London Zoo Darwin had stared into the eyes of their female orangutan,
Jenny, and decided, for all the hoo-hah, to leave 'man' out of it. But the
inference of *Origin* did not escape the press. It was not just Creation being
dealt a fatal blow, but man's special status. Despite *Origin* sidestepping
the human animal, our ape ancestry was the talking point as cartoons
lampooned a hairy Darwin swinging from the trees. Where Darwin
provided the evidence to remove the Maker,★ it was his arch supporter,
the biologist Thomas 'Bulldog' Huxley, who was happy to wield the knife.
Charismatic, eloquent and up for a show-down, Huxley's moment came
on 30 June 1860, in a room crammed with clergy and undergraduates
during a debate with Samuel Wilberforce, Bishop of Oxford. The two
showmen went head-to-head. In a taunt Wilberforce is supposed to have
asked Huxley if his gorilla ancestor was on his mother's or father's side.
To which Huxley responded (by his own account):

> If then the question is put to me whether I would rather have a
> miserable ape for a grandfather or a man highly endowed by nature
> and possessed of great means of influence and yet employs these
> faculties and that influence for the purpose of introducing ridicule
> into a grave scientific discussion, I unhesitatingly affirm my prefer-
> ence for the ape.[27]

Huxley relished the boxing ring of evolution. Another adversary was
a renowned anatomist and palaeontologist, Richard Owen, who had
singled out the *hippocampus minus* as a uniquely human part of the brain
which separated us from the apes. Until Thomas Huxley pointed it out
clearly in both ape and monkey brain. This became the gorgeous 'great
hippopotamus test' after the Reverend Charles Kingsley sent the episode
up in his children's book, *The Water Babies* (1863), in which Professor

---

★ Darwin categorically states in the introduction of the first edition, 'the view that each
species has been independently created – is erroneous'. Nevertheless, what was crucially
missing was Mendel's discovery of how hereditary traits were passed on.

Ptthmllnsprts (Put-them-all-in-spirits),* who 'held very strange theories
about a good many things', maintained that nothing could be depended
on but the great hippopotamus test. 'You may think that there are other
more important differences between you and an ape, such as being able
to speak, and make machines, and know right from wrong, and say your
prayers, and other little matters of that kind; but that is a child's fancy, my
dear.' For the hippopotamus was the difference, and 'if a hippopotamus
was discovered in an ape's brain, why it would not be one, you know,
but something else'.[28]

Origin had roughed up the comfort blanket of natural theology and
intelligent design. Our divinely ordained position in the world began to
seem precarious. We don't like the idea of no one at the controls. Even
Darwin, at the end of *Origin*, made his compromise: 'I should infer from
analogy that probably all the organic beings which have ever lived on
this earth have descended from some one primordial form, into which
life was first breathed by the Creator.'

By the time that Darwin published *The Descent of Man*, in 1871, even
the Church had given ground: for though physical man might have
evolved, immortal man and soul were safely inviolable – and that was
impossible to disprove.

LOVE, ACTUALLY

That the advance of science could drive religion into the wrecker's yard
was illusory. The spiritual reach is a deeply human impulse. It is our story,
and long has the transcendence in the experience of nature inspired
epiphanies of the soul.

Llewelyn Powys, born in 1884, younger brother of the writer John
Cowper Powys, lived his adult life in Dorset, with pulmonary tuberculosis.
That he was not expected to live long heightened his living moments.
He had a favourite dew pond which enchanted him, by which he would
sit, hoping some secret message would be delivered from its mysterious
heart. Here is what he wrote:

Always hoping for this hour of grace, I have loitered by the pond's
edge at every season . . . It was on a soft evening of this last

* A parody of both Richard Owen and Thomas Huxley.

September that there came to me the breath of the knowledge that I sought . . . All was silent, all was expectant. The messenger for whom I had awaited was at last revealed.

It was a hare. I saw her from far away and did not so much as venture to move a finger. She approached with uncertain steps, now advancing, now retreating, now frolicsome, now grave . . . Nearer and nearer she came. Was she actually intending to drink? Was it possible that I should see her lower her soft brown chin to the water ten yards from me? Surely if permitted to witness so delicate an operation, then at last I should receive the revelation I sought. The stillness of the evening was so profound that the fur of a field mouse's jacket brushing against the stems of its grassy jungle would have been audible, while against the sky, infinitely remote, the moon hung in utter calm . . .

I was suddenly awakened from my rapture. I had heard a sound, a sound sensitive and fresh as soft rain upon a leaf. It was the hare drinking.[29]

III: THE ANIMAL WITHIN

> Chuang Tzŭ and Hui Tzŭ had strolled on to the bridge over the Hao,
> when the former observed, 'See how the minnows are darting about!
> That is the pleasure of fishes.'
> 'You not being a fish yourself,' said Hui Tzŭ, 'how can you possibly
> know in what consists the pleasure of fishes?'
> 'And you not being I,' retorted Chuang Tzŭ, 'how can you know that I
> do not know?'
>
> Chuang Tzŭ on 'The Pleasure of Fishes'[30]

MY WORLD YOUR WORLD

I have one dog sitting on me and another panting down the back of my neck. Joanna Kossak is driving like a demon. We are 6 miles deep into the Białowieża Forest, still no sign of her house. Jonathan is laughing in the back seat. My relief is boundless for she is like a naughty schoolfriend and stories pour out of her in an English with Polish decoration. We swerve off the dirt road, stones spitting, through a gateway, and here we are in front of a tumbledown Goldilocks house. Wooden roof shingles curling, patches of corrugated iron sliding, pink paint flaking, long grass up to our knees. More animals come out to greet us. I am bewitched.

'The moment he could fly he spent every single moment tormenting every other creature.' Joanna dives behind the couch to emerge with a pile of photographs. 'Here, look! There, and there!' Korasek, the raven brought to Lech and Simona as a chick from a storm-damaged tree, was the only animal Joanna wanted 'to terminate'.

'He pecked the dogs' noses when they were asleep – this is a raven's beak, remember, a beak to break bones! Look, look at the poor donkey!'

We stare at Korasek pecking the donkey's rump and clinging on for the gallop.

'Because he liked going for a ride?'

'What else do you think? Here, riding the donkey. Here, fighting the cock. Look!'

The only creature who could stand up to Korasek was the stork, Joanna says, showing us a photograph of fully fledged battle, wings out, aggressive posturing.

'He drove Zabka to within an inch of insanity.'

'The boar?'

'Oh, yes. She never gave up trying to get the raven. He would pretend he was interested in some crumbs on the ground and she would charge, but just 3 inches from her open mouth he would fly up and jab her on the nose. Lech and Simona were his only gods, everyone else was to be terminated. I hid from him. He would attack from the air. Honestly! Look, there, I had to wear wellington boots and Lech's crash helmet to go outside.' A photo of young Joanna in a helmet wielding a big stick.

Korasek began to steal the freshly laid chicken eggs, not to eat, but for the fun of it. Every day there was a race to get the eggs. Any lumberjack working nearby had to watch their power saws or Korasek might steal the screws and sparkplugs or fly away with their packed lunches. He loved to divebomb girls on bicycles, he pecked at their heads and when

they fell off he sat triumphantly on the bike's seat watching the wheel spin round. The real trouble started when Korasek began hitching rides on the back of Lech's motorbike and discovered Białowieża village. Now he had a whole new world to torment.

Joanna laughs, 'Oh, the enormous attractions waiting for him! A lady was hanging her washing. He liked the pegs! For a simple person in those times a raven was an omen of evil, but Korasek of course didn't know this. He walked up and introduced himself. He would ruff his feathers and caw, *Korasek, Korasek!* He was a narcissist. The woman was scared so she hit him with a wet piece of washing and chased him away. A week later, the same woman was cycling to church and Korasek attacked her. She fell off her bike and broke her hip – in her 70s! Oh, my God. He was flying into the village, stealing from gardens, seed potatoes, milk churns, picnics, everything he could. One day he came home and Simona could see something in his beak. She told Lech to pretend not to see it, so Korasek dropped it to get Lech's attention. It was a purse with 700 złoty. Then, that was a lot of money!'

Joanna finds more photographs: Korasek's head down the meat mincer; Korasek on Lech's shoulder; Korasek lifting the lid of a milk churn; talking to a group of boys on a forest path in the snow, they are standing in a semi-circle around him and Korasek is holding court. His reputation spread fast. He stole a ring of 50 keys from the man in charge of all the forestry buildings. They never got them back.

Joanna disappears into the kitchen to fetch lunch.

Simona and Lech lived in the forester's lodge with the rhythm of the forest, the seasons, and the animals: the deer, the badgers, the fox, the martens, a black stork, a white stork, two owls, their chickens, Korasek, Zabka the boar and their donkey, Hepunia. And the host of injured and orphaned creatures brought to their door, who once nursed back to health were free to come and go, and some, like Zabka and Korasek, remained resident. Dziedzinka was part hospital, part home, part observational laboratory, and all of these at one of those rare times in life when everything comes fleetingly together. In Białowieża the locals became used to seeing Lech's bike whizzing by with Korasek gripping onto the back. When they saw Lech in the forest with Zabka walking by his side, Hepunia behind him and Korasek circling over the three of them, they called him St Francis.

Simona raised a small band of young deer who followed her on walks through the snowy forest. One day they were suddenly terrified, ears raised, hair standing up on the hind quarters; they would not enter the wood. Simona stopped, looked back, then went ahead to investigate, but as she did so a loud chorus of warning barks from the deer broke out behind her. She turned again, they continued to bark, standing tall and stiff. Simona pressed on slowly. A few feet away she came across fresh lynx prints and faeces. In that moment, as her deer warned her of danger, Simona felt she had crossed the bounds that divide the animal world from humans. That she was a member of the herd.

Hello, Brother Ox

> One morning, when Gregor Samsa awoke from troubled dreams, he
> found himself transformed in his bed into a horrible vermin.
>
> Franz Kafka, *Metamorphosis*

The German biologist Jakob von Uexküll (1864–1944) tried to imagine
the world of an eyeless tick, waiting on a stem of grass for the whiff of
butyric acid from the mammalian sebaceous follicles of a potential hairy
host. A wait that could be as long as 18 years, until she leaps onto a passing
meal of warm blood, embeds herself up to her neck (if she had one), gorges,
then lays her eggs and dies. Uexküll believed any organism that reacted to
sensory data should be judged a living subject and considered in terms of
their subjective world. To describe an animal's unique sensory surrounding
world he used the term *Umwelt*. This is not an animal's habitat, more their
universe, their particular reality, their unique sensory experience, taking into
account how a creature perceives the world: with sonic hearing, or zoom
vision, or swivel eyes, or infrared detectors, or electric field sensors, or blind
with a star-shaped nose, or deaf with three hearts, or with skin that can
change colour, or with long whiskers, or a lateral line of pressure sensors.
Short, tall, strong, fast, a tough hide or 3 feet of blubber. The permutations
are endless. In 1974, the philosopher Thomas Nagel concluded that a human
cannot know what it is like to be a bat. His point being that a human
imagining being a bat was not a bat *being* a bat. Without echolocation,
wings or fabulous ears, being unable to hang by our feet, without sleeping
upside down or catching moths, we were on a losing wicket. Dry moth
wings down human throats just don't elicit *Delicious!* Ludwig Wittgenstein
had already told us that if a lion could talk we would not understand him.
Not least for all the lion things we could never appreciate: stink, taste,
digestion, attraction to lionesses. Nevertheless, Uexküll's conception of
Umwelt inspired animal behaviour scientists to widen their scopes. Indeed,
as the primatologist Frans de Waal points out, Nagel could not have reflected
on what it felt like to be a bat at all had not an American zoologist named
Donald Griffin, in 1940, tried to imagine how it was to be a bat, and so
made the astonishing discovery of echolocation.

Four cows are standing upright on two legs, chatting in a field, when
the cow nearest the road shouts, 'CAR!' Two humans in a car then drive

past a field of contented cows grazing on four legs. Road empty, humans gone, the cows return to chatting on two legs. Gary Larson mocks our assumptions about animals, that they cannot think ahead or say goodbye to each other. Larson shakes humans upside down into his cartoon animal *Umwelt* where beasts play dumb on purpose or behave as absurdly as humans. When a family of spiders go for a Sunday drive, the sticker on their car has a smiley face with 10 eyes.

Charles Foster is a vet who shoved two fingers up to Nagel and Wittgenstein and just stripped off, crawled in the mud, ate worms, sniffed badger shit and slept underground in a sett on bracken bedding. He describes his intimate nose-snuffling, dirt-smelling perambulations on all fours as intense *localness*. It's fair to mention that he gagged on the worms. But in his committed experiment of *Being A Beast* while stuck with human shortcomings, he tried to get as physically close – not observing, but doing and being – as he could. In this way he thinks about the different elements of wild worlds. He is both a vet and an animal so he knows he shares the same nerve endings to transmit sensory data in the same way, like *That hurts!* for instance. 'What's an animal?' he asks. 'It's a rolling conversation with the land from which it comes and of which it consists. What's a human? It's a rolling conversation with the land from which it comes and of which it consists – but a more stilted, stuttering conversation than that of most wild animals.'[31] Dao philosophy, then.

All around us are the billions of individual selves acting out their dramas in their individual worlds, connecting to bigger and bigger worlds, to the higgledy-piggledy live jigsaw; the termite hill; the gannet colony; the auditory bubble of a female mosquito's rapturous wing-whine in the narrow range of around 380 Hertz (the only sound the male needs to hear); the infrared wavelengths of prey in the dark, detectable if you are a pit viper with sensory pit organs; to the mad hatter survival ploy of the *Uraba lugens* caterpillar who wears his moulted heads *with eyes*, one on top of the other, in a towering top hat to scare away predators.[32] Maybe the end of this *Umwelt* road for us is to imagine what it might be like to be them when we are doing things to them, like isolating them when they are social creatures who need to be with their herd or pack or mother, or locking them up when they are made to roam.

★

Konrad Lorenz was the Austrian zoologist who, in 1935, became a goose. To be precise, he became the surrogate mother goose to three goslings whom he hatched in an incubator. He began to bond with his goslings by speaking to them in their eggs, much as their real mother would. The day before they hatched they began to answer, *peep, peep*, and when they broke out of their shells he was the first thing they saw, and so they attached themselves to him as their parent in the phenomenon we call 'imprinting'. There is glorious footage of Konrad being chased by his gosling children, swimming in the lake together, following his canoe, peeping at his heels, tearing after him wherever he goes. His research showed that imprinting in ducks and geese happens in a critical period, between 12 and 17 hours after hatching, which has fed into research on brain development and the role early experience plays in shaping behaviour. Interestingly, a bird raised away from their own kind, who has imprinted on a human, will try to mate with a human, but *not* the human who raised them. Lorenz relates the bizarre example of a white peacock, the sole survivor of a clutch hatched at the Vienna Zoo, whose keeper put him in the warm reptile house with the giant tortoises. Throughout his life, he never looked at the lovely parade of peahens offered him: he only had eyes for enormous tortoises.

The many creatures Lorenz kept during his life flew and ran freely about his family home and garden in the village of Altenberg, on the River Danube. There were troops of lemurs, capuchin monkeys, and jackdaws in the attic. Songbirds decorated the furniture with their elderberry droppings, and a greylag goose slept in the bedroom, leaving by the open window in the morning.

I love the writing of this charismatic scientist, the tales of his free-wheeling menagerie-cum-home laboratory filled with constant drama and delight. He lived and breathed his belief that it was impossible to know the extent of an animal's intelligence, or understand any natural behaviour, if creatures could not move around freely and be themselves. 'How sad and mentally stunted is a caged monkey or parrot, and how incredibly alert, amusing and interesting is the same animal in complete freedom.'[33] The only cages Lorenz would allow in his house were what he called the 'inverse cage principle': to keep animals *out*, of his wife's flowerbeds, for instance. The only creatures contained were the fish in his aquarium.

In 1941, Lorenz acquired a pair of tropical West African jewel cichlid fish, renowned for their parenting skills. The male, in his nuptial ruby red and turquoise spots, cleaned up a smooth stone onto which the female obligingly spawned. Both fiercely guarded their brood, fanning the eggs until they hatched a few days later. At this stage the fry lie on the bottom in a little nesting hollow, until their swim bladders inflate. As they began to swim their parents escorted them around, returning them each night to the nest into which they sank by contracting their swim bladders. As they became more adventurous it was Father's job to collect the stragglers each night, which he did simply by inhaling them, swimming back to the nest and blowing them in. One day Konrad was late feeding the fish and arrived with a chopped-up earthworm just as the cichlid father was collecting the last of his errant young to put to bed. Mum was at the nest with the rest of the brood. Lorenz threw in the worm. The mother refused to come, though Father couldn't resist. But just as he began to chew, one of his babies swam past. He automatically swooped and inhaled it in. Then stopped. His cheeks blown out, mouth full of chewed worm and live young. Lorenz was riveted. The fish remained motionless. To chew or not to chew? The seconds drew out. It was one of the most astonishing things Lorenz was to witness in his scientific life. 'If ever I have seen a fish think, it was in that moment!' The same reaction a human might have, stopped in his tracks confronted with an unexpected conflict of interests. Father fish spits out both worm and baby fish. Both sink to the bottom. Father gobbles up the worm, all the while watching his baby. Finally having swallowed the worm, he sucks up the baby and takes him to his mother.

That cichlid's deduction might give us pause. What we think about fish tucked away in their cold tight armour, and what *they* might be thinking as they flicker by.

Inspired by Uexküll's principle of *Umwelt*, a new wave of field naturalists returned to study animal behaviour in relation to their natural environment. Konrad Lorenz is seen as the father of this science of ethology, as it became known (from the Greek *ethos*, meaning character). It is baffling that such a free-spirited, life-embracing man could ever join the Nazi Party. Yet, in 1941, the same year that he witnessed his male jewel cichlid decide his progeny's fate, Lorenz joined the German army as a medic. He believed understanding animal behaviour could have far-reaching implications for psychiatry and medicine and provide insights into the human mind. During his 'goose summers' in the 1930s, Lorenz had observed the loss of specialised instincts in domesticated hybrids and suspected a similar process might explain 'defective traits' in humanity. Unhappily, he confused genetic deterioration as a consequence of domestication with the Nazi terminology of the time. So goes our history: the confluences of humankind's relationship with animals that can hitch a goose to an ideology of racial purity, and by association the worst crime of the twentieth century. Posted to the Eastern Front, Lorenz was captured by the Red Army in 1944 and spent four years as a prisoner of war. Repatriated in 1948 – and touchingly allowed to bring home his pet starling – he resumed his life's work in the biology of behaviour, which could hardly have been less Nazi in impulse, goal or character.

It was during the Second World War that Lorenz's friend, Karl von Frisch, an Austrian with a Jewish grandmother, was saved from the Nazis by his honeybee research when he argued he could help fight the nosema parasite that was threatening bee populations and provide an important contribution to the maintenance of Germany's food supply. It was then that Frisch discovered one of the most mind-vaulting revelations of the twentieth century: the honeybee waggle dance. What a moment, the first spark of suspicion that an exchange of information was going on, to have the imagination for it. The rigour to test, and think, and test, and think, and slowly untangle it. Frisch set up two food stations at different directions and distances away from a single hive; he then painted a red spot on the bees who visited one station, and a green spot on the bees who visited the other. When the bees returned to the hive he noticed the green spot bees danced differently to the red spot bees. Both performed the figure of eight, but the orientation was different. This was the exquisite moment that would blow apart the assumption that only humans could communicate in an abstract way. How complex and precise this communication is, we are still discovering. Hives are vertically orientated, with the combs constructed to hang down in the direction of gravity. Gravity gives the bees a consistent marker: directly up (12 o'clock), which they use in their dance to represent the sun. If the bee dances straight up, then the other workers will fly towards the sun. If the dance heads off at a tangent of 90° to the right (3 o'clock), they will correspondingly fly 90° to the right of the sun. If the dance heads down (6 o'clock) the bees will fly away from the sun, and so it goes with all the intermediates. The amount of waggling represents the distance; a long waggle means the nectar is further away.★ Roughly, it's a second per kilometre of flying. It gets better. Headwinds are accounted for because a bee measures distance by the energy it takes them to travel there. Cloud cover doesn't faze them because they can see ultraviolet and polarised light, so can determine the exact position of the sun all the time, a perfect solar compass. Their other magic is their internal clock, which can estimate the new position of the sun as it travels across the sky, so a bee knows where the sun is even after being in the dark hive. They can also account for change in seasons or latitudes. A lot of humans I know can't even read a map. How do bees see the dance in the darkness of the hive? They sense air vibrations through their sensitive antennae, which the workers hold close to the dancing bee. For goodness' sake. It's the vibes! The pollen brought back to the hive gives the workers the scent of the nectar source. The honeybee's *Umwelt* ticks around the sun, gravity and the fragrance of flowers. The pattern and grammar of their dance communicates nest sites, water and probably landscape features, like trees. And it is likely we don't know a fraction of it.

We have been stealing honey for millennia. There are Neolithic paintings of humans taking honey using ladders and ropes. Which was a faff, so for convenience we took the bees and housed them in straw skeps. Then we made wooden hives fitted with frames for them to attach their comb – and for us to lift out the honey easily. One colony of bees will visit 500 million flowers in a season and make 90 kilos of honey, which

---

★ If the nectar is nearby they do a circular dance.

will see them through with a supply of food until spring. The problem is that we take most if not all of it.

Matt Somerville, a shipwright by trade, who made the wild beehive for our nature reserve, comes from a family of apple growers in Kent. In 1990 he planted his own cider apple orchard. He loved the sound of bees pollinating blossom, but one year there was no buzzing. The terrible silence turned his life around. Matt stepped into honeybee *Umwelt*. He saw the bees' resilience, long in trouble with systemic pesticides, mites, parasites, new diseases and reduced foraging, further weakened by conventional beekeeping. Compared to a thick tree trunk used by wild honeybees, a traditional hive is thinly constructed, close to the ground, without insulation and needs more bee-energy to keep warm. By feeding sugar and soya solution as a substitute for the honey we take, the bees lose the nutrients, enzymes and medicinal properties that keep them healthy. Matt thinks we need to rethink our relationship with honeybees if they are to survive in our degraded landscape. We have long reflected our own social systems onto honeybees, and the other way around, using bees as a model for how humans should work together. In Oman the different castes of the honeybee are known as the 'sheikh', the 'people' and the 'slaves', who, in the Oman system, are not the workers but the dark-coloured male drones! (Just sayin'.) The swarms that take up residence in Matt's insulated and hollowed-out log hives are subject to natural selection (notably smaller and darker), and produce stronger, healthier, more vigorous colonies that cope far better with varroa mite, viruses and other pathogens. Matt calls this method bee-centred.[34]

In 1973, the Nobel Prize in Physiology or Medicine was awarded jointly to Konrad Lorenz, Karl von Frisch and Dutch zoologist Niko Tinbergen (for his field studies of herring gulls and the ways behaviour is adapted to the ecological environment). That the prize went to the three pioneers of ethology was seen as giving the science of animal behaviour the tick of respectability. Then the attention turned to Lorenz's wartime record. 'None of us as much as suspected that the word "selection", when used by these rulers, meant murder,' wrote Lorenz after the award. Tinbergen, who had been a prisoner of the Nazis, supported him: 'I know that he turned round as soon as he saw what really happened.'[35]

CRIME & PUNISHMENT

Harry Harlow, born in Iowa in 1905, was by his own account raised by a cold mother. In 1932 he established the Psychology Primate Lab at the University of Wisconsin with a breeding colony of rhesus macaques. His interest was maternal deprivation. This is what he did. He removed infant monkeys from their mothers and isolated them in cages to record their behaviour, which was: staring, blinking, circling and self-mutilating. Harlow noticed how the infants clung to their cloth nappies for comfort, like a security blanket. This gave him the idea for his surrogate mother experiment. Harlow provided the infants with two surrogate mothers: one made of wire which dispensed milk from a bottle, and one of soft cloth, without milk. Harlow recorded how the infants spent more time with Soft Cloth Mother, going only to Wire Mother to feed. Now isn't that amazing. Next trick was to put a mechanical teddy in the cage banging a loud alarm. With no surrogate mother to cling to, the terrified infant huddled into a frozen ball. The photographs of the experiment are indescribable. But Harlow had only just begun. Meet Evil Mom. Harlow embedded spikes into the soft cloth mother which could shoot out to hurt the infant in a surprise attack. There were variations: Evil Mom who shook infant violently; Evil Mom who froze infant with cold air; Evil Mom spring-loaded to hurl infant away. When the 'cruel behaviour' stopped, the infants returned to try and make their soft cloth moms love them. Yes.

Harlow isolated some monkeys without contact *of any kind* for up to two years, observing them through one-way glass. He was interested to see what kind of parents these damaged monkeys would make. The problem was that they were so traumatised they would not mate. So he devised, to use his own term, the rape rack, on which he strapped female monkeys in the mating position so the business could be done. 'Not even in our most devious dreams could we have designed a surrogate as evil as these real monkey mothers were,' he reported. One mother chewed off her baby's feet, another crushed her baby's head. Harlow's next invention was perhaps his most sordid. The Pit of Despair was a vertical steel chamber with shiny sloping sides, like an inverted pyramid, fitted with a food box and water bottle, and a grate over the top. Its design was to instil the sense of hopelessness. The pit received monkeys after they had bonded with their mothers at three months old. Of course they all went insane.

INTERMISSION

The BBC's *Spy in the Wild* uses animatronic spy animals fitted with video recorders to infiltrate animal communities in their natural habitats to observe their behaviour. The animatronic model of a baby gorilla, brown eyes, shaggy black hair, can move his mouth and head, blink and make gorilla noises. He sits there. The young gorillas are curious. They look a bit like toy gorillas themselves, with their big tummies, their hooded brows, their perfect radical hair. Us, not us. Little ones, big ones, they are *so* curious. Then a silverback climbs down a tree. He thumps the ground, *What the hell is going on?* The other gorillas make way. Spy-gorilla sits his ground (he has to). The real gorillas crowd round. Spy-gorilla blinks, and through the eye-camera we get to see what he sees, all the gorillas staring at him. It's brilliant. A bit cheesy, but brilliant. Gorillas without humans, being gorillas. A conundrum is in their midst: one of *them* . . . or is it? I freeze the picture: a tier of seven gorillas looking, the young ones over the shoulders of the adults. They come closer in. Then leap back. Great clouds of flies, pant-grunting. Spy-gorilla baby averts his gaze. As the silverback chills out, the real babies begin to play and everyone relaxes. One of the youngsters returns to Spy-gorilla baby and beats his chest in an invitation to play. Spy-gorilla baby beats his chest back. Very clever. Real baby, excited, beats his chest even more. The other gorillas are still not sure. They are not stupid. Then a baby gorilla gets carried away and knocks Spy-gorilla over. Spy-gorilla doesn't get up (he can't). And then it gets priceless. Real baby peers at Spy-gorilla aghast, touches it, oops, no life, he backs away slowly. There is no question that Real-gorilla baby thinks he has hurt this strange baby, or worse. We know this because we can see it plainly on his face and by his body language. The baby searches for his mother and climbs on her back. David Tennant narrating says, *Best pretend it never happened.*[36]

The beastly burden

In the early twentieth century, behaviourism – the doctrine that animal behaviour is conditioned purely by stimuli in the environment – became a religion in animal study, with preachers and messiahs who inculcated their classrooms with its dogma and employed ridicule to squash dissenters. Their biggest weapon was a word. Anthropomorphism. Accused of that, you would be derided as sentimental, gullible or, worse, unscientific. The attribution of human behaviour or characteristics to an animal was some- thing to be very careful one didn't do. Non-human animals are, after all, *not* humans. Humans can't fly, breathe under water, lay eggs or hibernate; non-human animals can't write poetry or drive a car. But stick a pin in the vast majority of creatures and we will all yelp. Even an ant will lurch back from a stabbing pin. Who would stab an ant with a pin? Only we would. Yet for all the pin-pricking we have done, and all the years of living side by side, we can be conveniently resistant to attributing any personality traits, emotions or responses to animals that might be shared with us. If they can't say *Ow, that bloody hurts!* how can we be sure?

The burden of proof turns out to be the creature's burden, because we cannot use our common sense, or our shared mammalian experience. Animals could not feel anxiety, anger, or sadness, for this would endow them with human states. And certainly not reason, purpose, or self- awareness. That was taboo. We might share skin, hair, lungs, nipples, nerve endings and similar anatomies but we would never share minds. It seems behaviourists ignored the fact that we were mammals entirely. For them, the mind of an animal was a virtual blank slate onto which one could imprint behaviour through punishment and reward.* Its doctrine cowed imaginative minds and regressed our understanding of animals for more than half a century. Anthropomorphism was taboo but, paradoxically, that didn't stop us testing animals to learn about humans.

For 50 years Harry Harlow was given free rein and plenty of funds to discover that infants need the comfort of their mother's touch for their psychological development and physical health. Professional criticism of these grubby experiments is scarce and cautious. Outsiders do not under- stand. (Dead right.) Harlow's defenders cite the strict child-rearing beliefs of the time, which he was trying to overturn; his 'groundbreaking methods'

---

* Behaviourists were only concerned with observable behaviour that could be objectively measured by human-devised experiments controlled and quantifiable in the lab.

of gathering empirical evidence; how his findings 'rocked the scientific world' and 'fundamentally changed our biological understanding'; and even now consider his work 'a breakthrough in child psychology' that 'shed light on the nature of love'.[37] The American Psychological Association (APA) shut Harlow's work down in 1985 on ethical grounds.* He had died by then, in 1981.

Sadism requires a reason to enjoy the suffering of others. Most often this is personal power (for those needy of it) and requires the absence of the restraints that normally prevent its manifestation. Inhibitors like the law, cultural taboos, guilt, empathy, revulsion can be disabled. Looking down at a stricken monkey, godlike, letting the lifeline reel out, the ultimate power at one's fingertips, is an acquired taste that needs bigger and bigger hits to satisfy its hunger. An ideology can justify anything, even violence, as a means to a greater good. Maybe it is not the monkey going mad who should teach us anything, but the human animals in the experiment.

In 1975, the 20-year-old Stephen Pinker worked as a researcher in an animal behaviour lab. Among the lab rats was a runt whom the professor wanted to experiment on. Pinker was tasked with setting up a Skinner box** to deliver electric shocks every six seconds until the rat discovered a lever which delayed the shocks to every 10 seconds. Rats will normally catch on and learn to avoid the shocks entirely by pressing the lever every eight seconds. Clever rats. Pinker put the runt into the box to 'condition' her behaviour overnight. He set the timings and went home. Next morning, he opened the box expecting the rat to have behaved as expected. The rat he found was a contorted wreck shivering in the corner of the box, nowhere near the lever. Pinker rushed her to the lab vet, but an hour later she died. He describes the story as the worst thing he has ever done. For even if the trapped rat had learnt to press the lever every eight seconds he had subjected her to 12 hours of relentless, unknowable lever-pressing stress. Pinker knew it was wrong as he was doing it, but he did it anyway. He was told to do it. It was standard practice. His behaviour,

---

* One 1973 experiment was called 'Induction of psychological death in rhesus monkeys'. The University of Wisconsin resumed controversial research into maternal deprivation using monkeys in 2014.
** A box invented by Burrhus Frederick Skinner to experiment on rats and pigeons, equipped with a lever or button to deliver rewards or punishments – food, flashing lights, loud noise, electric shocks, etc.

he points out, mirrors the same behaviours that had those dark consequences in the worst episodes of twentieth-century history.[38]

Deep in Africa, another type of animal behaviour study was going on.
The palaeoanthropologist Louis Leakey, excavating fossils in Tanzania's
Olduvai Gorge, believed we would better understand early hominids by
observing the natural behaviour of our closest cousins, the great apes, in
their own habitat. Best suited to the task, Leakey thought, would be young
female humans for their natural empathy and calm, their greater sensitivity
to mothers and offspring, the lesser threat they posed to the males, and
importantly, unbiased minds without the conditioning of any scientific
doctrines. The strategy couldn't have been further from those American
psychologists peering at crazed creatures inside boxes; Leakey's protégés
would instead hide themselves in a 'box' to observe what the animals
were doing. His prescience would give us the great female primatologists
of the twentieth century.

REWARD

Beneath the skin, kin

Jane Goodall arrived at Gombe on the northern shores of Lake Tanganyika in 1960. Day after day, for 9 to 10 hours, she tracked and watched chimpanzees at a distance, until they accepted this skinny blonde outsider-weirdo who followed them around. Patience. At night she transcribed her field diary by typewriter, charting group hierarchies, facial expressions, gestures, vocalisations, the contents of faeces. When she gave her subjects names – David Greybeard, Flo, Fifi, Flint – irritation rippled through the scientific community. Naming was very definitely against the rules. Yet Goodall's detailed observations of chimpanzees using sticks to fish out termites from deep crevices shook up the world. Tool use was supposed to be uniquely human. 'We must now redefine man, redefine tool, or accept chimpanzees as human!' Leakey famously commented.[39] Goodall's discovery flipped our perception of who we are. In the same decade that we learnt that a monkey isolated in a steel box will go insane, a single-minded young woman discovered chimpanzees use tools, eat meat, live in socially aware hierarchical arrangements, read body language, cooperate, help each other, manipulate, deceive, employ cunning, grab power with divide and rule tactics, and have a complex culture with which they pass on information from one generation to the next. Goodall saw individual personalities and documented strong family bonds; she observed displays of affection, hugs, kisses, pats, tickling and reconciliation after a fight. She watched organised teamwork in the hunting of bush pigs and the killing of a colobus monkey. She recorded the dark side of violence and murder; a female stealing, killing and eating another's infant for dominance. She witnessed the grief of bereavement.

Then there was Bananagate. To enable closer monitoring of shyer chimps Goodall set up feeding stations. Fights erupted. Relationships began to fray. Were these the seeds of conflict that instigated an all-out war? Fingers would point to it . . .

In the shadow of Bananagate came a territorial war between two chimpanzee communities that lasted from 1974 to 1978. The first civil war of killings and land grabs ever observed in wild chimpanzees. When all the males of one community were killed, the victors expanded their territory but were themselves later ousted by another group. Animals at war? It hardly seemed possible. A kinship with our closest cousins was found in our grimmest capability. Ambushes, retaliations, kidnapping of females, childhood friends pitted against each other and bloodthirsty acts of murder. Goodall struggled to come to terms with the extent of the violence of the Gombe Chimpanzee War. Her reports were disbelieved. Human disturbance and the feeding stations were highlighted as the cause that must have distorted normal behaviour and got out of control. However, a 2018 study concluded that chimpanzee societies can and do indeed wage war, and that this instance was most likely the result of a shortage of fertile females during a power struggle between three top-ranking males.[40] In other words, power, ambition and jealousy; internal political conflict fuelled by competition for limited resources. The very things that incite lethal conflict in humans.

When the dominant matriarch, Flo, died in 1972, there was an obituary in *The Sunday Times*. Goodall recalled how Flo had taught her so many valuable lessons, 'especially the importance of a mother and child relationship'; which is bitterly ironic considering this was the area Harry Harlow was researching at exactly the same time, with his wire mother and his Pit of Despair.

★

Not far away, in the volcanic range of the Virunga Mountains in Rwanda, another war was raging out of control – between humans and primates. Dian Fossey, the second of Leakey's 'angels', had arrived in Rwanda in 1967 to study the mountain gorillas, one of only two populations in the world.★ The story goes that Fossey came to Leakey's attention by falling into his excavation site at Olduvai Gorge, breaking her ankle and throwing up on the fossil of a giraffe. Where Goodall was calm, composed and imperturbable, Fossey was all force, fury and fire. Only one thing worried Leakey as his researchers' work progressed: their emotional attachment to

★ Biruté Galdikas, who studied orangutans in Borneo, was Leakey's third female researcher in the field.

their subjects. He was right to be concerned: Fossey's passion for her gorillas would kill her.

Dian Fossey put her success at being accepted by the gorillas down to her previous work with autistic children; for two years she doggedly hung out, munching celery sticks, imitating their belching grunts, scratching herself, knuckle walking, keeping her eyes averted. After a year and a half, the gorillas began to approach out of curiosity. A 250-pound gorilla she called Peanuts was the first to touch his fingertips to hers. She was lying in the undergrowth as he watched her when he stretched out his hand and they touched. He then beat his chest and ran off in excitement. These enormous animals, so feared at the time, allowed Dian into their intimate and affectionate family world. She was not feeding the gorillas, nor helping them, nor imprinting herself as their mother; they were free to come and go. Silverbacks groomed her, mothers let her hold the infants. A gorilla with a bent finger couldn't keep away, flopping on his back and waving his legs in the air, inviting her to play. 'Digit' and Dian formed a friendship that had no contract other than wanting to be in each other's company. How could she *not* love them? She wrote to Leakey, 'It really is something, Louis, after all these years, and I just about burst open with happiness every time I get within 1 or 2 feet of them.'[41]

Dian's problem was poaching. Correction: the gorillas' problem was poaching, which became Dian's problem. Gorilla-hand ashtrays, $20; gorilla heads, feet, skins, bushmeat, and big bucks for live infants to sell to zoos. To capture an infant you had to kill their parents first because they would fight to the death. In 1969, Cologne Zoo commissioned the capture of two young gorillas; 18 adults were killed, with both infants injured in the process. Fossey was given the orphans, Coco and Pucker, to nurse, and though she tried to prevent their shipment it went ahead.* In the 1960s there were 475 mountain gorillas in Rwanda; by the 1980s the population had almost halved, not least because of agricultural pressure for the land.** Fossey became a vigilante force. She made enemies, destroying traps, burning out poachers' camps, confiscating herders' cattle. Here is what she wrote to a friend when the guards brought in a poacher:

---

* Both gorillas died at nine years old, having spent all their short life in captivity.
** Western lowland gorillas are more numerous and widespread, though threatened and declining. A survey in 2018 estimated a population of 361,900 in the remote dense rainforests of Central and West Africa.

We stripped him and spread eagled him outside my cabin and lashed the
holy blue sweat out of him with nettle stalks and leaves, concentrating on
the places where it might hurt . . . I then went through the ordinary
'sumu', black magic routine of Mace, ether, needles and masks, ended with
sleeping pills . . . That is called 'conservation' not talk.[42]

On New Year's Day 1978, Dian's beloved Digit was found beheaded with
his hands chopped off. Dian did not speak. Her silence set hammer hard.
In July, Uncle Bert the silverback, Macho, and their three-year-old son,
Kweli, were killed. Everything was different now. No rules, nothing she
wouldn't do, she *had* to stop it. She wore masks, cast malevolent spells in
KiSwahili. *I am the Goddess of the Mountain*, she hissed, swearing retribution for her murdered children.[43] Her kitchen was set on fire, her parrot
was poisoned, her dog disappeared. Her anger congealed. In the Virunga
range *she* was the live volcano. Her patrols found 987 snares and traps.
She fell out with colleagues; she fell out with the organisations fundraising
to protect gorillas in her name. Money was not spent to tackle poaching
as Dian wanted, but on education and films, on park vehicles and tourism.
Rwandan officials wanted her out. Conservationists wanted her research
station turned over to tourism. Dian fought tooth and nail; human presence would disturb their natural behaviour and the gorillas would be
susceptible to anthroponotic diseases against which they had no immunity.*
Her reputation nosedived: Dian kidnapping poachers, Dian drunk with
a gun, Dian firing at tourists. Her paymaster, National Geographic, cabled
to convey their concern.**

We know what happens. The dappled sunlit images of Dian with her
beloved gorillas ends with the attack on 26 December 1985 by an unknown
assailant with a machete that cleaves her skull. One name in her diary is
repeated again and again, *Digit, Digit, Digit*.[44] She was buried in her gorilla
graveyard next to him. Before her death Dian predicted gorillas would
become extinct within a decade; David Attenborough credits her as the
person who saved them. His own memory is inscribed for ever – and
recorded in the 1979 BBC *Life on Earth* series – with the young Poppy
who tried to undo the lace of his shoe. Yet, contrary to Dian's instinct,
tourism has made it possible for gorillas to survive. A revenue-sharing

---

* Pathogens that are capable of being transmitted from humans to animals.
** Louis Leakey had died in 1972.

scheme makes sure tourist money goes to the adjacent communities. The destruction of habitat for agriculture has ceased. The mountain gorilla population has quadrupled to over 1,000, though Covid presented new challenges with the dip in tourism, and if they are not careful, the disease itself. Poppy, it is good to report, was a long-lived mountain gorilla who knew her granddaughters.

In 1970 Dian Fossey reported a particularly curious and moving event. An old female gorilla with atrophied arms, whom she called Koko, lived with five younger males. Dian guessed she was about 50 years old. 'If it isn't being too anthropomorphical, the five males seemed to love her, and most group activities centred around this aged matriarch who would initiate grooming sessions.' Koko began to show signs of senility by wandering in aimless circles. The five males would sit and wait for her. Sometimes one of them, Rafiki, would give a soft hoot-bark to call her, and Koko would return to embrace him and he would hug her back. Koko and Rafiki would often share a sleeping nest where they curled up together like an old married couple. Then Koko and Rafiki disappeared. For two days they were absent. Then Rafiki returned alone. Though Dian retraced their trail, finding shared night nests, she never found Koko's body.[45]

MIRROR, MIRROR

'Can you row?' the Sheep asked, handing her a pair of knitting needles.

Lewis Carroll, *Through the Looking Glass*

Male bowerbirds build pavilions of wonder, extravagant as the Sydney Opera House. These rainforest-dwelling songbirds from New Guinea and Australia range from the size of a starling to a slim jackdaw. They weave twigs, leaves and grass into their stage set with mind-boggling building proficiency, then decorate it. Not just a few bright berries here and there, they go doolally. For these are theatres for courtship. For years Western Europeans assumed the structures were the confections of native forest people, so impossible was it to credit bird brains as their architects. The polygamist men of Papua New Guinea simply saw the bowerbird displays in the same way that they themselves must impress with a 'bride price' to acquire wives. Once the bower is built the treasure is laid out in front of it, each pile separated by colour and form — berries, shells, seeds, little bits of bone, clothes pegs, plastic bottle tops, anything that glitters or comes to beak. (Or stolen from a neighbouring bower.) The satin bowerbird constructs a twig avenue which he then dabs with natural pigment mixed out of chewed-up berries and charcoal that is loaded onto a sponge he has made out of plant fibre. Really. Different species have different techniques. A bowerbird must learn his trade. He visits the bowers of his elders and betters, carefully inspecting their technique before starting to practise. Early efforts are rudimentary affairs, and sometimes older males — not necessarily related — will show a young stud how it's done, prodding and weaving, patching up a weak spot.[46] While the bower follows the species' general form of construction, there is huge scope for individual variation. What kind of industry is this other than to show off individual personality . . . where, how many, what colour, how high?

Bowerbirds are, of course, not alone, fruit flies, bees, spiders, all vary from individual to individual. Some fruit flies go off 'roving', while others hang about the food. Some funnel-web spiders rear up and bare their fangs if they are disturbed, where others scuttle away. (Aggression and caution need to be finely balanced for survival.) Individual worker bees also display different predispositions. Lazier worker bees consistently don't venture far from the hive like other busy bees – and they live longer! Animals have friendships, rationality, empathy, and they have culture; humans have long been resistant to this idea.

In 2012 a group of neuroscientists signed the *Cambridge Declaration on Consciousness*, which stated that all mammals, birds and many other creatures, octopuses included, are conscious beings with the capacity to exhibit intentional behaviour. The extraordinary thing is that it took so long. More extraordinary still, perhaps, is that it needed to be stated at all. With the presence of neurological substrates, from brains to ganglia to nerve fibres, consciousness seemed highly likely (I mean, a world of unaware beings unresponsive to their surroundings seems an odd notion to me), but the onus has been on providing incontrovertible proof, and until then we had to remain agnostic.

Behaviourists also tended to ignore the golden rule of experimental science that absence of proof is not proof of absence. In his book *Descartes' Error*, the neurologist Antonio Damasio argues that reason actually *requires* emotion and feeling to guide behaviour and decision-making.* Yet we must still run the gauntlet with cautionary science if we want to talk about animal emotions. Pigs can be stressed, their corticosteroid levels might rise, but they cannot be unhappy. Darwin in contrast, observant and careful scientist that he was, was so at ease with the notion of animal emotion that he wrote a whole book about it, *The Expression of Emotions in Man and Animals* (1872). But what was okay for Darwin was not okay for Goodall. She already knew, from a whole childhood of close observation, that her dog, Rusty, was quite capable of rational thought, had a vivid and unique personality, and experienced an array of complex feelings. Anyone with a dog as a pet, she sagely pointed out, knows this. However, scientific orthodoxy was hostile; Rusty could leap around the garden doing somersaults all he liked, but he could not be happy without hard evidence. Any cattle farmer letting calves out of the wintering barn knows what happy looks like. Watch a wood pigeon swoop from the

---

* As demonstrated by patients who have suffered damage to the parts of the brain that govern emotional connections and are unable to function rationally with impaired decision-making ability.

canopy in the spring sunshine in a massive unnecessary ark; a seagull gliding in the thermals; a community of rooks playing broken umbrellas on a blustery autumn evening. Instinct? Come on, surely it must be bird flying pleasure. Moreover, if animals were so different why were psychologists studying them to learn about human behaviour? There were enough convenient similarities to electrocute, scare and isolate them, but not to admit to their feelings. Researchers in white coats kept their laboratory animals hungry to keep them responsive to rewards. After the experiments they switched off the lights and locked the door. As the German physicist Werner Heisenberg said in 1958: 'What we observe is not nature in itself, but nature exposed to our method of questioning.'[47] Behaviourism had wound the clock back to Descartes by treating animal as machines. Pigeons in boxes pecking discs, monkeys pulling levers. Test my memory after depriving me of food, sleep, friends and family before throwing me into a tank of opaque water with high sides, and you may find that I do not remember where the submerged escape platform is.* And if you dose me with enough radiation to make me feel sick while feeding me sweetened 'popping' water from a spout that flashes and clicks as it dispenses, like any rat I definitely wouldn't want that damn saccharin drink again.**

By the 1970s a new generation of psychologists were open to the similarities between animal behaviour and our own. In 1970 Gordon G. Gallup wanted to know if animals were self-aware. Could they visually recognise themselves? So he invented the mirror test. A very visual and egocentric test for animals who are far more interested in what their mating partner looks like, or more often, smells like. How, I wonder, would humans do in a pong test? Presented with their own smell, they might, or might not, recognise themselves. Nevertheless, the mirror test is how you test sense of self. The basic method is to put a cross on the animal's face (without them knowing),*** show them a mirror, and see if they either stare at the cross, or touch it. Very few species pass. Us of course; great apes; the odd elephant who doesn't smash the mirror with their tusks first; dolphins in aquariums race to the mirror to check themselves out; songbirds fail, they think the reflection is a competitor in their terri-

---

* The Morris Water Maze, standard memory test for rats, developed in the 1980s by Richard Morris.
** John Garcia and Robert Koelling's Bright Noisy Water experiment with rats to test 'conditioned taste aversion', 1966.
*** Usually by anaesthetising them (!)

tory and become cross (I mean display aggressive signals). But then in 2008 the Eurasian magpie passed; and in 2019 the cleaner wrasse passed by scraping his fish body on the side of the mirror. Manta rays also showed self-awareness. Cats don't give a stuff about mirrors, so they don't pass. As far as a self-awareness test goes, it doesn't seem watertight. A badger will know if he can get *his* body through a hole or not – so he knows how big and what shape he is. Same goes for a mouse. And is not defending one's territory an indicator of sense of self? *This is my joint, buzz off!*

For all our caution *not* to be anthropomorphic, we do have a penchant for anthropocentric tests.

For years, experimenters believed elephants incapable of using tools. The test had been to leave a stick on the ground with which to reach a banana placed outside their enclosure. Chimps passed the test in a flash. Elephants failed. It was baffling because elephants pick sticks up with their trunk all the time. Then someone stepped into their *Umwelt*. Unlike a chimpanzee's hand, an elephant's trunk is also his nose. So, when he is trying to sniff a banana, he doesn't want a stick in his nose. It would obstruct his sensitive smelling and feeling. So this bright someone hung a banana on a branch too high for the elephant to reach, but left a stool in the enclosure. The elephant moved the stool under the branch so he could reach the banana. Hey presto. A species-appropriate test.[48] Don't try throwing a ball for a rabbit to fetch. Animals can do what they *need* to be able to do. Like remember where the water hole will be in the dry season, and how long it takes to get there. Elephants are good on memory. So are animals who cache food. The Clark's nutcracker, a North American corvid, stores 20,000 pinecones in hundreds of different locations and remembers most of them. I can't remember which room I left my glasses in. Animals who roam have good spatial understanding. It works the other way too. Amazingly, kittiwakes *don't* recognise their young because all they need to know is where their nest is – because it's on a very narrow ledge inaccessible to predators, on which the chicks will stay put. By ignoring animal-specific skill sets we are blind to many of their cognitive abilities. Octopuses! Some carry coconut shells with them for shelter in case they need to hide. Isn't that tool use *and* forward planning? Rats cooperate, can regret their decisions, and hesitate before a difficult task. Fish shoals send out scouts to scope for predators. In Australia, black kites will pick up smouldering sticks from bushfires and fly them to dry grass to start their own fire to flush out prey. Indigenous Australians call them firehawks.[49]

We seem perpetually surprised when animals turn out to be more intelligent than we thought. One face recognition experiment for chimpanzees was to test them with photographs of . . . yes, humans! They scored low, like we might if tested on face recognition of sheep. When the Yerkes National Primate Research Center in Atlanta tested them with photos of their own species, they did better. Golden paper wasps recognise one another, for each one has unique face markings.

1995, Ohio. Sheba the chimpanzee is offered two bowls of candy of different amounts. She points at the full bowl. She is given the small bowl. This test is called 'reverse pointing'.[50] Sheba must learn to point at the small bowl to get the big bowl. Sheba fails this test of human perversion continually. How confused these apes must be by our weird behaviour. *You've taught me to point, now give me the ruddy sweets!* However, when the candy was removed and replaced with numbers, high and low, to represent the bowls of candy, Sheba learnt to point at the smaller number to get the large bowl of candy. When the ethologist and primatologist Frans de Waal asked if this experiment had been done with children, no one seemed to know. Yet apes are often tested against young human children to see who's cleverer . . . As de Waal points out, many of these tests require interaction with an examiner, eye to eye. The three-year-old human gets to sit on the lap of an adult of their own species who murmurs encouragingly, while the ape is brought in by someone in a white lab coat and kept behind a safe screen or bars.[51]

Another question that exercises scientists studying animal cognition is whether they possess 'theory of mind' (ToM), the term used for being aware of the mental states of others, for the ability to step into someone else's shoes and understand that their intentions and emotions are different from your own. We all might struggle on a bad day. Scientists believe chimpanzees, bonobos and orangutans have it. Some think corvids have it: the crow who hides his stash then re-hides it when the other crow who had been lurking about has gone. Would not most predator–prey relationships require theory of mind? Animals who engage in teamwork. Which way is he going to run? The ToM test devised for primates was to show them a film where one character (human) hides his bananas in a box and leaves the room. Another character (human in a gorilla suit) comes in and moves the bananas to a different box. The human character returns for his bananas. The researchers want to know if the chimpanzees can predict where the human will look. The first box, where he left the bananas? Or the second box, where the chimpanzees know they are? If they expect him to look in the first box this shows they have theory of mind, because they have stepped into his skin. The chimpanzees' eye movements revealed they anticipated the human would look in the box where he'd left the bananas because they understood he *expected* them to be there. This was about seeing if they understood when another holds a *false* belief.[52] The results challenged the idea that only humans were capable of this sophisticated skill, and validated ethologists like Frans de Waal — a perceptive animal who never forgets he is one — who sees a mental continuity between us and the great apes. That this 'King Kong' test of false belief was a big deal says more about *our* slow learning than about primates.

Let the animals speak for themselves – I

A chimpanzee walks down a track with her two young. The mother chimpanzee stops and looks back at her son lagging behind. Cat Hobaiter, a scientist studying chimpanzee body-language gestures in wild populations in Uganda, stops the video. She says it's like trying to decode alien communication because you are starting from scratch. Hobaiter learnt to look for and spot the thing that *stopped* a chimp gesturing — assuming he had succeeded in communicating his message. By studying what happened before and after each gesture, Hobaiter began to unravel meaning. 'Right there,' Hobaiter says, pointing to where the mother is showing the heel of her foot and giving it a little wiggle. This foot gesture is not very obvious (to us) but once Hobaiter had seen it a few times she worked out what it meant: *Hop aboard.* Each time the mother stopped and waggled her heel, the infant jumped on board. We watch it again, and now that we know, it's obvious.[53]

★

'Wolf kills often attract ravens by the dozen. Yet if humans put out elk carcasses, ravens generally ignore them. Ravens trust wolves. Ravens don't trust humans.'[54]

Crocodiles go bird fishing by balancing a stick on the end of their snouts. A bird comes down to perch, or get a stick for her nest, and bang.

Humpback whales blow bubbles to encircle schools of fish into a corral.

Green herons lure fish into their range by dropping berries, twigs, feathers or crumbs of bait.

Archerfish knock insects from overhanging leaves and branches by firing jets of water up at them.

Burrowing owls collect mammal dung to attract dung beetles.

Dolphins in Shark Bay, Australia, attach marine sponges to their beaks, so they can sift through sediment for food without hurting themselves.

Mice use markers to find their way home.

A carrier crab carries a jellyfish around like an umbrella in defence against predators.

Net-casting spiders gather a bit of their web to throw over passing victims like a net.

The bolas spider secretes a blob of sticky thread to make a ball, baits the ball with a pheromone that mimics her prey's mating pheromone, attaches the ball by a silk thread to a trapeze line, holds the ball in her mouthparts and front legs, waits until an amorous moth approaches, then swings the sticky ball and reels in her prey.

Caledonian crows carry their favourite handy tools with them.

Swifts and whales knew the world was round long before we had an inkling.

★

Meet Kelly. Kelly the dolphin learnt that whatever the size of the rubbish she collected from her pool at the end of the day she was rewarded. Paper cups, cans, cameras, glasses, all received a fish. Kelly began to find lots and lots of small pieces of paper, for which she was rewarded lots of fish.

When the pool was drained for maintenance, a stash of paper was found under a rock. Kelly had her future fish supply banked by multiplying her harvest then rationing the humans with torn bits. But Kelly was about to hit the big time. One day a gull flew into her pool; she presented the drowned feathered prize to her trainer, who gave her *several* fish. So Kelly saved some of the fish under her rock, and used them to lure other gulls into the pool to catch and get more fish. This involved planning *and* delayed gratification. She then taught her calf, who taught other calves, and gull-baiting in Mississippi's Marine Life aquarium caught on. In 2005, when the aquarium was destroyed by Hurricane Katrina, Kelly and eight other captive dolphins left to the mercy of the storm miraculously escaped the flooded wreckage into the harbour at Gulfport.★ Kelly kept the whole pod together at a safe distance for 12 days. A pity it wasn't further. Kelly was moved to the swanky Atlantis Resort in the Bahamas where the Dolphin Cay pool has the appearance of a large open space, but the dolphins are only released into cake-slice sections of the wider bay when it's time to do their tricks.[55] Dolphins have the disadvantage of a permanent smile, which we read as if they are happy.

The implication of rat laughter is chilling in our world, and even more so in a laboratory. In 1997, when the neuroscientists Jaak Panksepp and Jeff Burgdorf watched their laboratory rats play in rough-and-tumble bouts, they began to wonder if there was a sound component they couldn't hear. An ultrasound recording then revealed the rats were admitting high-pitched 50 kilohertz chirping noises, beyond our hearing range. Panksepp and Burgdorf described it as a '"laughter" type response'.[56] These laughter-type sounds came only with pleasant experiences. These were lowly rats. Laughter was our domain. It was one of the characteristics that set us apart. We laughed. Uniquely. Panksepp, who had been mulling it over, came into the lab one morning and suggested to Burgdorf, 'Let's go tickle some rats.' So they did. They tickled the rats and the chirping sounds increased to double the intensity they had recorded during play. The rats liked being tickled and began to seek it out. The sound of rat laughter being an important component of their play (as for children) was confirmed by deafening a control group of rats, who played far less

---

★ The metal roof sheared off and the amphitheatre was torn apart; aerial photographs show a scene of total devastation.

intensely. Which is a little depressing in an experiment that demonstrates the 50 kilohertz chirping as 'the sounds of social joy'. As usual, the burden of proof was the creatures' burden.

Thanks to ethologists like Frans de Waal, psychologists began to look beyond the limits of behaviourist theory and are even attempting to step into the *Umwelt* of an octopus. Sheep call out to photos of their friends. Crows recognise their capturers. Chimpanzees wipe the floor with us in computer-simulated memory tests where numbers flash up and disappear in milliseconds. A chimpanzee in Kyoto called Ayumi can replay sequences of nine numbers in 60 milliseconds, in less than a blink. Rhesus macaques are all over us at paper, scissor, stone. Horses can read *human* body language better than we can ourselves. Most humans can't tie a clove hitch. Even with a YouTube video and a pair of tweezers we couldn't build a long-tailed tit nest for love or money.

Do I sound like I'm tearing my hair out? I have just watched a fledgling blackbird watch another blackbird pick and eat berries off a tree in front of my window. She watched and then she had a go. So that was 'the mental transformation of sensory input into knowledge about the environment and the flexible application of this knowledge',[57] or cognition (information processing) from a bird brain, and it took a couple of seconds. We are more similar than alien, closer than far apart. When apes touch their lips together like kissing, it *is* kissing. When they put their arms around each other as *if* they were embracing, they *are* embracing. And when they are tickled, they laugh.

GOAT KING OF THE STRAW CASTLE

The rolled aluminium lid reveals the contents of its tomb. I am staring at a school of *sardinillas*. Twenty-two headless and tailless fishes in one sardine can. One of those tins that lives in the cupboard, bought in Spain years ago, opened during lockdown.

I try to imagine the super trawler sweeping them up, flipping silver, little gasps, eyes popping in air, the weight of them sliding across each other. Trouble is, now I know the stickleback makes a nest and fans his young; I know a jewel cichlid can think through the problem of a worm and fry in his mouth. I also know that a fish injected with vinegar under his skin (to see if he feels pain) will rub himself on the gravel. There are no protections for 'fish' because we assume they don't feel pain, even though we now have the science to tell us they do. I examine the lemon-yellow box: *SARDINILLAS en aceite de oliva, 20/22 Piezas.* There's a picture of a bright blue sea, gulls flying overhead, sailing boats in the harbour. *Product of Spain.* I doubt that. We mow the sea and expect it to grow like grass. It is the harvest for which we give nothing back. On the radio this morning a British minister was trying to persuade the fishing industry of the importance of marine protection areas, where there was 'no dumping, dredging, and restricted fishing'. Jonathan and I howled '*Restricted?*' in unison. Example after example shows that if you enforce strict no-take zones, the consequential nurseries seed the ocean around them. But we can't have our liberties taken away, so we plunder on. We deserve what we have got, but the fishes don't. Fish or fishes? Fish beings, not clones of one fish, so fishes. Every year 47 billion fishes are caught for recreation, and between 1 and 2.7 trillion to eat.[58] Our non-attitude to them allows us to leave them to die, however long and painfully it takes. A fish out of water. The moral question that entitled an animal to consideration for the eighteenth-century philosopher Jeremy Bentham was not, can they talk, nor can they reason, but can they suffer? Wishful thinking that they can't.

The silvers of sardines and other flashy fish (I know) are created by guanine crystals in the scale cells called iridophores, which act like minute mirrors that gleam and flicker. By reflecting light in front of them to match the light behind them, silvery fish can just disappear. They blend into the water and light, then twist off the vertical and you see them cascade, bright and glinting. Sardines need each other. They swim in harmonies by watching their neighbours. They have pressure-sensitive pores along their body which can feel the furrows in the water of their nearest schoolmates, so they can gauge exactly how close they are. *Swooossh*, they turn at once. A ripple through the school: Danger. The fishes at the front have seen a porpoise. The message travels back in movement, light, water and speed. Now everyone knows. They swim tight together, heightened attention, faster, bodies synchronised (parallel), moving as one. Stick out, and you could be the next target. It's faster, safer and more economical swimming together. Get into the right spot and the slipstream spinning vortexes from your schoolmates' tails will give you a speed boost – like cycling behind your husband (yes, okay) – while the front guy gets a shove by riding the bow of the wave. Together they can swim up to six times further. Human aeronautical engineers have studied the dynamics of racing schools to exploit the turbulence created between the individuals for the layout of vertical axis wind turbines to increase their output up to tenfold.[59] A sardine school is made up of individuals, each with their own proclivity, some braver (or hungrier) swimming at the front, some trailing, some who tend to keep their left shoal-eye on their mates and swim on the right side of the pack where they can multi-task by keeping their right eye on the lookout, and vice versa. If it's your turn to scout you must leave the throng to check if the nearby predator is in hunting mode. Fish life is far more complicated than we'd realised, and a fish's cognitive abilities are far greater too.[60]

When I ate the 22 *sardinillas* on a piece of toast, with no consolation to them, they stuck in my gullet.

In coral reefs there are cleaning service stations run by wrasse. Queues form, big fish, little fish, waiting for their clean. All day the wrasse performs his duty, removing unwanted sucking parasites while feeding himself in the deal. The occupation requires cooperation and a good memory. The wrasse knows his clients. Who he needs to be careful of (grouper with sharp teeth) and who he might take advantage of. If he can get away with it, he will munch on a bit of skin and mucus for its sunscreen properties. Sometimes he has to apologise by doing a sort of shimmying massage. Cunning and diplomacy are alive and swimming in these complex, newly discovered reef relationships. Groupers go hunting with moray eels. They fetch them out of their holes by doing a little dance, *Hey, Moray! Fancy a spin round the reef?* The eel's snout appears. And off they go. The grouper can even point to where prey has disappeared into a crevice by standing vertically on his tail and shaking his head rhythmically towards the spot. The eel goes in and either nabs it or flushes it out for the grouper to have a go. Both do well out of the partnership. It's a bit of a surprise. Flexible brain work that can adapt and respond to the challenges of a sophisticated social life. Not the zombie world we imagined, glug, glug, aimless swimming, food in, waste out, big fish eating little fish and little fish swimming very fast.

Fish were not included in the 2012 *Cambridge Declaration of Consciousness.* It is not surprising that fish feel pain. Why wouldn't they? Pain is a limiter. Without it, what is there to stop them bashing themselves on a rock, or tearing their lips getting algae off barnacles? Rainbow trout tested in the lab show nerve cells that sense noxious stimuli: acid, bee venom and burning.

Ah, but how do they perceive pain? At the Roslin Institute scientists injected acid into the trouts' lips and they rolled around on the bottom of the tank rubbing their mouths. The trout injected with blanks didn't do this. When they were injected with morphine the trout stopped rubbing their lips.[61] What they couldn't do was squeal.

★

At breakfast at our lodgings on a walking holiday in Transylvania we watch some goats in the field outside. Two goat kids are playing a game. It's very familiar. We know the rules. One kid is standing on top of a stack of straw bales butting at another who tries to jump up. He's up! Now he butts the first one off. And so it goes. Round and round, up and down. Goat King of the Straw Castle.

IV: DUMB

The language of birds is very ancient, and, like other ancient modes of speech, very elliptical: little is said, but much is meant and understood.

Gilbert White, *The Natural History of Selborne*

boobooboo   *Short-eared owl*

John Bevis, *An A–Z of Bird Song*

'I was just thinking,' said the parrot; and she went on looking at the leaves.

'What were you thinking?'

'I was thinking about people,' said Polynesia. 'People make me sick. They think they're so wonderful. The world has been going on now for thousands of years, hasn't it? And the only thing in animal-language that *people* have learned to understand is that when a dog wags his tail he means "I'm glad!"—It's funny, isn't it? You are the very first man to talk like us. Oh, sometimes people annoy me dreadfully—such airs they put on—talking about "the dumb animals".'

Hugh Lofting, *The Story of Doctor Doolittle*

LET'S SPEAK TO THE ANIMALS

'Language is the Rubicon which divides man from beast, and no animal will ever cross it,' declared the Oxford professor Friedrich Max Müller in 1861. Since we started walking upright, what we had, and what other primates did not, were long pharynxes and larynxes low in our throats.★ The arrangement has allowed us to shape air into a vast array of sounds with our mobile, muscular tongues. With our big brains and big mouths we abstract objects and thoughts into words which we arrange in a

---

★ We are the only mammal not able to breathe and swallow simultaneously. (And consequently, the only species who can die choking!)

myriad of ways to encode and decode information. Language (and then writing) changed everything for humans. Information was able to transfer not just across genetics, or generations, but across centuries. We could cache knowledge and pass it on. Now each one of us can propel words across the world at the speed of light. Collaboration elevated us, all the way to the moon.

Yet, whale song carries across oceans; rooks talk together as they fly overhead; dolphins have individual signature whistles to identify themselves; honeybees waggle their abdomens; vervet monkeys have specific warning calls for *Snake! Eagle! Leopard!*; prairie dogs have a different call for *Man!* and *Man with gun!*; and I can tell you, having witnessed it many times, blackbirds have a very particular short note chant for the presence of a stoat in the garden. Wolf staring is what Barry Lopez called 'the conversation of death'. Animals communicate by sound, smell, chemicals, body language; there are animal sounds we cannot hear, colours we cannot see. In 1901 a horse called Clever Hans was so expert at reading the minute flickers of body language that he fooled his German owner and audiences worldwide that he could spell and do addition. This astonishingly sensitive and clever horse had learnt that the cue to get his carrot reward was to stop tapping his left hoof (the way he was taught to respond to questions) the moment his questioner's posture unconsciously relaxed – invariably when Hans had reached the correct number of taps. This observer-expectancy phenomenon, when uncovered, became known as the Clever Hans effect.★ As far as we know, animal communication is about territory, food, mating, danger, aggression, dominance, group socialisation, calling for Mum, that kind of thing. The syntactical range, flexibility, poetry and vast capacity of human language is our particular pleasure, and probably our pain. Animal language is understood to be biological, inborn, whereas 'true' language is supposed to be transmitted culturally. This belies the observed breakdown of song culture reported in 2021 of the Australian regent honeyeater, a bird now so rare that the young males are unable to find adults from whom to learn, thus the once long, complex song has become diminished and less attractive to females. And it forgets how whale songs change year by year like a hit parade; and that British wrens have different dialects depending where

---

★ Discovered by Oskar Pfungst, part of a commission sent to investigate in 1911. Pfungst, so impressed with Hans, suggested we humans should rethink our treatment of horses.

they are from – Geordie wrens, Cornish wrens, Essex wrens – all distinguishable by regional accents. When musician Joe Acheson slowed down the wren song it sounded like the swooping morning calls of gibbons echoing across the rainforest.[62] It is a very complex song that can throw out 740 notes a minute and is one of the loudest bird songs in the UK. When the BBC's *Springwatch* presenter Megan McCubbin speeded up her voice to wren-speed, the result was completely unintelligible. When she slowed it down, this is what she said:

Hi. I'm a male wren. Look at me on top of this hedgerow singing my heart out. I am exposing myself to all the elements just to show you how fit I am. Now if you are a male coming by, I want you to know that I am here and this is my territory. But if you are female, why not come and look at one of the six nests that I have already pre-made for you to have a look at. Our offspring would be so successful. So ladies, this breeding season I've got the nest for you.[63]

Exactly.

For me, the most astonishing recordings are of captive beluga whales imitating human speech, much as we mimic birds or howl like a wolf. The belugas sound just like a human speaking through a trumpet.[64] They are good at blowing raspberries too.

The linguist Noam Chomsky believes language is innate or pre-set, an inheritable evolutionary development unique to humans, and that learning comes only partly into it. He thinks this because of the relatively modest parental input in relation to a young child's super-fast linguistic competence. As Frans de Waal points out, no trait, not even language ability, evolves without antecedents. Indeed, songbirds and humans share at least 50 genes connected to vocal ability, including the gene FOXP2. Mice also have a version of FOXP2, active in the ultrasonic peeps made by the pups. For all the uniqueness of human language, speech is the sum of many parts: genes, brains, neurological wiring, fine-tuned motor control, anatomical architecture, and of course, something to talk about. Humans do, nevertheless, have a godly and fiendish faculty for language. And thereby hangs our tale.

But do we really *want* to talk with animals? If our dogs and cats could talk would we love them as much as we do? It could be embarrassing. Or even boring. Spaniel: *Throw it again! Throw it again!* Do I really want to talk to the thrush I love to see on the lawn? We are the grammatical creatures; it is how we experience the world. We replay the past and project into the future. Animals are more present. More now. What is the thrush singing? I think it's *Glory, Glory, Hallelujah*. Poetry, after all. We don't know about other animals' interior monologues, but it is hard to believe from Goodall's observation of Flo's son, the chimpanzee who sat for days by the stream where his mother died, that he did not have an interior monologue. Or our dog, Frida, yelping while chasing rabbits in her dreams.

From the horse's mouth

In the early 1700s the world was still a place of unexplored lands and unknown creature inhabitants. As Linnaeus refined his classifications, the Irish clergyman and prose satirist Jonathan Swift shipwrecked his sailor-surgeon, Lemuel Gulliver, on a series of ever stranger shores.[65]

The first creatures Gulliver lays eyes on in Houyhnhnms Land (discovered 1711, south of New Holland, Australia) are uncomfortably like himself but degenerate 'brute' versions. Bearded, hairy, filthy, bare-bottomed without tails, bounding around on all fours, the female's 'dugs' almost trailing on the ground. Gulliver is utterly repulsed. One of these 'brute' humanoids confronts him curiously, but Gulliver slaps him away. The creature's roar brings the herd charging over, forcing Gulliver to back up against a tree. The savages leap into the branches to 'discharge their excrements' on his head. The fabulous drama is interrupted by a horse, at whose arrival the defecating beasts flee. The horse looks Gulliver up and down but greets his outstretched hand with disdain. And so we enter a world where the horse, the Houyhnhnm, dignified, honest and rational, is master; and the human degenerate, the Yahoo, provides the labour. Another horse arrives. The two horses confer in neighs so articulate that Gulliver's ears can scarce believe. In turn, the horses can scarce believe the appearance of this well-dressed, shoed and stockinged Yahoo. Gulliver is escorted to their home, but to his horror he is put out with the 'detestable' Yahoos. On closer inspection Gulliver sees their human form is no different from his, but for the length of their nails, the profusion of their hair and filth of their skin. A mirror of his beast self. Gulliver's civility, cleanliness and curious clothes confuse the Houyhnhnms, whose language Gulliver begins to learn. What the Houyhnhnms want to know is how Gulliver learnt to *imitate* a rational creature.

Jonathan Swift turns the tables on us. Annotated editions of *Gulliver's Travels* ascribe the name Houyhnhnm to come from the sound of a horse's whinny; but accidentally or on purpose, the phonetic version of *Human* in a horse accent plays on the role reversal too.★ There's another dig at human hubris when the etymology of the word is described as 'perfection of nature'. Later the Houyhnhnm will comment to the clothed Gulliver on the contradiction of why nature had taught those of his kind to conceal parts of their bodies that nature had given them. Not that the Houyhnhnm is impressed with the human creature: unable to feed itself without lifting its forefoot to its mouth; its flat face and limited vision range, with eyes unable to see either side without turning; feet too soft; nails useless; eats without hunger; precarious on two legs. Not surprising to the Houyhnhnm the natural antipathy that every other creature has for the Yahoos.

Gulliver distances himself from the Yahoos and engages in conversation with his host and new master, describing the human world he has come from. To the rational Houyhnhnm, the ways of humans sound illogical and shocking. He cannot credit a world where Yahoos are the governing animals, or that horses could be in their employment. It gets worse . . . When the Houyhnhnm asks about the lives of his brethren in this far-off land, gullible Gulliver tells all. He describes riding, saddles, harnesses, horseshoes, carriages, coaches, spurs and whips. The Houyhnhnm is incredulous. How could a puny Yahoo get on board a strong, large, dignified Houyhnhnm in the first place? Why not shake the Yahoo off? Or roll on him and 'squeeze the brute to death'? Why indeed. Gulliver awkwardly explains. That human power over horses comes from training them young, beating them for misbehaviour and *castrating* the males 'to render them more servile'. We wince at every detail. Swift puts us behind un-human eyes to observe ourselves with their incredulity. Gulliver even reveals the horse's final destination when he is lame or spent. As dumbfounding are Gulliver's descriptions of drunken sailors, fighting, drinking, debauchery, whoring, gaming, murdering, robbing, poisoning, perjuring, cheating, forging, raping, sodomising . . . (no holding back, this eighteenth-century man of the cloth). For a creature pretending to reason to be capable of such odious behaviour, the corruption of reason seems worse than the brutality itself, where in contrast the Yahoos had no pretension

---

★ As Gulliver plays on gullible.

but purely followed their natural laws. How monstrous to give males
different educations from females, for instance; and how irrational to say
'the thing which was not' (a lie). Gulliver, as Swift's mouthpiece, gets
stuck into the human race as we listen with the Houyhnhnm's wide
eyes and ears: lawyers, drunks, syphilitic toffs so idle and stupid they lose
their fortunes and marry for money, producing 'scrofulous, rickety or
deformed children'; backbiters, scoundrels, ranters, tedious talkers, high-
waymen, 'bawds, whores, pimps, parasites and buffoons'. And with them
a multiplicity of maladies that occasioned enemas and purges of dubious
nature – and contrary to nature by changing the direction of the orifices,
'forcing solids and liquids in at the *anus*, and making evacuations at the
mouth'! Gulliver contracts 'such a love and veneration' for the noble
quadrupeds that he now sees his fellow man as detestable, a creature
who has made no use of his pittance of reason except in multiplying
desires he spends his miserable life trying to satisfy by his own inventions.
Gulliver doesn't want to go home.

The problem of the Yahoos is a much-debated matter for the
Houyhnhnms. An invasive, non-indigenous species, should they be exterminated? The Houyhnhnms wonder if Gulliver's description of castration
could render the Yahoos tractable until they should happily die out.*
That's telling us. Alas for Gulliver, it is decreed he must leave the island
lest he incite the other Yahoos. And so he is forced to make a canoe (of
Yahoo skins), in which he paddles away.

Ever the priest, Swift had pulled off a manifesto for animal rights-
cum-diatribe against fallen humanity. A master stroke to choose fiction
to lay out the horse's lot not to a rational man, but to a rational horse
in a human's earshot (ours). Everything sounded just *so* diabolical. What
if they *were* sentient, intelligent beasts with personalities? And yet as
fictional Gulliver described the sins of man in his treatment of the horse,
cargoes of human slaves were filling the holds of ships to shore up
empires. It is a Portuguese ship that rescues Gulliver. When he finally
reaches home he finds both his fellow man and family repugnant and
intolerable, so he buys two horses – never to suffer harness – and sets
them up in stables where he spends most of his time. The Houyhnhnm
is what the human is not, which is the point, I think, and curious in
light of Swift's vocation.

---

* When they would train the more amiable ass to do their service, hmm . . .

Signing over, signing off

It wasn't until 1953, when psychologists Keith and Cathy Hayes failed to teach their human-raised chimpanzee, Viki, to talk after six years of trying, that we humans grasped that non-human apes cannot physiologically produce the human vocal range because of anatomical differences. The logical next step was to try sign language. In 1967, psychologists Allen and Beatrix Gardner began to teach a young female chimpanzee American sign language, ASL, in their home, using it exclusively with each other. Her name was Washoe. Although she was two years old when her training began, she learnt 350 signs and invented her own terms, like 'water bird' for swan, and 'candy fruit' for watermelon. But what if a new-born chimpanzee was brought up as a human? Could an ape form a sentence? And what might they tell us? The possibilities of the most ambitious ape-language experiment to date preoccupied Dr Herbert Terrace, professor of psychology at Columbia University. It would be like talking to aliens. (Except chimpanzees are not aliens.) In 1973 Terrace acquired a male infant removed from his mother at two weeks old from a breeding facility in Oklahoma, and gave him to an ex-lover psychology student of his to rear as one of her own and teach ASL. Stephanie LaFarge had a number of children, and was happy and hippy enough to breastfeed the infant chimp. A considerable flaw in the plan was that neither she, nor indeed any member of her large, well-heeled bohemian family, was proficient at sign language. The baby chimp was given a joke name: Nim Chimpsky, a pun on – and dig at – Noam Chomsky, the American linguist who had famously concluded language was uniquely human.

With no chimpanzee contact, Nim was embraced into the bosom of his human family, but chimps develop faster than children and no sign language was used until he was three months old. He wore human baby clothes, sat on a lavatory, ate with cutlery, played with baby toys, whizzed round the garden on a bike and just about crawled across the ceiling. It didn't take long for Stephanie to realise how powerful her ape baby was. Way stronger, tougher, wiry, more agile, faster, more mobile and more demonstrative than her own brood, and impossible to subdue. She was nevertheless happy, as a loving mother, for Nim to express his chimpness. Terrace, however, had a more human destiny for Nim in mind. After 18 months Nim 'outgrew' his Upper West Side brownstone family home and was moved to a rambling estate in the Bronx belonging to Columbia

University, where Herb Terrace installed Mum No. 2. Laura-Ann Petitto
was a research student, 18 years old, and having an affair with Dr Herb,
20 years her senior. When the affair floundered Petitto fled in tears and
Nim was motherless once more. For three years Nim was allocated a
stream of language teachers who collected 24,000 'utterances' in sign
language, and after four years a vocabulary of 125 signs which he could
use in multiple combinations. But Nim, growing fast into chimp adoles-
cence, was not easy to manage. He was sweet and tactile, and violent and
vicious. This was the 1970s, but hey, surely it was not a good idea for
Nim to smoke pot and drink beer?★ Nim became unpredictable and
began to bite. A researcher needed stitches. Funding was drying up. Terrace
called a meeting with Nim's carers and teachers. He had ample data
thanks very much, and now the project was over. Everyone was shocked.
Nim was four years old and since leaving his mother had never seen a
chimpanzee, but when he woke up after being tranquillised he was in a
cage back at the breeding facility in Oklahoma where he had come from.
Nim's one piece of luck was called Bob Ingersoll, an undergraduate
psychology student at the University of Oklahoma, who befriended him.
Nim didn't bite Bob; Bob said, some people have 'bite me' written all
over them.

Terrace analysed his data and had an epiphany. It was a sham. Nim
was not signing spontaneously but only when prompted by a teacher;
nobody had a *conversation* with Nim. Nim learnt to sign words by imitation to get rewards. 'He just said, "gimme, gimme," and then he got. But
he didn't say "thank you" or "this is an interesting-looking orange" or
anything like that.'[66] Hadn't Terrace picked this up before? He claimed
he was too intent on watching Nim. Not good enough, Mr Psychologist.
And we might ask: would individuals of the innately strong hierarchical
chimpanzee species naturally *initiate* conversation with us, clearly the top-
dogs? A subordinate chimpanzee would not naturally pick up a banana
while being seen by a dominant chimp. Chimpanzees pant-hoot. Maybe
Terrace can tell us what pant-hooting means, *exactly*? Terrace said Nim's
first approach was to try and grab what he wanted; I don't know why
we would assume a first-generation human-reared chimpanzee could leap
into an evolutionary spaceship and *ask* for a banana when everything in
his chimpanzee make-up is screaming that he can just grab it instead?

★ Nim appears in *High Times* magazine in 1975 as a pot-smoking chimp.

The process of domesticating animals did not produce readymade sheep dogs. Terrace is still trying to figure out 'what all this meant' and 'why we talk and they don't'. Shame on him that no primatologist was overseeing this project.

James Marsh's 2011 documentary, *Project Nim*, shows Nim with his human family, with Terrace in a car, Nim drinking, smoking a joint. Another study more illuminating about our own species. Other participants in the project say Nim used signs in combinations and that there was ample evidence he initiated conversation. Nor is it the case that signing in apes is a form of imitation, for in the wild signing is one of their own powerful forms of communication. Moral philosopher and Princeton professor of bioethics Peter Singer asks the nagging question, 'Was it possible that Terrace had taken such a negative view of Nim's abilities because dumping a language-using, humanized ape back in a cage with non-language-using chimpanzees would be worse than doing that to an animal without the ability to use language?'[67] At best this was a sloppy experiment, at worst an arrogant and profoundly immoral one.

Nim and Bob Ingersoll became friends; Marsh's film shows them roughing about together, playing and signing. In 1982 Nim was sold to New York University's Laboratory for Experimentation and Surgery in Primates to be used in trials testing drugs for hepatitis C. Bob kicked up a media stink and Nim was rescued by an animal rights activist and sent to a sanctuary where Bob was able to visit Nim until he died of a heart attack, aged 26. In Marsh's film Terrace says, 'I never regarded him as a child, I regarded him as a scientific project. Nobody keeps a chimp for more than five years.'

'We did a huge disservice to that soul,' says Joyce Butler, one of Nim's carers, 'and shame on us.'[68]

Mike did it

A 300-pound gorilla looks through a camera at a mirror and clicks. Koko has taken her own picture. A self-referential image rich in double-double meaning. Here's you, looking at me looking at me. It made the cover of the *National Geographic* in 1978. The story of Koko and her guardian, Penny Patterson, was a love story that transgressed the bounds of scientific research. Koko, born in 1971 and separated from her mother after an illness, was lent to Patterson by San Francisco Zoo for a four-year PhD project to teach Koko to sign. When the time was up, the idea of returning Koko to zoo life was unconscionable for Patterson. She persuaded the zoo to let her buy Koko and set up The Gorilla Foundation as a funding organisation. Koko entered the human realm where she would live her 45 years as not human, nor free, and no gorilla life to speak of, but with a devoted carer who was in it for the distance. She had human playmates, famous visitors and a pet kitten, even a suitor, Michael (Mike), borrowed in hope he might become a mate, yet there is no escaping that Koko's home was a trailer behind heavy wire mesh. Penny and Koko were inescapably trapped by each other.

It's hard to get the raw video data on Koko's language abilities. What there is suggests prompting, projection and interpretation by Patterson, but whatever *was* going on didn't fall into the catch-all of the Clever Hans effect either. A gorilla is not a horse. There was clear signing, comprehension, self-awareness and two-way interaction that was not all about reward. Koko's funniest joke was to sign 'Mike did it' when she was asked what happened to some missing food from the fridge. The controversy of how much apes can understand and how much is projected and ascribed by us continues. It was as if Koko had to be more than she was, and yet she was more than enough. She blew raspberries and play-acted talking on the telephone. She stole Robin Williams' glasses, tried them on and looked in the mirror at herself. They tickled each other. She laughed. She was not imitating him. She was an expressive, demonstrative gorilla with obvious emotions and an inner life, whose gorilla *Umwelt*, had been taken away. Patterson did not forsake Koko. She knew you cannot just sever the bond without consequences on both sides. It does not follow that because animals do not have complex syntactical language they cannot think, as is often assumed. Koko's cognisance of human syntax or lack of it is not the point here. It's the dysfunctional relationships, inevitable when we pull wild creatures into our world and remove them from theirs. It rarely works out well, and we should stop doing it.

In 1981 Thomas Sebeok, an outspoken critic of ape-language studies, organised a three-day New York Academy of Science conference on interspecies communication entitled 'The Clever Hans Phenomenon: Communication with Horses, Whales, Apes and People' to call for a halt to teaching animals human language. Human researchers have stopped teaching chimpanzees ASL. The last chimps with sign language ability are reaching the ends of their lives. There seems to be a consensus that no good has come of it. It has taken us a long time, but we are now more interested in the chimpanzee or gorilla's own world, in their own methods of communication. What they might be saying to each other in their own way. Horse whispering, which involves more attention and acute 'listening', is altogether a happier place.

When a group of Dutch chimpanzees were moved to Edinburgh Zoo in 2010, it took three years before they communicated in the same way with the Scottish chimps. It was discovered they had a completely different vocalisation and gesture for apple.

LOVE, ACTUALLY

Alan Rabinowitz did not speak as a child. Whenever he tried to put words into his mouth, it was as if they constricted his ability to breathe. His whole body would spasm. This was no regular stutter; this was a meltdown. The New York City public school system put him in a class for disturbed children.

Alan discovered there was one situation when he could talk without any stuttering. When he talked to animals. Every day, after his special class, he came home and shut himself in a cupboard with his hamster and gerbil and told them his thoughts. He identified with animals, because he saw they had feelings which, like him, they were unable to express. Alan could see that because animals couldn't speak, people thought they were stupid. He felt he lived in two completely different worlds: one in which he couldn't speak with humans, and another in which he could speak with animals. One day, his parents took him to the Bronx Zoo. He remembers staring into the face of an old female jaguar. He looked at the bare walls of her cage and wondered how she had got there. He identified with the big cats as kindred spirits in their isolation. Alan made a pact that if he ever found his voice he would speak for them.

Alan got through school, passed exams, but he never spoke a whole sentence out loud to another human being. His parents tried psychologists, hypnotherapy, drug therapy, but nothing worked. In his last year at college his parents sent him on an experimental programme in Upstate New York where he learnt a mechanical method of channelling airflow and controlling his mouth. After 20 years of silence, he finally spoke. But this breakthrough did not change his inner turmoil of being alienated from humanity. Alan was good at science and began to study medicine. But he couldn't see it through because of the pain he felt on behalf of the laboratory animals. So he applied to the University of Tennessee to study zoology and wildlife biology. Just before he got his PhD he was invited to go to the small Central American country of Belize to study jaguars by the preeminent wildlife biologist, Dr George Schaller. Alan's task was to catch jaguars, fit them with radio collars and collect data. It must have seemed as if everything was against him. He was nearly killed in a plane accident, then one of his workers was bitten by a fer-de-lance snake and died; and as fast as he could gather data the jaguars were being poached. He knew he had to use his voice to fight for the jaguars and fulfil the childhood promise he'd made.

Six months later, Alan was standing outside the office of the prime minister of Belize. Ironically, for someone who avoided speaking to people, he had managed to get through to the highest level of government and been given 15 minutes to speak to the cabinet. He could not afford a single stutter. He needed to persuade this impoverished country, with no animal protection areas and little tourism, of the importance and economic benefits of saving jaguars. His 15 minutes stretched to an hour, and the prime minister agreed to set up the world's first jaguar sanctuary.

A month later, deep in the Belize jungle, Alan found the tracks of a large jaguar and followed the trail until it began to get dark. He turned around and right behind him, less than 15 feet away, was the large male jaguar he'd been tracking. He knew he could not run, so he squatted down facing the jaguar. The jaguar sat down. They watched each other. Alan was transported back to his five-year-old self looking into the eyes of the female jaguar at the Bronx Zoo. But *these* eyes were strong and powerful, and it occurred to him that he too was quite different from the young boy he had been. He suddenly felt terrified. He stood slowly up and took a step back. The jaguar stood up, turned and began to walk back into the forest. After about 10 feet he stopped and turned to look back at Alan.[69]

Alan Rabinowitz dedicated his life to saving wild cat species. He established extensive wildlife reserves in Myanmar and Thailand. He conceived and implemented the Jaguar Corridor across their entire range from Mexico to Argentina. He co-founded Panthera, an organisation dedicated to the conservation of the world's 38 wild cat species and their ecosystems. He mentored young scientists in the field. He made a difference. Alan died in 2018. He is loved and remembered by his colleagues as 'The Indiana Jones of Wildlife Protection'.[70]

LET'S SPEAK ABOUT THE ANIMALS

The gobshite, the gum-sucker,
the scare-the-man, the faith-breaker,
the snuff-the-ground, the baldy skull
(his chief name is scoundrel)
                    Seamus Heaney, 'The Names of the Hare'

The cow who

In 2016 *The New York Times* ran the headline 'Cow Who Escaped New
York Slaughterhouse Finds Sanctuary'. That 'Who' caught the eye of Peter
Singer, and he was pleasantly surprised. The perspicacity of a cow's dash
for freedom from the jaws of death appears to have thrown the reporter
and said cow out of the newspaper's stylebook. Could this at last reflect
a change in policy? Not according to the paper's Editor for Standards,
Philip Corbett, who held to the Associated Press's guidelines: that 'person'
pronouns are only given to animals with a name or where the sex is
specified. Otherwise, tough. Cows are its and thats and whiches, and that's
it. So the 'who' was a blip then. Nevertheless, the story was reported
variously across the media as the cow 'who', and the cow 'that'. Technically
this cow was a steer, a castrated male – which demonstrates the way we
talk about animals does not encourage paying attention – and now he's
got a name, Freddy, so it doesn't count. Freddy went to the Skylands
Sanctuary in New Jersey to live out his natural life. Which is so typically
contrary of humans. This sanctuary seems to do a roaring rescue service
for escaped slaughterhouse cows and steers. Brianna, for instance, who
fell off a cattle truck on her way to the slaughterhouse, was rescued and
gave birth to a female calf who will now 'never be without her mother',
said Mike Stura of Skylands. From meat line to the sanctuary in one
imaginative leap, Brianna earned herself a lifespan. March 2019, same
thing: a calf escaped the slaughterhouse, ran loose down the street, became
a TV star, earned his 'person' pronoun, and was rehomed at Skylands. At
writing, Skylands has 70 escapees. If livestock have the wherewithal to

not become deadstock, we humans can relate to them, and with our divine whim we name them and grant their reprieve.

England, 1998. Two five-month-old ginger Tamworth pigs, brother and sister, escaped from a Wiltshire abattoir while being unloaded from the lorry. They were chased through the streets of Malmesbury, but they dived into the River Avon and swam to the other side. They were on the run for a week. Top slot on the news desk, the nation fell in love. A celebrity tried to buy them. Donors offered 'silly money' for their safe retirement. Sanctuaries across the country competed to give them a home. By the time they were captured the boar had been named Sundance, and his sister Butch. Their owner, Arnaldo Dijulio, said they were worth 40 quid each and he wasn't prepared to discuss it. Whatever money passed hands, Butch and Sundance ended up at a rare breeds farm in Kent where for 13 years they were the main attraction.

Third-person singular

What is 'it' about? It: '*pronoun*, the neuter of **he** or **she** and **him** or **her** applied to a thing without life, a lower animal, a young child . . .'[71] It's a lot for such a tiny word to carry. Is *it* serviceable? Perverse? Anachronistic? It's unscientific, surely? It's demeaning. Literally. Some animals, like parrot-fish, begin as females then change into males, and with wrasse it's vice versa, and some animals are difficult to gender, but there are so many we can. Cows, for instance. A handsome pair of bull's castanets will put us on the right track. Antlers, a peacock's tail, a lion's mane, a blackbird's song. Nevertheless we still use 'it'. To call a person an 'it' is the height of insult, *Oh, look, it's arrived.* Pet owners will bridle if, after meeting Daisy the dog a few times, we persist in calling her a he, or an it, so we make an effort. We now graciously extend personhood to chimpanzees and gorillas, *who* can be *he* and *she*, but that's it; the rest get to be 'it' with the pronoun 'that'. The gorilla who, but the dog that. What is the point of grammar if not precision? Scientists with the strongest commitment to precision are prescribed a grammar unfit for purpose: '. . . a dead female guillemot with a fully formed, perfectly coloured egg in *its* uterus . . .'[72] Why, for Darwin's sake, if he or she has a demonstrable gender, is he or she an it? *It* is so entrenched that if you question *it*, you are being sentimental at best, or insanely politically correct.

Of course, animals won't know what we call them, but language directs

how we think about them. Spiders are 'it', even when triple the size of
the male of the species. We make exceptions for the *femmes fatales* to
whom we give humanoid pronouns on the basis that they eat their
husbands after copulation. Hence the black widow spider gets her name
(unpack that for a bit of not-so-covert racism and sexism). Follow the
logic of this pronoun exclusion zone, and we can only ask 'What' questions of animals and 'Who' questions of ourselves, when surely today
there is more occasion to wonder *who* the bear is, rather than *what* 'it' is.
The *Oxford English Dictionary* permits 'who' to be used for an animal
'with implication of personality' . . . But *who* will decide this?

★

In 1935, in a garden on the Greek island of Corfu, the ten-year-old Gerald
Durrell made a discovery. He was watching a lacewing on the roses,
admiring the delicate insect's glass-green wings and golden eyes. No
third-person singular nonsense for him. She was a she from the instant
she lowered her abdomen.

> She remained like that for a moment and then raised her tail, and
> from it, to my astonishment, rose a slender thread, like a pale hair.
> Then, on the very tip of this stalk, appeared the egg. The lacewing
> had a rest, and then repeated the performance until the surface of
> the rose leaf looked as though it were covered with a forest of tiny
> club moss. The laying over, the female rippled her antennae briefly
> and flew off in a mist of green gauze wings.[73]

An unkindness of ravens

As slaves were slavish, animals were brutish, and still are. With our meta-
phorical language we do things we hardly notice we are doing. Like all
powerful tools, words can be used for good and ill. Names foul in the mouth.
They can obscure as much as they can reveal. We can use them to pervert,
poison and play with our minds. Language structures our consciousness.
Changing language changes views, because language is loaded.

Bitch.

Pig.

Snake in the grass.

The Taliban entering Kabul in August 2021 were reported by news commentators as showing their 'sheep faces', although it was expected that they would soon show their 'true nature as wolves'. Even the word *wild* has connotations: out of control, unkempt, mad even. The opposite of civilised, cultivated, sophisticated. What makes us human, or like an *animal*? Words project thought. And insult.

Vermin. Scavenger. Pest.

Let's blame the animals.

Ape /eɪp/ noun
1. a large primate that lacks a tail, including the gorilla, chimpanzee, orangutan and gibbon.
2. an unintelligent or clumsy person.
verb (**apes, aping, aped**)
    imitate (someone or something), especially in an absurd or unthinking way . . .
PHRASES
    Go ape (go crazy), God's ape (a born fool), go ape-shit . . .[74]

If someone called you Lizard Lips, would you like it? Perhaps not. Nor do we like being accused of weaselling out of something, of worming our way in, or of toadying up to someone. We hang adjectives on creatures which they cannot shake off. Sly. There is nothing *deceitful* about a fox, for he must eat and he must feed his cubs. In the same way that we dehumanise our human enemies before we ask our young men to kill them, we en-mean animals. Vermin implies vicious, wicked, detestable; say 'vermin' and you are absolved. From vermin to varmint to the near extermination of the wolf. That is how far words can stretch.

Euphemisms are the cloaks we employ to protect our sensitive souls. Control. Manage. Harvest. We know we do it, but we do it anyway. Other words can write things off. Wasteland. Quagmire. Swamp. Wilderness. It is empty. There's nothing there. Language is so deeply ingrained in our psyches that we succumb to it even though we understand the mechanics of metaphor. Even our efforts to be politically correct produce 'non-human animal' – a negative term for what is not us. Non-rhinoceros? Non-pigeon?

Other words shoot blanks. Like by-catch. A tiny word for a very big thing. A shrimp trawler throws 80–90% of the sea animals in its nets, dead or dying, overboard. For every pound of shrimps, 26 pounds of sealife must also die. That's dolphins, turtles, sharks, whales. By-catch. I've mentioned 'biodiversity' (unless you skipped the introduction). Such a flatliner for *everything* that has life; the incalculable species of insects and fungi and slime, let alone a wren's beating heart. 'Biodiversity' squashes the luminous zing, the sleek glory. Wild living community? Planet inhabitants? It's not easy. But we need more to care about, more *magpie garble*, more moose who loom.[75]

In 2007 the *Oxford Junior Dictionary* contentiously removed the English words for adder, beaver, boar, heron, herring, kingfisher, lark, leopard, lobster, magpie, minnow, mussel, newt, otter, ox, oyster, panther. As if they were no longer necessary in a modern world.* The writer Robin Wall Kimmerer explains how, in the language of her people, the Potawatomi Nation of North America, a 'bay' is a noun only if the water is dead and stilled between its shores. Whereas the verb *wiikwegamaa*, to *be* a bay,

> releases the water from bondage and lets it live. 'To be a bay'
> holds the wonder that, for this moment, the living water has
> decided to shelter itself between these shores, conversing with
> cedar roots and a flock of baby mergansers. Because it could do
> otherwise – become a stream or an ocean or a waterfall, and there
> are verbs for that, too.[76]

For the Potawatomi all beings are persons – the Beaver people, the Bear people – and trees are the standing people. Kimmerer calls it the grammar of animacy, that every sentence reminds us of our kinship with the

---

* The writer Robert Macfarlane and artist Jackie Morris responded with *The Lost Words*, an illustrated book of spell-poems to bring back the excised creatures.

animate world. Humans are considered the beings who must look to the teachers all around them. Had the human people learnt from the council of animals not to interfere with the sacred purpose of another being, the eagle would look down on a different world and salmon would be crowding up the rivers.

Let the animals speak for themselves – II

> But where a passion yet unborn perhaps
> Lay hidden as the music of the moon
> Sleeps in the plain eggs of the nightingale
>   Alfred, Lord Tennyson, 'Aylmer's Field'

The female *Photuris* firefly is as duplicitous with language as we are; she flashes the female semaphore of another firefly species, the *Photinus*, to lure in the smaller *Photinus* male. Then she eats him.

Many songbirds use alarm calls of pure tone, making their whereabouts difficult to locate, but for an elaborate song to impress they tune to a wavelength that will make themselves easy to locate by their own kind. The flourish of the nightingale's virtuoso performances, his piping, fluting and limitless extravagance, with the added anticipation of every pause, might make *us* pause to wonder how he is composing as he goes along, and what he is singing. Remember, too, that most of the song's complexity is beyond the scope of our hearing. The thrush in our garden will sing far longer in autumn than necessary to assert himself as the territorial king, appearing totally absorbed, falling under his own spell. The writer Richard Mabey sees birdsong as not necessarily a language 'but it is expressive – expression*ist*, if you like. It conveys a bird's emotional tone, be it proprietorial, angry, sexy, contented, sociable, exuberant – states of mind we intuitively understand.'[77]

Consider what the enigmatic giant cuttlefish is flashing with her surround-vision mantle, a living billboard of technicolour signalling. Rippling of stripes, blotches, flares, necklaces of glowing pearls, a cloud passing, dots, jags, washing rainbows. The displays are brain-activated through neural pathways to muscles that contract and relax to reveal or obscure the particular pigment held in each chromatophore – of which there are millions. In the layer below, iridophores reflect and bounce light like a stack of mirrors, filtering into blues, greens, violets and silver-whites. To this electric colour show add a skin with papillae that can sculpt itself into an array of shapes and textures; then add the gestures of eight independent arms (and two feeding tentacles). A show of horns, hooks, clubs, a flinging of arms aloft; the male will flatten his fourth arm into a flat blade in a show of aggression.

For goodness' sake, what do *we* know? The cuttlefish has a potential banquet of signalling variations of which we can only dream. The combinations of shapes, gesture and colour has the potential (if not the need or realisation) of a language as complex as our own. Peter Godfrey-Smith, writer/philosopher/diver, once observed a cuttlefish from above flash a *passing cloud* on her right side to another cuttlefish, while her left side remained unchanged and camouflaged. What *was* she saying?

Bewilderingly, these magicians of colour are supposed to be colour-blind.

With only one kind of photoreceptor cell (we have three), the cephalopod should be unable to respond to different wavelengths of light. But that conclusion is hard to accept from an animal who can trigger displays which appear highly intentional – for camouflage, defence, mating. So what's going on? We don't know. The suspicion is the answer lies in the unusual off-axis shape of their pupil, and by exploiting something called chromatic blurring to decipher spectral information. Cephalopods have been observed going through tremendous choreographed repertoires on their own, for no apparent purpose. Of course, it's unscientific to suggest they might be practising, or doing the visual equivalent of whistling, or occupying themselves for their own – dare I say – pleasure? We are having to change our minds about the cephalopods, the mollusc family that includes squid, octopus and cuttlefish. Once thought of as unsocial creatures, we now observe gatherings of octopuses where behaviour and radiant expression seem to exceed any biological function. It's possible, yes, that these shimmering shows are just manifestations of electrical activity. More likely, there is more going on out at sea than we imagine.

V: DAM NATION

I consumed his heart, and then
on his antlers hung my raincoat.
Nina Cassian, 'Sacrilege'

PARADISE

The Eurasian jay, who can fit up to nine acorns in his gullet, and works
10 hours a day selecting the best for a winter larder from mid–September
to November, should be credited with planting the fleets of ships which
sailed the oceans to settle lands and conquer nations. For though the jay
needed most of the acorns himself, there were enough to grow into great
oaks and medieval wood pastures to build the hulls to cross the seas and
feed the next generation of jays to plant the acorns to grow the oaks to
feed the jays to plant more ships to sail the seas.

One German study found that 250 jays planted 3,000 kilos of acorns
in 20 days. In the spring, when the acorns have begun to grow, the jays
will tweezer out the sapling's juicy cotyledons, which the oak – now
established with a tap root – doesn't seem to mind, for the jay and oak
have a mutually beneficial partnership which humans muscled in on.
Licences to exterminate jays are still given to humans who call them
pests, even though they planted our galleons, the rafters of our churches
and the barrels for our brandy.

In front of me, the delicate wash and ink line of a cartographer's pen:

A CHART OF PART OF NEW ZELAND OR THE
ISLAND OF AEHEINOMOWE LYING IN THE SOUTH SEAS
BY LIEUT: J. Cook. COMMANDER OF
His Majesty's BARK the ENDEAVOUR.
1770 —.

The mouth of the island yawns open into the Bay of Plenty; a pimple rash of mountains flares down the east coast of the North Island to Cook Strait. Names mark the human biography of place: Cape Runaway, Poverty Bay, Hawke Bay, Cape Kidnapped, Cape Turnagain. Naming will seal its fate. *Country here fertile and well Cloath'd with Trees* is written above clusters of tiny cauliflower florets that dot the chart like parkland oaks on wallpaper. It is the unnamed 'emptiness' of the interior of this map that beguiles me, pulling me into a deafening cacophony of birds. Jonathan's family live opposite the *Hen and Chickens* and *Barrier Isles*; Little Barrier, unmarked, is now a refuge island of dense native bush cleared of rats, cats and people for New Zealand's flightless birds. On a calm day you can hear the soundscape of Little Barrier drifting miles across the sea, the booming kakapo, the bellbirds, the ethereal fluting of the kokako, the loud call of the kiwi. The chorus is a vestige of what creatures this map once held in its unmapped heart.

In the early 1300s Polynesian navigators first pulled their canoes ashore onto an untouched land whose inhabitants had thrived in isolation for 80 million years. Frogs, geckos, skinks, beetles, giant wetas,★ a thousand species of land snail, earthworms over a metre long, and birds, birds, birds. Birds dominated the ecological niches. Without mammal predators there was no need to fly, and some had just carried on growing. A striking anatomical feature of the enormous ostrich-like moa is that they have no trace of wings, no upper arm bones, not even a vestigial structure. A female moa could stand 12 feet tall and weigh 250 kilos. Almost twice as tall and three times as heavy as the male, whose bones were classed as a different species until 2003. With their long necks they could browse high and graze low. Their only enemy was the Haast's eagle, the largest eagle that has ever lived, at a whopping 15–18 kilos borne on a 10-foot wingspan, fast, strong, with huge musculature and bone-crushing talons as big as a tiger's. She could fly through the forest at great speed, swooping at 40–50 miles an hour. Her huge size, a direct evolutionary correlation to the size of her prey in absence of competition, is called *island gigantism*. The arrival of humans invariably marks disappearances. With no time to evolve defence against tricks or trapping it took less than 200 years to see off the moa, which was in turn the nail in the coffin for the Haast's eagle. The eagles, losing their dinner, turned briefly – so they say – to

---

★ Huge flightless crickets.

hunting Māori children; programmed to hunt large creatures on two legs, they were certainly capable of it. Isolated and tucked away in the southern seas, these were the last of the giant species to go extinct. In the 1800s sealers, whalers and missionaries would bring their cats and rats to this terra incognita. Then axe and fire.

★

In 1833, William Forster Lloyd, a political economist at Oxford University, wrote an essay on the unregulated grazing of common land. His thesis was that in a system of open access, where a resource is shared, individual users will take as much as they can, contrary to the common good. So, before the other guy eats it, fishes it, hunts it, gathers it, I'd better get as much as I can, because if I don't, he will. This concept became known as 'the tragedy of the commons' after biologist and philosopher Garrett Hardin wrote an article in 1968 extending this idea of the 'commons' to all shared unregulated resources: the sea, rivers, fish, atmosphere. Free access to common resources tends to collapse from overuse or over-harvesting, except when small populations manage to cooperate, or self-regulate their use of the resource prudently, like the Native Americans and the buffalo, and some fishing communities before industrial fishing.

The world's creatures lived peaceably, eating each other – but only what they needed – for millennia. Our needs were never-ending. We had wars to win, soldiers to feed, bodies to clothe, desires to satisfy. Head out to sea and we can light up the world: oil, spermaceti, blubber. Tons of it. Fifteenth-century engravings celebrated the harvest, great leviathans rolling over beneath a hail of harpoons. By the seventeenth century some species were near extinction, like the right whale, named for being 'the right whale' to kill. To picture the abundance of life in the ocean before industrial fishing, in his book *Oceans of Life*, marine biologist Callum Roberts points to a painting by Flemish artist Frans Snyders (1579–1657), *The Fishmarket*. A fishmonger's stall overflows with giant cod, whopping crabs, lobster, ling, eels, steaks of salmon as big as his head. These sizes were normal. In early photographs, Grimsby pier is covered end to end in Goliath-sized cod. In 1939 the portly mayor of New York City stands dwarfed beside a 300-pound halibut twice his size. 'For every hour spent fishing today, in boats bristling with the latest fish-finding electronics, fishers land just 6% of what they did 120 years ago.'[78] And that was from sail craft open to the elements. Roberts likens the relationship between politicians and the fishing industry as 'a doctor assisting the suicide of a patient'.

Atlantic cod was exploited to levels that reached evolutionary suicide. The fishing-out of large individuals induced selection towards earlier sexual maturation, and a fish less than half the size. Since 1960 Newfoundland populations have fallen by 99% – that's more than 2 billion reproductive individuals. When the Newfoundland cod fishery closed in 1992 scientists were confident stocks would quickly recover. So far they haven't. Gone are the days of the 25-year-old, 5-foot-long, 100-pound fish. There are multiple problems. One is that juvenile cod, eggs and larvae are easily predated by the fish who were once prey to the larger cod. Without big cod to keep these prey fish in check, their numbers increase with more pressure on the cod juveniles. Another is that smaller cod are less successful in mating. The bigger cod courted females more vigorously, produced more sperm of higher velocity that were able to fertilise more eggs, and so produced more offspring to replenish their population.

★

Here's the trail of a bountiful marine mollusc with a glossy, egg-shaped shell. The Money Cowrie, as the name suggests, was adopted as currency (since around 1200 BCE) for the properties of its shell – being small (3 centimetres long), uniform, lightweight yet durable and hard to break. The female Money Cowrie broods her eggs until they hatch and the tiny molluscs are swept away in the current before finding a place to grow up in. In the Maldives (where cowries were harvested in eye-watering numbers for centuries) the islanders wove coconut-leaf mats to float in the sea to attract them. When the millions of molluscs that had attached themselves to the mats reached the right size, the mats were hauled onto the hot sand to dry out. The shells were loaded into baskets, each one holding 12,000 shells, then packed off on a ship to sail to India, where they were exchanged for rice and cloth. So plentiful and cheap were they that Arab merchants carried them on camel trains across the Sahara, while Portuguese, British and Dutch trading ships filled their holds with cowries as ballast. From the London and Amsterdam trading houses they were shipped to West Africa, where they became the local medium of exchange – for enslaved humans. What a journey. Clink, clink, from humble marine

snail in the warm waters of the Indian Ocean, to hard currency. A trans-
action of mollusc exoskeleton for human muscle. As Europeans bought
up cowries in such vast quantities, so the monetary systems and economies
crashed. In 1520 an enslaved adult man fetched 6,000 cowries; in the
1700s you'd need 25,000; by 1770 the price was 160,000 cowrie shells.
As the eighteenth century saw the peak of the slave trade, so it was
mirrored in shipments of cowrie – British fleets imported 40 million a
year. As marine biologist Helen Scales points out in her book *Spirals in
Time*, the billions of shells pay testament to the humble cowrie's reproductive proficiency against our rabid exploitation.[79]

The never never ending

I was done for from the moment I first laid eyes on the monumental
landscape paintings of the American Sublime. The never-ending golden-
green horizons of forests, the copper light reflecting on a distant bend
of a river, the smoky purple ranges and sparkling cascades, rainbows arcing
from great bruising skies, the misty indigo deluges of faraway rain, bolts
of light blasting through stacks of cumulus and crashing into rocky crags
and snow. An Arcadian wilderness, bountiful and untamed. Yet frightening
too, and easy to be unnerved by vertigo and one's own shrivelling. I tried
to bore my gaze into their heart. What I was thinking was: lurking bears,
prowling mountain lions, waterfalls tumbling with bounding salmon,
rainbow glitters of fish, running deer, browsing elk, rocks howling with
wolves. These paintings thrilled my eyes and imagination, but I should
have looked more closely. In *Lake George* (1869) by John Frederick Kensett,
there is a tiny canoe paddling across the glazed surface of the water in
the dark shadow of a willow thicket. In 1869, when the Hudson River
School was at its height, every pristine river and creek was studded with
long trap lines of iron jaws set wide open. If not, it was because trappers
had already cleaned the place out.

The fur trade was a force like no other in the course of American
history. From 1600 to the twentieth century it spurred colonisation,
displaced indigenous nations, decided settlements and determined the
course of empire. The land was mapped by its fur- and feather-rich river
courses, those vascular paths of water budding across parchment with
names that spoke of a country patterned by its inhabitants, *Beaver River,
Otter River, King Fisher Lake, Bear Hill, Antler Mountain, Moose Woods, Swan*

Lake, Elk River . . . almost everything is animal. Wherever there was water
and wood there once were beavers. From the Atlantic to the Pacific, from
the deltas of the Arctic Ocean to south of the Rio Grande, beavers had
been hunted for centuries. European rapacity, however, would change
both the ecology and the shape of America.

By 1670 the vast watershed flowing into the Hudson Bay – 1.5 million
square miles then known as Rupert's Land – had become the commercial monopoly of the Hudson's Bay Company (HBC), which functioned
as the area's de facto government for an astonishing 200 years.* Forts,
factories and trading posts sprang up along the waterways all the way to
the Rockies. The indigenous peoples, the 'Indians' who had traded beaver
pelts for beads, blankets, knives and cooking pots, now shared the rivers
with white trappers. North America was made by beaver. The unit of
legal tender became equivalent to the value of the animal itself – the

---

*   Granted by charter by King Charles II after the English had ousted the Dutch from
    what is now New York in 1664. (The East India Company operated in India during
    the same period in a similar way.)

'Made Beaver' or MB; one adult beaver (winter) pelt = 1MB. Brass coins were stamped with the MB value on one side, and the HBC coat of arms and motto, *Pro Pelle Cutem* (a skin for a skin) – whatever that was supposed to mean – on the other. A hunter in 1733 who came to a trading post with a black bear skin would get 2MBs; a trapper with two otter pelts, 1MB. For 10MBs the trapper could buy a gun, or a blanket and a gallon of brandy. Ships loaded with pelts plied their way to London and Canton. The demand was inexhaustible. As the rivers became depleted, the response was to expand. In 1821 George Simpson, the HBC's governor, wrote, 'an exhausted frontier is the best protection we have against the encroachments of rival traders'. He targeted the Snake River, the largest tributary of the Columbia, 'a rich preserve of Beaver . . . which for political reasons we shall endeavour to destroy as fast as possible'.[80] As the French emptied the St Lawrence River networks, the English worked their way through Rupert's Land and the Yankees came up from the south. Every dam was broken, every lodge cleaned out. River by river, beaver populations that had sustained North America's First Nations in food and fur for 20,000 years disappeared.

Beavers are big rodents who fell trees. It can be an uncomfortable idea for animals like us who believe *we* are the stewards of the land. When you see your first 2-foot-diameter beaver stump surrounded by a halo of curling beaver-toothed wood shavings, it's hard to take in. The Scottish writer Jim Crumley describes beavers as nature's architects. They are hydrologists and engineers, too, redirecting water, heaping up sediment, a bit of felling here, digging a channel there, a dam, a splash pool and a very snug house for the winter. Ernest Thompson Seton, a naturalist/artist/writer raised in Canada at the end of the nineteenth century, credited them with inventing reinforced concrete in the form of mud mixed with sticks and stones. Crumley even wonders if Frank Gehry watched them. Gehry's phrase 'liquid architecture' sums it up: 'It's like jazz – you improvise, you work together, you play off each other, you makes something . . . .'[81] Gehry's buildings look like they are mid-grow.

There is a beaver dam in Northern Alberta, Canada, which is half a mile long. It was spotted on Google Earth by landscape ecologist Jean Thie in 2004 while studying the rate of permafrost melt. With the help of NASA World Wind images, Thie estimated beavers had been working on the dam for half a century. The reason it's so long is because the water run-off from the higher ground of the Birch Mountains creeps slowly down a gently sloping alluvial fan through wetlands over a wide area, which forces the beavers to keep lengthening the dam to corral the water. Beavers take the long view and each generation takes up where the last left off. Their landscapes are dynamic and changing. The life-giving power of water is the expertise of beavers. Remote and undisturbed, 118 miles north-northeast of Fort McMurray, the dam which Jean Thie spotted lies inside the Wood Buffalo National Park, whose staff had no idea of its existence until a BBC film crew gave them a call, having seen Thie's web page. Park staff flew over, but the ground swamp water was too shallow to land a floatplane. An hour's flight from the nearest town and part of a series of dams in a major beaver wetland system, they could hardly wade in. So Parks Canada took aerial photographs and on 16 September 2020 wrote: 'Wood Buffalo National Park staff have flown over the location but because of its dense forest cover they have been unable to land and do a physical survey of the dam.' It's a miracle. Somewhere so difficult to get to that we don't go. They are on their own, doing their beaver thing. That thought brings me back to the way I once saw the paradise paintings of the American Sublime. Aerial images of the dam show a long gash in the quilted green canopy – a defined raggedy line where the dam arcs against the dark indigo water that mushes out into pale swamp before the trees give back the green. The closeups reveal a mosaic watery world without us. I'm looking at the lodge, a humped lozenge of brash rising from the surface of the black-blue lake. You can see the backbone skeletons of drowned trees. This is the second-largest protected area in the world at 27,800 square miles and one of the largest inland freshwater delta ecosystems; wild whooping cranes breed here; wood bison roam. A northeast view shows forest wetlands marching into the curvature of the Earth.[82] When I feel down about the world, I look at these.

While hunter-gatherers were napping flint to sharpen their spears, beavers were logging forests. The beaver's tools were their tooth chisels, their dextrous front paws, and multipurpose tails: flipper, rudder, warning slapper and ballast when walking with armfuls of mud. They can stay underwater for 15 minutes; they keep a clean house; they play, wrestle, fight. They might be rodents, but they don't breed like them for they bond for life, with two to four kits in each yearly litter. What did for them was the quality of their fur – a coat to die for. Up to 23,000 barbed hairs per square centimetre interlocking into a dense waterproof felt. It took three

beaver pelts to make one exceedingly fashionable, water-shedding top hat.
Wander through any national portrait gallery, past Vermeer, Rembrandt,
Gainsborough, and the black hats you see on any important head, indeed
every gentleman until the mid-nineteenth century, are likely to be pure
beaver. Stove pipes, the Paris Beau, the Regent, the Wellington, the D'Orsay,
Napoleon's bicorne, Nelson's naval cocked hat. Chaucer's Flemish merchants
wore 'bevere' hats in the 1300s – although these would have come from
the European beaver.★ Firm, smooth, waterproof (no umbrellas until the
late 1600s) and durable. The other valuable commodity was the yellowish
aromatic liquid produced by the beaver's castor glands to mark territory:
castoreum, a seventeenth-century wonder drug prescribed for gout, epilepsy,
toothache, even madness (pity that didn't work). The perfume industry
(Chanel, Givenchy, Lancôme, Guerlain, Paloma Picasso) continues to use
it for 'notes of leather'. To this day, the US Food and Drug Administration
(FDA) approves the use of castoreum as a 'natural' ingredient to enhance
strawberry, vanilla and raspberry ice cream flavours.

★

Importation dockets tot up the wild lives taken from North America's
rivers, lakes and forests – every fur-bearing creature that breathed the
North American air. In 1840 Edward Taylor, Broker, imports for sale in
London: 55,432 beaver, 198,236 muskrat, 4,923 black bear, 35,845 lynx,
56,860 marten, 8,185 wolf, 8,636 otter and similar eye-watering numbers
of mink, racoon, wolverine, fox, badger, isinglass (gelatine from sturgeon),
'Sea Horse Teeth' (walrus tusks), goose and swan quills, oil and bed feathers.
Ship after ship, year after year. A thousand wolves here, 20,000 martens
there. Among the 'Goods for Sale by The Candle at THEIR HOUSE, in
Fenchurch Street', the HBC lists also include porpoise, whale bones, white
bear, eider down, buffalo wool and crow quills. In London's Bishopsgate
stands the last 'Hudson's Bay House' built in 1926, its weathervane – a
golden beaver, on top of its cupola – turned no more by London's smog
thick breeze, but by gusts that roar round the towering glass edifices that
deal not in fur but plain money.

Naively, I had thought *wild* fur was a thing of the past, but no one today

★ Once numerous and widespread, by the twentieth century there were only about
1,200 beavers left in Europe and Asia. The beaver was extinct in Britain by the sixteenth
century.

sells more wild furs than NAFA, North American Fur Auctions, in business since 1670. In their 2014 May auction it took 2 hours 17 minutes to sell 157,207 wild pelts: beaver, otter, timber wolf, bears and lynx. Wolves, with their black nose pads pointing towards the ceiling warehouse, are hung just high enough so their bushy tails don't sweep the floor.[83] NAFA's July sale offered 18,483 wild coyotes, 259,951 wild muskrat, 50,769 wild red fox, 3,933 wild lynx, 19,778 wild beaver, 1,988 wild otter and 67 wild bears. In March 2018, the city of San Francisco passed a bill to ban fur. Just like that.

Cry me a river
Population estimates of beaver before Europeans set foot on American shores range from 60 to 400 million, and more than 25 million dams with ponds, wetlands and meadows full of life. In 1797 surveyor David Thompson reported a dam a mile long and wide enough to ride two horses abreast. Yet when American naturalist John James Audubon travelled 1,860 miles up the Missouri River in 1843 to collect specimens for his illustrated book on the mammals of North America, he did not see one single beaver. By the beginning of the twentieth century less than 1% remained. The loss of the beaver affected the hydrology of the land, and a whole host of North American species. Creeks dried up; beaver mead-owlands became tinderbox dry and fires began to rage. North America stood for a fast-speed ecological demonstration of what not to do. Let's see how quickly we can wipe out the buffalo, extirpate the wolves, empty the rivers, stop salmon, starve orcas, contaminate waterways, scorch the earth, dry the place out and blow away the soil. It was like putting a continent into a laboratory.

In the march of American 'progress' in the eighteenth and nineteenth centuries, a few souls could see the road to disaster. One was George Perkins Marsh, an American diplomat who described himself as 'forest-born', and how as a young boy 'wild animals were to me persons, not things'.[84] On a trip to Egypt in 1851 he was struck by the scars on ruined landscapes in the wake of the history of agriculture, all barren hills and worn-out soil, in contrast to America, where everything still seemed so new. Humboldt had written about the despoiling and deforestation caused by mines and sugar plantations, and Marsh could recognise the same environmental problems happening in America as forests were felled, rivers polluted and settlers set upon the great grasslands with industrialised machines powered by steam. In 1860 Marsh began his opus, *Man and Nature*, to warn of the dangers in man's avarice. Marsh spelt it out. His book became a lodestone for the generation of conservationists who campaigned to preserve surviving areas of wilderness. In 1872, Yellowstone became the first national park signed into law by President Ulysses S. Grant.

★

When Eric Collier first saw British Columbia's Meldrum Creek in 1922, it was on fire. Flames leapt across the old channel towards a stand of 60-foot spruce. Beyond the crumbling remains of an old beaver dam, a thin stream trickled towards the trees. All that was left of the beavers' old lake was a stagnant spit set in the parched land. It was common practice to burn last year's wild meadow hay to dispose of the matted under-swarth in spring, and that acres of forest were engulfed in these fires didn't seem to matter at a time when timber had such paltry value. When the fire reached the spruce it flared through the resinous needles, sparking like a bundle of firecrackers. When Eric Collier took a last look through the smoke he knew that if just one pair of beavers had been allowed to remain 'the gate would have been shut' and instead of inflammable dry grass there would have been water stopping the flames and the spruce would have been spared. When he turned his horse and rode off to safety, something about the place planted itself in his mind.

Five years later he met Lillian, the girl he would marry, at a place called Riske Creek. Lillian's 97-year-old Indian grandmother, Lala Ross, born in 1830, remembered Meldrum Creek when the beaver ponds were teeming with enormous trout, when thousands of honking Canada geese settled there during migration, when so many ducks lifted from the marshes at sunset that they obliterated the sky. Mink and otter sunned themselves on the banks, and evenings were punctuated by the slap of beavers' tails on the water. 'S'pose you take all beaver,' she told him, 'bimeby [by and by] all water go too. And if water go, no trout, no fur, no grass, not'ing stop.' Then she said, 'Why you no go that creek and give it back the beavers? . . . S'pose once again the creek full of beavers, maybe trout come back. And ducks and geese come back too, and big marshes be full of muskrats again all same when me little girl. And where muskrats stop, mink and otter stop too. Aiya! Why you no go that creek with Lily, and live there all tam', and give it back the beavers?'[85] Eric and Lillian went to Meldrum Creek and built themselves a cabin, but the beavers were long gone and their efforts to fix the old dams did not hold water in a storm. Making a living was a struggle until a ranger gave Eric two pairs of living beavers, and Meldrum Creek began to flourish once again.

But tap Riske Creek, the home of Lala Ross, into Google today and photographs from 2017 of fierce wildfires raging across 98,000 hectares will come up. Red-outs, lone waterjets and impotent hoses defending ranches and houses. Fir trees light up like roman candles.

Aldo Leopold, an American forester later regarded as the father of wildlife ecology, recounts a wonder-filled river journey that he made by canoe with his brother in 1922 through the myriad of green lagoons of the Colorado Delta. Travelling through deep emerald waters that were flanked by walls of mesquite and willow, the river flow carried fleets of cormorants and avocets, huge flocks of mallards and widgeons. There were so many egrets in one green willow that it seemed covered with snow. They came across a bobcat balancing on a log, a paw poised for mullet. Raccoons waded the shallows munching on water beetles. They saw coyotes and burro deer, doves, quail, and geese by their thousands. They felt the elusive presence of cougar. Their impression was of a forgotten and unchanged secret land.

Don't, for God's sake, tap Colorado Delta into Google images today.

In 1948, Idaho Fish and Game Department embarked on a unique beaver reintroduction programme in the state's backcountry, into what is now called the Frank Church – River of No Return Wilderness.★ They para-chuted them in. An ingenious warden, Elmo Heter, knew there was a surplus of parachutes left over from the war, so he designed a beaver-dropping box that would open when it touched the ground, but only after the parachute canopy had collapsed, allowing the fastening ropes to slacken and fall away. Elmo had already discarded his first idea of putting the beavers in a box of woven willow for them to chew their way out, after realising how dangerously fast they could chew. Elmo's test parachutist was a long-suffering beaver called Geronimo, who was put through dozens of jumps. Geronimo's reward was to be the first released into his own

★ At 2.367 million acres, the largest contiguous managed wilderness in the US outside of Alaska.

paradise, with three young females. In its simplicity, the idea was utter genius, and so typical of humans: wipe out 200 million beavers, and then design a special box to parachute them back. The cost of dropping four beavers from a plane in 1948 worked out at $30. Out of the 76 para-chuting beavers, all made it but one. They and their descendants have created the 'amazing habitat that is part of what is now the largest protected roadless forest in the lower 48 states'.[86]

Beaver have come back. In North America there are now over 10 million. In British Columbia beaver ponds have made perfect habitats for juvenile salmon with natural protection from predators. In Europe, Eurasian beavers are coming back to their old places too. George Eustice, a British DEFRA (Department for Environment, Food & Rural Affairs) minister in 2014, wanted to catch the beavers who had mysteriously appeared on the River Otter in Devon and put them in a zoo because they could be carrying a disease. Too bad, for the River Otter locals had become attached to them. Tourists were coming to see them. We went to see them. We didn't see them, but there was a frisson just knowing they were there.

THE BLOODY TOURNAMENT

Horse power

As we've seen, there is a lot to admire in a horse: speed, strength, agility, alertness, endurance and those springy toes. Their feet work like pogo sticks. Their biceps contain collagen, which gives an elasticity that boosts power, like a catapult. The slingshot action propels the foot at a hundred times the punch of a non-elastic muscle. Elastic cartilages and cushions reduce concussion, which allows fast recovery. They also expend less energy in locomotion, which gives them great stamina over long distances. They're curious and smart. As herd animals with strong social hierarchies, they are biddable. But if we isolate them away from company, as you sometimes see, then we have not understood their nature, for it is their misery. That is when they hang their heads between their shoulders, or sway their necks, or kick you. For 6,000 years we have bred horses to eat, haul, plough, ride, hunt, charge, jump and prance; from black to white, small to enormous, light to heavy, nimble, fleet, docile and fiery.* Because there are few stallions in proportion to mares, the male lineage has little genetic variation on the Y chromosome, whereas the maternally inherited DNA is broadly diverse. For thousands of years horses were the fastest form of land transport, and our assault vehicles. No sooner had we invented the wheel than we put it on a chariot and went to war. It would be hard to imagine a more peaceable animal shackled to such violent purpose.

In the Americas, the horse had been extinct for 10,000 years until the conquistadors dropped anchor off the Yucatan coast in the early sixteenth century and their Iberian steeds swam ashore. The locals had never seen horses, let alone ironclad men glued to them like mythical beasts. The Spanish put their victories down to God – synonymous almost with the psychological shock of an animal one has never seen. Horses descended

---

* Paradoxically, if we had not domesticated the horse, it is likely we would have hunted them to extinction.

from these Spanish mounts spread via native trading networks through the Americas. In North America they became both the cause of war and the means to wage it. The Plains Indians could now hunt buffalo on horseback. The Comanche became formidable riders. Their power (and culture) depended on horses, raids, buffalo, daring, defence. But they had no defence against the invisible enemy of smallpox. By the 1870s their population had shrivelled to a few thousand. Along the Texas frontier early maps mark *Droves of Wild Horses*. A million mustang strong.

Cut to Marilyn Monroe's stricken face.

'Eighteen hundred, two thousand pounds,' the dealer calculates.

He is talking the weight of horseflesh for 120 dollars. Clark Gable, hobbling a mustang, bends the horse's legs double with a rope.

'Eight hundred on the stallion,' he says to the horse dealer.

Cut back to Monroe. The dawning realisation. She throws her head back, spins around and runs into the expanse of desert behind them.

'Killers! Killers!' she screams.

The men turn round.

'MURDERERS! LIARS! All of you! LIARS! You are only happy when you see something die! Why don't you kill yourself and be happy? You and your God, country, freedom . . .'

The screaming gets so hoarse and wild and desperate that I cannot untangle the words. She's angry. Let's say that she is beside herself with fury.

'She's crazy,' the horse dealer says. 'They're all crazy.' (He means women.)

Monroe in the distance, screaming like a banshee: 'LIARS!'

A free foal walks up to the trussed mustang on the ground, nuzzles him and tenderly paws his neck.

John Huston's *The Misfits* was a flop on its 1961 release, but it exposed the truth about the mustang hunts, or 'mustanging', for pet food that sprang up in the 1950s. For cattle ranchers, the roundups were about maximising the number of cattle on the public range. But not for rancher Velma Johnson, who became known as 'Wild Horse Annie' and led a 27-year crusade to protect mustangs on public land. The issue was unusual. Having descended from the conquistadors' horses, these were feral non-native grazing competitors with cattle, yet the mustangs also had iconic status. They were the heartbeat of the Wild West and the stars of all those cowboy films, part of American heritage. The beloved horse, after all, had put wings on human feet for millennia. Johnson instigated a nationwide letter-writing campaign to senators. In 1971, the Wild Free-Roaming Horses and Burros Act was signed into law without a dissenting vote, mandating that wild horses and burros would be managed humanely as integral to the landscape as 'living symbols of the pioneer spirit of the West'.*

Horses have carried men to kill each other for centuries. The word 'chivalry' comes from *chevalier*, meaning a nobleman with horse, the medieval mercenary paid in privilege and land. But a weighed-down warhorse with a knight and 75 kilos of steel armour was not efficient when you needed speed and agility. Which is why pitched battles were avoided and sieges became the thing. Otherwise, the charge was the initial shock attack, then the cavalry dismounted to fight, clank, clank.

The disastrous 'Charge of the Light Brigade' in the Crimean War, on 25 October 1854, was blamed on a miscommunication and remembered for the valour of the men. The Russians thought the British soldiers must have been drunk. For it was a headlong mile and a half charge of 670 British cavalry troopers down a valley surrounded on three sides by 20 battalions of Russian infantry, into a battery of guns. The eyewitness account of Godfrey Morgan describes headless bodies still in the saddle, limbs flying, sabres flashing, smoke, flames, shells, explosions, screaming, 'the noise of the striking of men and horses by grape and round shot was deafening'.[87] Men following orders. Horses following orders.

In Michael Curtiz's 1936 film *The Charge of the Light Brigade*, for special effects 125 horses at full gallop were tripped by wire. Twenty-five of them died. The actor Errol Flynn was so enraged by this he had a go at Curtiz and they had to be pulled apart. The public furore caused Congress to pass laws to protect animals in motion pictures.

The writer Michael Morpurgo is recalling the genesis of his novel, *War Horse*, about a farm horse who is sold to the army and sent to the Western Front. In 1976 Michael and Clare Morpurgo founded Farms for City Children, a charity which offered city kids the opportunity to stay on their farm in the Devon countryside. Billy, a boy in foster care from Birmingham, was among the children staying for a week. His teacher told Michael that

---

* Even so, more mustangs are in captivity than run wild, and some still are being 'mustanged'.

Billy didn't speak, so he shouldn't speak to him and expect a reply. Billy had not spoken for two years and the teacher didn't want him to run away – so please, Michael, don't ask him questions. Michael observed the silent Billy all week. It was November, and on the last day Michael went over to read the kids a story by the fire. As he walked into the stable yard with his book he saw Billy standing in front of a horse who had his head over the stable door. Michael stopped, unnoticed, and watched Billy reach up and stroke the side of the horse's face. Billy began to tell the horse what he had done that day and how he wanted to take him home.

Michael says the horse was listening, that Billy talked without hesitation, and he could see there was a connection between them. That both horse and boy, in that moment, shared a kind of comfort. The horse understood he was needed. It was watching this private moment that gave Morpurgo the confidence to write *War Horse*. He said it dispelled the struggle to believe in the sentient connection between a human and a horse.[88]

Rider and horse are the enduring legacy of our old relationship. When I was 12, I lived on a scruffy pony who could fly. We were both at a moment in our lives when our vitality was boundless. Tuned to the slightest muscle quiver, we could read each other. And something ancient struck my young animal soul.

One spring morning, as we came down a deep sandy path, his hind legs sliding beneath him, our excitement criss-crossed flesh to flesh and our energies knotted together. No one around but us. As the path flattened out I leant forward, lifting my weight over his shoulders. His nimble feet picked their way. At the bend his pace quickened. We both knew what was coming. I loved this horse. Every cell primed as the wide grassy track stretched out in front of us. We were off! His horse-heart pumping. His mane splaying, wiry and rook-black. My body inscribed along his body. As close as I could get. To become horse. Breath mingling, sweat bonding, slewing our oxygen from the sky. The wind ripping tears from our eyes. Two animal bodies helixed by blood and history. Nothing would stop him, even if I wanted to. His rhythmic panting, the pounding of hooves; my nose drank in the muck sweat of him. I kissed the whorl of chestnut hair on his neck, like a thumbprint. The demon-wild soared in me. I knew enough to hold the moment and whisper, *Remember this.*

All the fashion

I run my thumb and forefinger along the shaft of a collared dove quill. Its solid nib and ivory lustre. You'd think it would be uncomfortable having those sharp points embedded into your skin. Such a weightless thing, yet with so many uses. Run a crow feather over your top lip, then run it under the tap. Its smooth, watertight engineering, its interlocking barbs. Pen, from Latin *penna*, feather, and of course the French, *plume*. The *Magna Carta* and the Declaration of Independence were signed with a quill, a primary wing feather. Feathers have fletched our arrows, lured fish from the depths and stuffed our beds. If you want to know how light they are, cut open an old eiderdown and try to sweep them up. How light, how warm. How beautiful. We'll have some of those, we thought. And by God we did. The fashion craze for feather hats decimated bird populations all over the world. Not just feathers, necks, heads and beaks lolled over brims; disembodied wings unable to take off, or whole stuffed birds hat-bound, flocks of them. Earrings of hummingbirds. A muff and tippet made of two herring gulls. In 1886, an ornithologist spotted 40 different species upon 700 hats he saw one day, walking the streets of New York City. Finches, parrots, birds of paradise, vultures, egrets, storks, herons, grebes, spoonbills, flamingos, ibises, pelicans, terns, albatrosses, lyre birds, Himalayan pheasants . . . the whooping crane was almost extermi-nated. In 1880 the value of ostrich feathers, pound for pound, was equal to diamonds. There were six London feather auctions a year. In the 1907 June sale, the hammer came down on 20,615 kingfishers, bang. In February 1908, 18,000 sooty terns, bang; 10,700 crowned pigeons, bang. Two Chinese traders exported 12,000 birds of paradise every three months. Toucans, cock-of-the-rock, finches, wings of 'sea-swallows'. Just another face of colonialism. Hiring the natives to hunt out their ecosystems and send them to us.

So hats off to the gloriously named Etta Lemon, an Englishwoman who had a habit of going to church, noting the ladies who wore plumed hats in the pews around her, then penning them an educational descrip-tion of the feathers' previous owners. She would include the graphic details of their procurement, not forgetting the starving chicks in the nest. Etta Lemon fought to stop the feather trade for 30 years. In 1883 she founded the Fur, Fin and Feather Folk; within a year membership was nearly 5,000. In 1891 it merged with the Society for the Protection of Birds, to become the RSPB (with the prefix, Royal) in 1904, with Queen

Alexandra banning osprey feathers at court. It was Etta who was the driving force behind the Importation of Plumage (Prohibition) Act introduced to Parliament in 1908, although it would be another 13 years until it was passed into law.

Birds of peace, birds of war

Dove is a tender word. A close-up of the immaculate dusk-pink fog-grey livery of the collared doves in the garden is a vision of awe. I watch two of them land in the ash tree. Ballerinas of pure loveliness. The gorgeous scallops that knit together the perfect ziplock of their feather vanes. Their plumage weighs two to three times as much as their skeleton, which is hollow and contains bubbles of air. This summer they fought viciously, screaming round the garden, pecking each other at the nape; leaving so many feathers everywhere that I thought the hawk had been through. White doves are released in heavenly flocks as envoys of peace, where carrier pigeons have been the unlikely candidates commissioned as messengers of war. Hannibal used them to send word across the mountains; Cyrus, King of Persia, used them to communicate across his empire in the sixth century BCE. In *Animal Farm* the pigeons were sent to mingle with animals on neighbouring farms to spread the story of the Rebellion. Pigeon Post was the fastest method of communication for centuries – astonishingly, the last service, in Cuttack, India, only closed in 2008. Before the telegraph, stockbrokers used it. Communications from behind enemy lines were kept open by carrier pigeons in both world wars in all theatres; in the early days of Occupation, pigeons were the only connection between France and Britain. Shooting homing pigeons in the First World War was punishable by six months' imprisonment or a £100 fine. In the Second World War, British pigeon breeders donated 200,000 pigeons to the services, and 16,554 were parachuted into Europe. The first report of the Normandy landings was delivered back to Britain by pigeon Gustav – due to the radio silence of the fleet – on 6 June 1944. He arrived at his loft on Thorney Island in Chichester harbour after being released by a Reuters news correspondent on board an Allied tank landing ship 20 miles off the coast of France. In March 1945 Kenley Lass was parachuted into enemy-occupied France with an agent, to be released 12 days later and fly 300 miles home with her message in less than seven hours. Messenger pigeons could draw extra rations of corn and seed. It was dangerous work, for the

Germans employed marksmen and falconers along the French coast to take them down. Commando, bred by Sid Moon, was awarded the Dickin medal of gallantry for completing 90 missions. The Dickin, the highest award for animal valour, has been awarded to 32 pigeons, 18 dogs, 3 horses and a ship's cat (rat duty and raising morale). However, the biggest accolade must go to First World War hero Cher Ami, dispatched into enemy fire on 3 October 1918 by Major Whittlesey, whose 194 surviving men were surrounded by Germans and being shot at by Allied troops who didn't know they were there. Cher Ami was shot down but managed to take off again and fly the 25 miles to HQ in 25 minutes to pass on the coordinates of Whittlesey's position. She had been blinded in one eye, shot in the breast, and her leg was hanging off. A medic saved her life, but not her leg, so she was carved a wooden one.

In 2017 Argentinian police shot a carrier pigeon running drugs into jail. The downed bird was carrying a backpack containing 7.5 grams of cannabis, 44 pills of Rivotril sedative and a USB stick.

In the West End of Derby lives a working man
He says I can't fly but my pigeons can.
And when I set them free
It's just like part of me
Gets lifted up on shining wings.[89]

'I had a grizzle I used to go down to Lake District and I would always take it with me,' says pigeon racer Clive. 'Let it go and he always go back. He was about 17 when he died. I had that pigeon al'time. He was a belting pigeon, him.'[90]

Pigeon racing is the sport of gentle folk bewitched by the magic of flight. I get racing pigeons. To stand looking up into the sky, waiting for your pigeon to come home, and then you see her, your heart lifts as she folds her wings and drops for home. Relief, pride, wonder. Love. Love for their bright eyes, their pink feet, the sweetness of their purring in the rafters, the glory of their iridescent neck ruff shimmering like the pearled cheek of a mussel shell. There are the fast flyers and the endurance racers. To navigate they use the direction of the sun above the horizon, and visual markers, particularly linear features like roads, rivers, railways, hedges, the edges of woodlands and towns. Like seabirds, olfactory cues help to steer them home. Pigeon-keeping world sounds utterly engrossing; waiting for eggs to hatch, watching the chicks grow, watching them fly and come home; learning everything you can about them. The tie that binds is their freedom to fly. It is their choice to come home, for that is their way. For the human – a metaphor of family; for both, a trust across species.

Clive says his wife of 22 years said to him, It's the pigeons or me. So that was the end of her.

Symphonies in the sky

In Ancient China, during the rule of the Song and Qing dynasties, circling below the clouds were great flocks of pigeons with delicately carved gourd whistles on their backs, their mouths open to the breeze. The orchestra they made arced its avian pathway, rising then falling, vanishing and returning in a pattern of undulating sound. A gossamer wind-music of sky flutes. How skin-shiveringly lovely is that? Each whistle, carved by a master, took on a different shape to produce a different aeolian tone. Reminiscent of netsuke but more curious, some were the shape of a peeled tangerine crowned in a bouquet of stubby fingers, others like tiny clusters of chimneys; they are as light as a leaf and can fit in the cup of your hand. A thousand years ago, in battles between kingdoms, the sound of the wing-borne whistles was a signal of attack. Not so long ago, in old Beijing, you might be woken by these flying symphonies each morning, for every street had maybe two or three keepers of pigeons with whistles.

I tap into YouTube and find Baotong, a master whistle-maker from Beijing. He goes to his loft, lets his pigeons out, and whoosh. A flush of sweet silver notes across the sky.

Ex-miner and steeplejack Pete Petravicius, born in Nottingham, raised his first pigeon from a nestling he scrumped from a wild nest in the roof of a terraced house when he was 12. He just climbed up and got it.

'Come on then. Come on. As a young lad I used to dream how fantastic it would be to fly. I wouldn't say I wanted to be a pigeon, but I must admit I wish I could do what a pigeon could do. You know, just take off and fly around. Come on then . . . . Come on . . .'

A rattling. A cooing.

'Come on then. They want to. They don't. Can you see? Come on then. You're silly. Come on then, come on then.'

A clatter of wings like a rainfall of feathers: off they go. 'Pigeon Pete' is the only man in England flying birds from a mobile loft strapped to the back of his scooter. It's not easy to train pigeons to come back to you, wherever you are, and not go home.

'Time, patience and . . . I don't know how to put it in words, you have got to have feelings, and you have to get the trust of the birds,' he says.

The musician Nathanial Mann met Pete when he was composer in residence at the Pitt Rivers Museum in Oxford in 2013. He was intrigued by the old Chinese pigeon whistles in the collection and imagined recreating this flying soundscape. To do that he needed a pigeon fancier, so he found Pete and they worked together. Nathanial designed the whistles; Pete attached them to the birds' middle two tail feathers.

'Come on, Irish. Are you in the mood today?'

'Fingers crossed.'

'Come on, my little beauty.'

Pete launches the birds. One after another. The whistling keens and dips. Trial run after trial run, whistle after whistle. They hone and hone it. Until Nathanial's imagining comes true and at last their orchestra takes to the skies.[91]

THE VAST INTANGIBLE BEAUTY

Breath taking

I'm in the Natural History Museum and I'm in a bad mood. I've come to London especially to see some tropical bugs, butterflies and moths. I'm not particular, anything will do. I wanted to trawl my eye across metallic green carapaces and wire antennae, over iridescent wings and tiny claws and nature's miniature magnificence. As I remember doing in childhood, lost beneath the reflections of the glass cabinets, transfixed with wonder by insects who looked like leaves, by butterfly wing eyespots, by legs, mandibles, goggle eyes. Now, you have to interact with a screen. And *know* what you are looking for. Or get a headache. I click away but get more data forms and more Latin, another order, another family, another genus, another species; I just want to *see* the darn thing! I don't care which one, any one, and then I want to see another one, and another one (like the voracious collector himself, more, more). I try butterflies, Lepidoptera, class, order, higher classification, country, continent, higher geography, catalogue numbers, collection code, sub-department, donor name, JUST SHOW ME THE BLOOMING BUTTERFLY.

I go to the Information Desk. They direct me to the Darwin Centre's Cocoon. It is in the Orange Zone. Shows how long I haven't been here. I might have fared better had I been an architecture student, for the 'cocoon' is an eight-storey white concrete £78 million torpedo set at a jaunty angle in an airy atrium with wide stairs (easy traffic flow) and a few darkened rooms at the top, where, granted, there are a couple of stick insects artfully displayed in a modern glass diorama with the odd butterfly among some twigs and leaves. The cocoon is the exhibit. Spacious, modern, tasteful, with lounging areas and computer monitor stations, and tantalising views into hi-tech labs where you can't go, but where you can glimpse live scientists among *their* 3.3 kilometres of cabinets of insects. The cocoon won an award at the Concrete Society Awards in 2009. I am possibly the only person in the country (in the world!) who has gone off the Natural

History Museum. Because I didn't come to look at an award-winning staircase. Or have lunch, coffee, cakes, drinks, sweets, in one of the many cafés, or buy gifts in one the *many* gift shops. Taking. Up. Valuable. Display. Space. Maybe I am an old-fashioned child, but this futuristic rocket would not have flicked the nature switch on for me. If I want to meet Archie, the 8.62-metre-long giant squid, or Darwin's octopus, I can go on a behind-the-scenes tour in the 'Zoology spirit building'. But booking is essential and there are only 'various dates' and it costs 15 quid. And when they say spirit, do they mean anima, soul, psyche, ghost, quintessence, inner being, or formaldehyde? I am told I missed the Sensational Butterflies exhibition, when you could see butterflies hatching from their cocoons, and there are models to show how dragonflies fly, but. I want to see *real* insects, who I *know* are here, and there is virtually nothing on display. Correction, everything seems to be virtual.

The Natural History Museum has the most important entomology collection in the world: over 34 million insects and arachnids; 80,000 drawers of nearly 9 million moths and butterflies (Lepidoptera); 22,000 drawers of beetles (Coleoptera); in Sir Joseph Banks' eighteenth-century collection there are 10 million specimens alone. 'Of the Museum's 80 million objects, only a tiny fraction ever go on display.' The museum's bird-skin collection numbers 750,000 specimens over 8,000 species. That averages nearly a hundred of each. Beaks pointing up, cotton wool eyes, fallen feathered soldiers in dark drawers; labels inked with Victorian writing tied to thin stick legs. The slender godwit, neck folded back into his body, his long beak stitched to his wing, Captain FitzRoy's bird, brought back on the HMS *Beagle* in 1836. Any creature in the nineteenth century who hopped in front of a naturalist would be lucky not to be bopped on the head in the name of science. It was good business. Baron Rothschild's bird-skin collection numbered 300,000. There are more than a million eggs, one of which is the emperor penguin egg collected by Apsley Cherry-Garrard in July 1911 on the fateful *Worst Journey in The World*, the British Antarctic 'Terra Nova' Expedition of 1910–13. I stomp across the stone flags and think of the community of death in this great mausoleum replicated across the world in every big city. Half the butterfly species in Singapore have disappeared. Of the multitudes, how many were taken for science with positive outcome? How many for pure excess?

As the breeze blew Spanish, Portuguese, French, Dutch and British ships across the oceans to distant lands, they took both the military *and*

scientific aspirations of their European nations. Expeditions carried cartog-
raphers, astronomers, meteorologists, geologists, botanists and naturalists.
We wanted to expand our minds *and* our territories. This wide remit put
Europe at the helm of global domination. For modern imperialism and
scientific progress were inextricably entwined. Discoveries drove break-
throughs in medicine, navigation and technological innovation. To rule
the waves was to rule the land and bring home the booty. Global trade
networks criss-crossed the sea with cargoes of gold, spices, fur, feathers
and worse . . . How often 'progress' in one area meant asset-stripping
another. For the flora and fauna and peoples of these lands, the arrival
of European ships would seed ecological disaster.

Alfred Russel Wallace was initiated into the wonders of beetles in 1844
by Henry Walter Bates (who later wrote the first scientific paper on
animal mimicry); they scoured the Charnwood Forest in Leicestershire
together with their killing jars, fired by reading Darwin, Humboldt, and
expeditions to foreign lands. In 1848 they set off for the Amazon with
plans of making great discoveries '. . . with a view to the theory of the
origins of species',[92] to be financed by doing the thing they loved, collecting
specimens. Though they parted company, Wallace remained in the Amazon
for four years, awestruck by the sheer magnificence of the animals he
saw – encounters that invariably concluded with a bullet and the collection bag. The 'splendid blue and purple chatterers' (toucans and parrots);
frigate birds, wood-ibis, cranes, a monkey who comes 'too near for its
safety'. He has great admiration for the shy umbrella bird 'expanding and
closing its beautiful crest and neck plume',' though 'being very muscular
will not fall unless severely wounded'.

It is not all adventure. Wallace falls sick and his visiting brother dies
of yellow fever. Then the brig which he sails home on – with his thousands of specimens, dead and alive, caged, skinned, stuffed and wrapped
in swaddling rags, plus all his notes – ends up at the bottom of the ocean
after fire breaks out in the ship's hold. Wallace survives in a leaky rowboat
with nine others for 10 days, until rescued by a cargo ship. He calculates
the loss of his specimens at £500 (£60,000 today). A loss to dishearten
most men, but two years later, in 1854, he ups sticks and off he goes to
the Malay Archipelago.*

---

* Now Malaysia, Singapore and Indonesia.

Here he is considering his emotions in the Aru Islands with nature's 'choicest treasures' – birds of paradise:

> I thought of the long ages of the past, during which the successive
> generations of this little creature had run their course – year by year
> being born, and living and dying amid these dark and gloomy woods,
> with no intelligent eye to gaze upon their loveliness; to all appear-
> ance such a wanton waste of beauty.

And yet he fears if 'civilised man' were to 'bring moral, intellectual and physical light into the recesses of these virgin forests' the finely tuned relationships would be destroyed and the birds lost for ever. Wallace is in a quandary. For he *can* see that creatures were not made for man. That 'their happiness and enjoyments, their loves and hates, their struggles for existence, their vigorous life and early death, would seem to be immediately related to their own well-being and perpetuation alone, limited only by the equal well-being and perpetuation of the numberless other organisms with which each is more or less intimately connected'.[93] And yet, he would have liked to have bagged more. He loves the pursuit and the capture. To gaze upon his breath-taken prize. In New Guinea he spots a new species of bird-winged butterfly. He must have it. It flies high into the canopy and eludes him. For two months he searches and begins to despair. Then he nets one. As he opens the glorious wings his heart begins to beat violently and blood rushes to his head, he is sure he will faint, 'to feel it struggling between one's fingers, and to gaze upon its fresh and living beauty, a bright gem shining out amid the silent gloom . . .' There you have it. The struggling beauty between his fingers. To want, to hold, to *must* have. So intense was his excitement that he had a headache the rest of the day.

The whole orang

Paradoxes coalesced in Wallace, the passionate naturalist, the empirical scientist, the collector and the conservationist. The ironies are stark when it comes to orangutans. These strange and marvellous primates 'which at once resemble and mock "the human form divine"' fascinate him, but his fingers are always twitching on his gun. He is a lousy shot. In Borneo, in 1855, he has to hire 'Chinamen' to cut down trees in which injured orangutans hide out of reach. The 'collecting' is relentless, but one day Wallace is confronted with the fruit of his labours. What he calls, in the malignant language of his time, his 'half-nigger baby'. He has killed a female orangutan, and this was in her arms. He washes, brushes, nurses and feeds the young orangutan on sweetened rice-water, who in return clutches his beard so tightly that Wallace has difficulty extricating his little hands. He muses how the vice-like grip serves to hold fast while his long-haired mother ('poor creature') is 'running about the trees like a mad woman'. He even provides his charge with a monkey for warmth and company. 'When the monkey wants to run away, as he often does, the baby clutches him by the tail or ears and drags him back.' Wallace is love-struck: 'my baby is no common baby, and I can safely say, what so many have said before with much less truth, "There never was such a baby as my baby", and I am sure nobody ever had such a dear little duck of a darling of a little brown hair baby before'. It was never going to end well. He had imagined them living together in England and grieves for his loss, 'its curious ways and the inimitably ludicrous expression of its little countenance'.[94] When preparing the little one's skeleton and skin, Wallace discovers he had broken an arm and leg in the fall with his mother. Wallace receives £6 for *his* pains from the British Museum, and in the meantime succeeds in shooting a full-grown male.

To put this in context, as Wallace cuts his deal, in Tasmania (then Van Diemen's Land) there was a thriving market for Aboriginal body parts, preferably skulls and whole skeletons. Assumed soon to become extinct, and considered the most primitive human on the planet, the remains of Tasmanian Aboriginals fetched the highest price on world markets. Native people were mercilessly hunted off their lands. In 1830 an open season sanctioned by colonial governors offered a bounty of £5 for an Aboriginal adult and £2 for a child. The idea they should be captured alive was undermined by paying the same bounty for the bodies.[95] Convicts and settlers were given guns and ammunition. Public museums and collectors took advantage of the Black War, as it was known, with departments of palaeoanthropology making bids and unscrupulous individuals placing orders. In 1838 a reforming government gave Aboriginals a legal status akin to that of children,[96] but the trade continued until the 1940s – yes, *nineteen* forty – as bone collectors raided burial grounds, hospitals, prisons and asylums for Aboriginal remains, exporting them to Europe and America, declaring them through customs as kangaroo. Truganini, believed to be the last full-blooded Tasmanian Aboriginal, died in 1876. Against her wishes for a respectful burial, and confirming her fears, she was displayed in the Museum of Tasmania until 1946.

In the 1850s, gorillas and orangutans were relatively new on the white man's scene. The Javanese name, *ourang outang*, meaning 'man of the forest', became used generically for anything human-like, wild, hairy and dangerous-looking. Orangutans were rumoured to be the missing link, they slept under blankets, used fire, buried their dead. The Javanese said they could talk but were not inclined to in case they were forced to work![97] It was Jenny, the orangutan at London Zoo (dressed in human clothes and taught to drink tea), who persuaded Darwin in 1938 that the difference between us and other animals was 'one of degree'. He witnessed her passions, her rage, her affection, her sulkiness and astonishment. He wrote in his notebook, 'Man in his arrogance thinks himself a great work, worthy the interposition of a deity. More humble and I believe true to consider him created from animals'.[98] The geologist Charles Lyell, who once ridiculed the notion, conceded to Darwin, 'that we must "go the whole orang"'.[99] Queen Victoria visiting Jenny's successor in 1842, also called Jenny (zoo lifespans didn't last long), was both fascinated and repulsed, and declared of her: 'He is frightful and painfully and disagreeably human.'

Perhaps the queen had read Edgar Allan Poe's *The Murders in the Rue Morgue*, published in 1841. Poe liked to fizz topical matters into spine-chilling tales. Our detective, Auguste Dupin, is intrigued by horrifying newspaper reports of the double murder of a woman and her daughter in a fourth-floor apartment. And pretty gruesome it is. Great clumps of grey hair are found in the hearth, pulled out by the roots and clotted with fragments of scalp; the daughter's corpse is wedged head-down up the chimney(!), while the mother's body has been thrown from the window into the rear courtyard below with her throat so badly cut that her head fell off when they moved her. The witness testimonies of neighbours woken by the screams describe hearing a peculiar shrill voice as they ran upstairs, to all ears an alien tongue in a language no one could identify. Dupin is permitted to inspect the 'wildest disorder' of the murder scene and put his analytical powers to work. He assembles the peculiarities: the strange voice devoid of 'intelligible syllabification'; that nothing had been stolen; the ferocious brutality and super-human strength required to commit such a 'grotesquerie in horror absolutely alien from humanity'. This was no ordinary assassin. Dupin discovers a tuft of tawny non-human hair. He consults Cuvier (the French palaeontologist we briefly met in Part II) and matches the hair and the victim's lacerations to Cuvier's description of the Ourang-Outang of the East Indian Islands. What could be more current than the talk of apes undermining our unique identity? Here is our 'foreigner'. The 'man-like' wild animal from the jungles of Borneo. Civilisation's villain. Or imposter. Hideous, depraved, a degraded version of ourselves. The orangutan becomes a stand-in for man's inhumanity. The wild beast within. Their size, strength, all beast and brawn, a living embodiment of the dark side of ourselves. Poe mines the anxiety of our relationship to apes, which of course is that they are too close for comfort. In a letter to a medical student, Poe wrote that it was the heart – 'that principle in the human breast' – that makes one human 'without which man would be a brute or a God'.[100] We could argue that we became both.

The beginning of the world

In my palm, a guillemot's egg. I cradle its stillness. Air light, vessel of emptiness. Tiny blow holes in the chalky sarcophagus that plumbed the speck that never flew. No wing tip in the breeze blown, no pinion curling over a sparkling sea. Fragile. Exquisite. Gone.

I stroke it on my cheek. Guillemot eggs were coveted for their calligraphy and the shape of their thick shells. And oh, they are beautiful. Rounded at one end, narrowing to a point. The Romanian sculptor Constantin Brancusi's *The Beginning of the World* couldn't be more guillemot. His obsession with the ovoid was his meditation on the pure form. Something Brancusi held as eternal, infinite and absolute, that grasped the vibrant current of the universe. The perfect beauty of his ovoid *Sleeping Muse* would occupy him for 20 years. The origin and mystery of *human* life cradled in the egg of a bird. Alfred Wallace supposed that without predators, guillemot eggs were an expression of unchecked pure exuberance. The diversity is endless and distinctive to each bird, like a fingerprint. The dark nose cap; the pepper pot; the shorthand dash; the blotch; the scribble; the scrawl; the spatter; on sky blue, blue-green, white-white, pink-red, red-brown. In the guillemot's gallery you will find every colour, every mark ever seen on an egg across 10,000 other species, somewhere . . . A seductive thing of wonder so exquisite, so varied, the entranced turned their pockets inside-out to own them.

I pass the guillemot egg back carefully to its owner; he gently lays it on cotton wool in its case with 11 others. A flock of nothingness. I think of their beginnings, of the parent birds facing in towards the cliff face on their narrow ledges, the cacophony and jostling of sisters and aunts, salt spray on their tongues. The dangling man, a spider on a thread with a canvas bag strapped to his waist. *Clink, clink.* A shouting, screaming, swirling of birds.

Summer 1935. George Lupton is standing anxiously at the edge of the 350-foot Bempton cliffs of Yorkshire. Banknotes limp in his pocket, he can hardly wait to see what comes up. Lupton is a well-known customer with more than a thousand guillemot eggs in his collection. His mind's eye holds them bright, for his obsessive-compulsive passion is all about beauty and arrangement: a row of black ink blots on bright white ground, a sequence of peppering, a trajectory of constellations. For, although he might not call himself one, George Lupton is an artist. Seabird eggs had been taken for food from the world's coastlines for centuries, but here in the 1800s it became all about possession. The craze for egg collecting supplied every curiosity cabinet and natural history museum in the land. Along the clifftop, gangs hauled up their men, canvas bags emptied to the satisfying chink of eggs loaded into baskets as buyers crowded round. There were collectors, dealers, and there were the 'climmers', as they were called.* Climmers knew long before biologists that if they took an egg, the bird would lay again a fortnight later. And again. And maybe even again. Each egg with her same signature markings as the one before, laid in exactly the same spot, a 7-inch stake on a precipitous ledge. As with many seabird species, only one guillemot egg is brooded. At 12% of her body weight, each one is a huge investment for the bird. The rare reddish-brown Metland egg (named after an adjacent farm) was taken every year from 1911–38 from the same bird from the same spot. In 26 years, after all that flying, diving, fishing, in a life devoted to the creation of new life, the Metland guillemot never raised a single chick.[101] There were many like her. For guillemot eggs were golden.

Guillemots breed crammed together on the sloping ledges of steep cliffs without nests. Generations and generations, preening, crying, greeting. The elongated shape of their eggshell was a conundrum that inspired theories and perplexed humans until very recently. The assumption had been that a pyriform (pear-shaped) egg spins in a tight circle or rolls in an arc if it is knocked. But it is the *blown* egg that spins, and the arc is too wide to save the egg from falling.[102] All you really have to do, dare I suggest (I do), is think of choosing to put an apple or a pear on a sloping, uneven ledge. The long edge confers stability, resting on the slope

---

* Yorkshire dialect for climbers.

more securely, and easier for the parents to manipulate when they change over. The ornithologist Tim Birkhead points out another selection pressure in the shell's evolution aside from rolling off the ledge: faecal contamination. 'The shape keeps the blunt end out of the muck,' he explains. Which is important, as the blunt end is where the air sac is, where the embryo's head develops and where diffusion of air through the shell is crucial.[103]

By the 1860s the Bempton climmers were taking everything they could reach, around 80,000 eggs a year.[104] A pillage exacerbated by the Victorian craze of shooting seabirds from boats. Panicked birds scattered; eggs rained from the ledges. An appalled vicar from nearby Bridlington persuaded Christopher Sykes MP to put to Parliament the sorry seabird's tale. Sykes explained how the birds showed the fleets where to cast their nets; and how the gulls' cries warned merchant sailors of the rocky shore in foggy weather. Indeed, at Flamborough Head the seabirds had earned the name 'the Flamborough pilots'.[105] The Protection of Seabirds Bill of 1869 (though sold as being to our benefit) was one of the first pieces of environmental legislation in the world to protect wildlife for its own sake – rather than as quarry.* It was another 85 years before the Wild Birds Protection Act in 1954 made it illegal to collect wild birds' eggs in Britain, which put an end to climming.

The nagging question for *this* story is that collectors such as Lupton did not weigh the cost of the plunder in their eye.** Egg collecting is history, yet the exquisite remains, in endless rows as ordered as the Great War cemeteries, speak volumes to our contrary relationship with our fellow creatures. Yet these collections can tell us other things about a nation's relationship with the natural world. We can measure the thinning of eggshells since the Industrial Revolution, due to acid rain that caused a reduction in the abundance of snails to supply the calcium. And the effect of DDT accumulating in the food chain that, by 1960, caused the eggshells of birds of prey to be so weak they were crushed by the weight of the incubating bird.

---

* Protection benefited 35 species with a closed season from 1 April to 1 August.

** It is not hard to detect the addiction involved in many collections. Psychologists suggest the practice of organising and arranging calms our anxieties and highlights the controlling side to our personalities. The psychoanalyst Werner Muensterberger thought collecting satisfied the 'notion of fantasied omnipotence', while others look back to the collecting of nuts and berries by our hunter-gatherer ancestors. Freud believed collecting came from unresolved toilet training (the collector trying to regain control over their possessions . . .)!

The guillemot chick will peep inside his shell before he hatches. The parents reply. Their individual voices begin to imprint and a bond forms between them so they will know each other the moment he breaks out. And that, when you picture it, is an impossibly lovely thing.

When he's ready to fledge he's called a jumpling. Smaller, downier versions of their parents, the jumplings shuffle around on the ledge, looking over doubtfully.

I watch footage of the guillemot chicks' first heart-stopping leap. Chicks are jumping from every cliff ledge. Legs outstretched, baby unformed wings flapping like crazy, 400 feet! It's terrifying. Splash! They are all jumping. Splash! Splash! Parents glide down to join them. One small chick is left behind. He looks down. A crow is taking an interest. The parent returns and calls to encourage him, *Aaa-aaaar, aa-aar.* He's so tiny. The parent continues to call.

Then a razorbill flies in aggressively, pushing the chick off the ledge! The camera pans his fall. He's heading straight for the rocks.

He hits a rock and bounces off, a parent in hot pursuit, *Aaa-ar, aaaa-aar.* It's hard to watch. He's disappeared into the foaming swell. But here he is! This tiny thing, struggling out of the sea onto a slippery rock, his little wings outstretched, his white tummy, his dark eyestreak, he scrabbles swaying like a penguin up the seaweedy slope. He looks around. All in one piece. Dad is waiting on the rock. The jumpling waddles over. And there they sit, together.[106]

FLAMBOROUGH—EGG GATHERING. THE END OF A HARD DAY.

DAMNED NATIONS & ISLANDS OF SHITE

Karl Marx thought the vast quantity of human excrement pumped into the River Thames in the 1840s a national scandal. As people migrated to the city, effluent began to fill the streets, inundate basements and seep up through floors. What once was valued as Black Gold in the cycle of fertility, was causing an expensive and dangerous problem, let alone an abominable stink. Solids piled up along the banks 6 feet deep. Cholera outbreaks took thousands of lives. We soiled the river. We soiled the streets. We soiled ourselves. Everything except the soil. In the new intensive system of British farming the food went to the cities, but the natural fertiliser didn't come back. If you take, take, take, you don't get more, more, more, you get a humungous problem. In the 1840s soil depletion was the humungous problem. The world population had reached 1.2 billion and needed food.

Justus Liebig, the German agricultural chemist who had identified nitrogen, phosphorus and potassium as the essential nutrients needed for plant growth, described British 'high farming' as a robbery system.[107] Karl Marx went further, considering the Thames contamination as an example of 'externalising' the costs of a capitalist economy. The amount of human waste generated in the city was a by-product of the labour required for the industrialists' profit, yet the expense of disposing of it was free to the employer, but costly to nature and society. Sound familiar? For Marx, the consequence of human alienation from their surrounding land represented a rift between man and earth. Soil exhaustion was the result of an imperialist mechanism that took away fertility in exported crops, from country to town, and from nation to nation. And marked a further disassociation between the human and animal worlds.

The quick fix for Britain was more robbing further afield. As the ruling classes retired to neighbourhoods away from the stench of working-class areas, the smell in business noses was of opportunity. Deals were done for bones rich in phosphorus and calcium, harvested from the Napoleonic battlefields of Waterloo and the catacombs of Europe, and taken to the bone grinders of Doncaster. But in 1842 William Gibbs, of the trading house Anthony Gibbs & Sons, set his sights 6,000 miles away.

The Chincha Islands, 13 miles off the coast of Peru, had been the stopping place for seabirds gorging on the gigantic shoals of *anchoveta* gorging on the plankton in the nutrient-rich Antarctic waters that run up the west coast of South America. Since the granite had risen out of the sea, since *achoveta* could swim and seabirds could fly, splat by splat, layer by layer, compacted by its own weight and baked in the hot sun in the dry, almost rainless, climate, undisturbed for millennia, the pungent, potent excrement of millions upon billions of Guanay cormorants, boobies and pelicans had built up a mountain of 'white gold'. The chemical analysis of a sample collected by Alexander von Humboldt in 1804 revealed it rich in nitrogen, phosphorus and potassium.* The deposits of guano in some places were 200 feet deep. Sixteenth-century voyagers had reported flocks around the Chincha Islands so enormous they blotted out the sun. Here was their parting gift for the taking.

And here we are again, when the plunder of nations irrevocably alters the fine-tuning of animal worlds. Fur, feathers, the conquest of land, the pursuit of science, the stealing of beauty, the commerce of shit. In 1841 William Gibbs struck a deal with the Peruvian government and exported 1,700 tons of guano. By 1847 it was 220,000 tons. That year he negotiated an exclusive monopoly for European and American trade. By 1850 guano was fetching a staggering $76 a pound ($167 per kilo) – the equivalent of $3,000 today. Ships poured into the islands' harbour – a bristling of masts from England, Holland, France, Germany and the USA. The Incas had known the value of this natural fertiliser and guarded the islands during the seabirds' breeding season. European demand paid no such heed. There was no attempt to protect the asset, or the indentured Chinese brought in to wield the picks and shovels in the acrid fumes and dazzling sun. Seabirds were harassed and killed, the seas overfished, and the populations collapsed. In 1856 the US passed the Guano Act inciting citizens to seize any uninhabited islands with guano deposits anywhere in the world; over a hundred were claimed, but nothing compared with the quantity or quality of Chincha guano. By the 1880s the Chinchas were guanoed out. Almost 13 million tons

---

* When he measured the temperature of the cold current of water that would bear his name.

of nutrients from one ecosystem had been transported across the globe. It was time to exploit another territory.

One unlikely source turned up in the desert at Beni Hasan, 100 miles from Cairo. In 1880 an Egyptian farmer struck a vein of cat mummies: 180,000 of them. These were not the beloved kitties of royal households, as one might imagine, but felines bred specifically for mummification and sold to pilgrims to offer as sacrifices to the gods Isis and Osiris. Six months old, tightly bandaged, battery kittens. Nineteen tons were shipped to Liverpool, pulverised and sold as fertiliser at £4 a ton. The Liverpool auctioneer used a cat skull as his auction hammer.

It didn't take long to ask whether guano was anything more than digested fish. So we cut out the seabird. Oceans of fish were processed into fishmeal to make into fertiliser. Then in 1909, the German chemist Fritz Haber magicked fertiliser out of thin air. Well, air, water, natural gas and *a lot* of energy. By 1913 synthetic nitrate fertiliser (and explosives) was being manufactured on an industrial scale.★ An inexhaustible (but not free) supply of fertiliser was the detonator for the human population to explode (50% of the nitrogen in human tissue originates with the Haber process). Fritz Haber changed the course of human history and cast the dice of consequence on animal history.

Meanwhile, the far-flung islands claimed for the US under the Guano Act, with the guano deposits taken, were found a different purpose. Some became national wildlife refuges. Some became military bases and sites for nuclear weapon testing. And some, contrarily, became both. Like Johnston Atoll, bang in the middle of the Pacific Ocean, designated as a bird refuge in 1926. In 1934 Roosevelt gave it to the US Navy, who blasted the coral to create a seaplane landing strip. The lagoon was dredged for a parking area and causeway. Three more landing strips, barracks, a steel control tower and an underground hospital were built (for the birds?). In the Second World War Johnston Atoll National Wildlife Refuge became a submarine refuelling base; in 1958 it became a nuclear test site for high–altitude nuclear explosions in outer space; in 1962 a warhead was detonated at 250 miles high with a 1.4 megaton explosion visible in Hawaii; on 3 November 1962, a nuclear armed missile was detonated that was too bright to view until an orange disc transformed into a purple doughnut. Plutonium contaminated the atoll

---

★ The Haber-Bosch Process combines nitrogen from the air with hydrogen from natural gas (methane) into ammonia (NH_3) under high pressure and 450°C heat. The ammonia is the base of synthetic nitrogen fertiliser used today.

and radioactive debris was dumped in the lagoon. The once pristine 46-acre wildlife refuge was dredged and filled in to become 625 acres, sporting a 25-acre radioactive landfill and two new imaginatively named islands: North (25 acres) and East (18 acres). In 1964 a new tick-borne virus was discovered in the nest of a common noddy (a large tern), named Johnston Atoll Virus. In 1964 the atoll became a biological warfare test site. Pathogenic bacteria were released downwind of the atoll and upwind of barges loaded with rhesus monkeys. Bacteria which cause things like rabbit fever (pneumonia, ulcers, fever, enlarged lymph nodes) or Q fever (flu, diarrhoea, profuse perspiration, joint pain, nausea, vomiting, liver enlargement, confusion). In 1970 Johnston Island National Wildlife Refuge served as a chemical weapon storage site: rockets, bombs, mortars, projectiles, mines, bulk containers of Sarin, Agent VX, vomiting agent, blister agent, mustard gas. Stockpiled weapons from the Second World War were shipped in from Japan, and later Germany. In 1972 more than 2 million gallons of Agent Orange arrived; the barrels leaked into the lagoon. In 1985 construction began on a chemical weapon disposal facility. Operational by 1993, chemical munitions were incinerated over the next seven years. In 2003 the disposal plant was demolished and the site was 'covered in crushed coral to help maintain the protected environment'. I kid you not. The US Army's Chemical Material Agency points out that, because of the area's 'sensitive ecological systems', environmental concerns were 'a major priority'.[108] Thus the island was 'environmentally remediated' for its green turtles and Hawaiian monk seals, its petrels, shearwaters, boobies, frigate birds and whales. Which is a big history for a tiny atoll. All because seabirds once landed to crap on it. I wonder what Marx would have thought of the costs to society and nature of all of that?

Space cadet

The ambitions of nations can atomise a pristine coral reef into a purple nuclear doughnut in the march (or myth) of human progress. Whatever it takes. As nuclear warheads blasted off from Johnston Atoll, the US man-in-space programme, Project Mercury, was accustoming Astronaut No. 65 to his chimpanzee pressure suit. On 31 January 1961 he was strapped into the spacecraft's contour couch, hooked up to various telemetric devices, then launched into suborbital flight. Only when the recovery vessel landed safely was No. 65 given a name: Ham. (An aborted mission carrying a chimp with a name to eternity would not have been good public relations.) Ham, an anagram of the institution that launched him, Holloman Aero–Medical, is also, more recognisably, the name of Noah's second son. He is the son accused of a heinous sin (we don't know what) by his drunken father, Noah, after Ham found him naked in his tent sleeping off his hangover. Remember the vineyard Noah planted after the Flood . . . In good Old Testament logic, Noah curses Ham's son, Canaan, saying: 'a servant of servants shall he be unto his brethren'. Known as the 'curse of Ham' it was this line that Christian slave owners conveniently took as a justification for the enslavement of black Africans, who they decided were descendants of Canaan.★ So the name Ham was an unfortunate choice for a chimpanzee whose servant role was to be a human substitute. It turns out our 'understudy for man in the conquest of space'[109] had already been given a name after he was captured in French Cameroon in 1957: Chop Chop Chang. Of course it was. Chop Chop Chang had been trained to pull levers in response to flashing lights with rewards for success and 'corrections' for failure. Whatever Darwin taught us, the Great Chain of Being is deeply ingrained in the human psyche. After his space mission 'Ham' retired to the Washington Zoo. He died aged 25, and is buried in the International Space Hall of Fame in New Mexico.

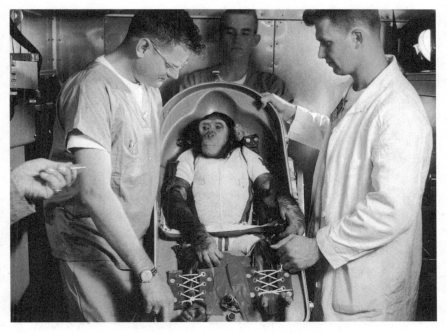

★ There is no mention of skin colour in Genesis 9. Scholarly speculation on the sin Ham committed ranges from laughter to sodomy.

LOVE, ACTUALLY

Flea Girl

Charles Rothschild, banker, naturalist, entomologist (brother of Walter Rothschild – zebra tamer and private museum owner at Tring), had the largest collection of fleas in the world, 260,000 of them. It was Charles who discovered the vector of the bubonic plague, the Oriental rat flea, *Xenopsylla cheopis*, which he collected in Sudan in 1901. Charles was the original rewilder. His Northamptonshire estate, Ashton Wold, was managed for nature and became the perfect classroom for his daughter, Miriam, who at four years old was engaged in breeding ladybirds and never went to school. Her idyllic childhood came to an abrupt end at 15 when her father, suffering from encephalitis, shot himself. In filial devotion Miriam took up with his 'beautiful' fleas to become a world authority: 'My father was a flea man and I became a flea girl, it's as simple as that.'[110] From him she learnt the need for precision and an open mind. Her father's rat flea can jump over 100 times its own height, 600 times an hour, for 72 hours without pause. Miriam worked out that the mechanism of the flea's huge jump, as fast as a rocket, was down to the rubbery protein resilin, which can trigger a sudden release of elastic energy into the legs.[111] Miriam spent 30 years cataloguing her father's flea collection into the six-volume *Catalogue of the Rothschild Collection of Fleas*. She was over the moon that her *Atlas of Insect Tissue* had the vagina of a flea on the cover. I tap 'vagina of a flea' into Google immediately. Up comes: Flea in vagina! (Maggot-infested labia, if you want to know.) I eventually find Miriam's cover: a section of a mole flea vagina, embedded in wax, cut into a ribbon of thin slices, then passed though fluid dyes to reveal the Paisley tissues of the cellular maze. She likened examining stained sections under the microscope to smoking marijuana. Another of her passions was the biochemistry of insect communication and the heady world of aromatic pyrazines. 'Squeeze a ladybird very, very gently,' she urged, 'and its characteristic aroma will be on your fingers for days.'[112]

After her father died, Miriam spent more time with her eccentric uncle, Walter, whose private museum held his own enormous collection of animal specimens – 300,000 bird skins, 200,000 bird eggs, 2 million butterflies, 30,000 beetles and much more. Working for Walter was hazardous (remember the groom), one of his specimen hunters died of dysentery, another from typhoid, he lost three men in the Galapagos and his ornithologist had his arm bitten off by a panther. One of Walter's many obsessions was giant tortoises. Over his lifetime he owned 144; Rotumah, originally a gift from the King of Tonga to a trader called Alexander MacDonald, came from a Sydney lunatic asylum and was rumoured to be 150 years old. Walter liked to ride his giant tortoises dangling a lettuce leaf in front, like a vision out of Lewis Carroll.

Witnessing Walter fall prey to blackmail by his mistresses and seeing the damage it wrought motivated Miriam to campaign for the decriminalisation of male homosexuality (particularly vulnerable to blackmail). In 1957 her testimony provided the Wolfenden Committee★ with scientific evidence that homosexuality was a natural occurrence throughout the animal kingdom. She deplored cruelty to all living things (uncles included). Her perennial footwear was white wellington boots in protest against leather and the abominable way we treated farm animals. She thought every child should visit a commercial slaughterhouse as a responsible part of their education, and that slaughter should be done on the farm by those who had reared the animals. Miriam funded her scientific research by farming the Ashton estate she inherited from her father, nurturing its wild side – long before it was fashionable. It turns out that Ruary Mackenzie Dodds, the dragonfly expert who inspired us to

★ Commissioned to enquire into the discriminatory law on homosexuality which outlawed any homosexual act – whether a witness was present or not.

dig a lake at Underhill, restored his first lake to create a dragonfly sanctuary at Ashton because Kari, his wife, conveniently, was Miriam's niece. Miriam's vision as a scientist was dynamic, rigorous and unconventional, her imaginative curiosity meshed with compassion for living things, and indeed love. At the imaginary dinner party in the sky the person I would choose to sit next to would be Miriam Rothschild. When an injured vixen she had saved, nursed and released returned with her cubs to her garden, she said she felt 'crowned'.

VI: IF YOU KILL IT, YOU HAVE TO EAT IT – 1

In my position – lucky, singled out
By death and beauty for the blessèd kill,
assenting to the creature's dumb assent
John Burnside, 'The Fair Chase'

I am the Spirit that Denies!
Goethe, *Faust*

KILL JOY

It is 1963 and I am perched on a maroon velveteen seat in the Embassy cinema in Fareham. I am six years old. My mother sits to my right. I am deep in a forest where squirrels leap from bough to bough, butterflies flutter, bluebirds twirl around each other and animals talk (albeit in very odd American accents). I am entranced. A doe and her faun lie together in a glade; a loving look is shared between them. But winter comes. Bambi and his mother are in a clearing nosing the snow for grass. The music takes an ominous turn. The doe looks up, alert. She has heard something. There is an alarmed look on her face, her ears prick up, her eyes widen, sensing danger.

'Bambi! Quick, the thicket! Faster, faster, Bambi! Don't look back, keep running.'

Bambi bounds ahead, obeying his mother's words.

A shot rings out.

'We made it, Mother,' Bambi says, reaching safety. He turns to face the direction he has just come. 'Mother, Mother!'

But Mother doesn't come.

Like many children in the 1960s and '70s, I was traumatised by Walt Disney's cartoon film, *Bambi*. Dustin Hoffman claims he still suffers from the shock. I waited throughout the whole film for Bambi's mother to return, hoping beyond hope that it was just an adult trick and that magic would prevail. I was old enough to understand that a cartoon story was made up (which made it all the more perverse), but that made no difference. I was emotionally invested, in Bambi and his mother, and in their forest world. Then danger waltzed in. And that danger was us. Somehow, I knew *this* was the real world. Alas, now I had entered it.

My young self had no idea we'd been let off lightly. Earlier drafts of the film and indeed Felix Salten's original book, *Bambi: A Forest Life*, written in 1923, were so much darker. Salten, an intellectual Hungarian Jew who lived in Vienna, had hunted with Habsburg aristocrats. He drew the world of his young buck's forest from observation and a passion for animals. In his book the harsh realities of life and death keep coming: crows peck at a baby hare, a ferret swallows a mouse, a wounded squirrel dies and is eaten by magpies, an injured fox lingers with an infection in the snow. But nothing scales the terror or agony inflicted by the creature the animals of the forest call 'Him', whose scent is 'heavy and acrid', who is pale and erect with a terrible leg that sprays sudden death. 'No one can escape Him . . . He kills what He wants.' Salten's *Bambi* is a tale of man's tyranny that in 1936 would be banned in Nazi Germany as a political allegory about the treatment of the Jews. Bambi's mother is killed in a game drive, Bambi is shot in the shoulder and learns caution after being nearly fooled by a hunter with a doe-call imitating his mate, Faline. Fate for the hunter's dogs was a life of servitude, fear and adulation. It's strong stuff; and an early contender for ecofiction. *The New York Times* raved, 'He has given us the life story of a forest deer, and Felix Salten's comprehension of the entire universe as well . . . read *Bambi*.' Nobel laureate John Galsworthy wrote, 'I particularly recommend it to sportsmen.'[113]

Enter Walt. Walt Disney had grown up on a Missouri farm where he'd spent many hours in spring watching rabbits with his elder brother, Roy. One afternoon Roy arrived with an air rifle and shot a big buck in front of him. It wasn't a clean shot and Roy had to break the rabbit's neck to stop him thrashing. The memory haunted Walt. When he bought the rights to *Bambi* in 1937, his animation studios had begun to push the envelope from short slapstick humour (the loathsome Mickey) to more sophisticated cartoons; he was no longer interested in two-legged mice

and ducks with human hands dressed up in gaudy costumes. He wanted
to tell a meaningful story and convey the essence of a real deer. He was
emphatic. No rubbery human body language. The problem for the anima-
tors was *how* to express emotion in the rigid face of a deer without
anthropomorphic gestures. They spent months watching films of deer,
examining every tiny movement, a flick of an ear, a tilt of the head, a
shift of weight, those big brown eyes. Disney set up a zoo at the studios
with rabbits, ducks, owls, skunks and a pair of fauns. Armies of inkers,
tracers and copiers were set to work. The first rushes were rejected because
of tremors in Bambi's legs. Back to the drawing boards. The story treatment was also going through the hoops. Great chunks of Thoreau and
rhyming couplets on man's savagery found their way in, then out. The
writers, aware of the gathering storm of Hitler's Germany, had Bambi's
mother shot mid-leap as she was entreating the young fauns to believe
in the possibility of human love. They sent 'slave' dogs through the forest,
praising man and shredding any cornered animal to bits. Bambi's friend,
a hare, was shot and died murmuring, 'I don't understand.' Fire raged
through the forest to Salten's denouement, as a great stag appeared, Bambi's
father, to show him the body of a poacher caught in the fire. The message:
humans die as animals do and are not the all-powerful immortal gods
they pretend.

It would never do. Salten's story was simplified further and further, until
all that remained was Bambi's mother's ominous reply to his question
asking why they all ran: 'Man was in the forest.' The grit from Salten's
original was replaced by the offscreen tragedy and intricate drawing to
enchant young eyes with a dreamy world where autumn leaves swayed
gently to the forest floor, where deer were fleet and fluid and full of grace.
There were ample Freudian devices to keep older titterers happy. The
bunny who stimulated Thumper into explicit convulsions by stroking his
upright ears, and the climax of flower petals to conclude Bambi and Faline's
love-dance would not have passed the censors in another guise. Faline's
virgin blue eyes mysteriously became the same chestnut brown of Bambi's
mother. Disney was a complex character who on one hand was prudish
about anything overtly suggestive in Hollywood films, yet happy in his
office to run a reel of spliced wildlife shots of *innocent* animals copulating
to select company.

As Disney sweated over *Bambi*, MGM Studios were not so timid. For
Christmas 1939 they released *Peace on Earth*, an extraordinary eight-minute

cartoon directed by Hugh (Looney Tunes) Harman.★ It's snowing, the camera pans across the blue-white landscape as carol singers sing 'Peace on Earth, Good Will to Men', but all is not as one might expect of a Christmas cartoon. The gable end of a church is a ruin, barbed wire sticks out of the deep snow, and what might at first be mistaken for tortoise shells are German steel helmets. The carol singers are three young squirrels, and the helmets have been niftily converted into squirrel homes. Grandpa Squirrel arrives at his family's door singing the last line of the carol to his grandsons.

'What are *men*, Grandpa?'

'Well, *heh, heh, heh*, there ain't no men in the world no more, Sonnies. Nope, no more men. As I remember the critters, they were like monsters, they wore great big iron pots on their heads and walked on their hind legs and they carried terrible looking shootin' irons with knives on the end of them . . .'

The two baby squirrels, now on Grandpa's lap, are glad there 'ain't no more men around'.

'I never could figure 'em out,' Grandpa says. 'They was the ornery-est, cussedy-est, dag-nam tribe of varmints I ever did see. Why, they was always afightin' and afoolin' and ashootin' one another. They'd no sooner get one argument settled than they'd find somethin' else to fuss about.'

The shadows of tanks roll by. Grandpa Squirrel acts out the Second World War with whistling bomb imitations and rattling machine guns.

'And then what, Grandpa?'

'It was terrible. They fought and they fought and they fought, until there was only two of them left.'

No offscreen inference here. To the menacing beat of a drum we watch the last two soldiers shoot each other and sink slowly into the mire.

'And that was the end of the last man on Earth.'

From the war-torn landscape life begins to return. A squirrel pokes his head out of a hole in a dead tree; a rabbit pops up, another squirrel, a bird . . . the forest animals come out of hiding and converge in the ruined church where the wise owl reads from a book they find there: 'THOU SHALT NOT KILL'. And so the animals rebuild their home from the wreckage and enjoy Peace on Earth.

★   The only cartoon ever to be nominated for a Nobel Peace Prize.

Bambi's 1942 release enraged the hunting community. *Outdoor Life* hunting magazine journalist, Raymond J. Brown, was particularly angry: *Bambi* was the worst insult to American sportsmen *ever*. Hunters *preserved* the wild. Killing a doe in spring, the idea! The *Audubon Magazine* disagreed; nature writer and naturalist Donald Peattie attested to the forests being full of maimed animals and badly shot deer, does *and* fawns, adding that spring or fall made little difference to the deer. Suspicions were raised about forest fires that began on the first day of the deer season. Hogwash, the hunters responded. This allegory of man's destruction and animal innocence was partisan and deeply felt on both sides. Pitting Bambi killers (malign and repellent destroyers of the natural world) against Bambi lovers (sentimental sandal-and-sock-wearing lily-livered lefty brainwashed nature ponces, or vegetarian feminists, or nudists or both), Bambi's name weaponised both camps. It still does. In 2007 *Time* magazine listed *Bambi* in the Top 25 Horror Movies of All Time, claiming it still haunts those who saw it 60 years ago. Maybe in my deep unconscious I'm still waiting for Bambi's mother to reappear.

If the forest is the stage on which we re-enact our primal past, none has been more prized for its theatre of man vanquishing beast than Białowieża.* Once part of the endless horizons of ancient woodland that covered northeastern Europe, these had been royal pleasure grounds since the Middle Ages, protected by law. Poach a bison in 1538 under Sigismund I, King of Poland and Grand Duke of Lithuania, on pain of death. The grandiose hunts were legendary and could last a fortnight. Killing by day, feasting by night, they required elaborate preparation. Peasants were given their liberty for their services as foresters, game-keepers, guards and beaters. A moving wall of 1,000 men swept the forest, driving everything in its path – boar, bison, elk, deer, wolf, hare, fox, badger, bear – to a fenced-off area up to 5 kilometres square. On the prescribed day, the crazed menagerie was funnelled into a special hunting garden in which a pavilion had been erected to afford the royal guests a splendid view. Hunters on horseback charged to stick a 1-ton bison or a mighty boar to the roar of the spectators. In they poured, elk, wolves, foxes, hares. The forest kept giving, the hunting horn sounding

---

* The name Białowieża means White Tower, and comes from the white hunting manor enjoyed by Władysław II Jagiełło, King of Poland, 1386–1434.

the *mort* as each beast fell. So plentiful was the mountain of meat that it was smoked, salted and fed to the armies defending the kingdom from Teutonic knights and Tartar hordes.

With the invention of firearms, the king could just sit in his armchair and shoot a matchlock from the stands. It was the mark of courage to kill a bear, with rituals becoming more elaborate through the ages. In 1752, Stanisław Poniatowski (later King of Poland after his affair with Catherine the Great of Russia) describes an astonishing hunting garden known as the 'Polish Versailles' that boasted a bizarre contrivance of wooden scaffolding that ushered the wild beasts along a raised path to a 30-foot-high bridge over a canal. The hilarious thing was to see the animal plunge through a trap door set over the water and to take potshots at whatever fell. Poniatowski's diary records a bear surviving King Augustus III's hail of bullets before she capsized the royal boat.[114] Alas, a rifleman dispatched the bear, and the king lived to host the night's drink-fuelled shooting jamboree with geese, monkeys and hares dressed in Harlequin costumes as targets.

By the time Białowieża was swallowed by the Russian Empire in 1795, the beavers and otters were all but gone, foresters had been turned back into serfs, and a railway had been constructed to bring hunting parties in and lumber out. In 1860 Tsar Alexander II blamed the low numbers of elk and wisent (European bison) on wolves, bears and lynx, and ordered their eradication. When the German army seized the forest in 1915 they slaughtered the remaining bison and logged 4.5 million cubic metres of wood to sustain the Reich's war machine. Between the two world wars and the brief period in which beleaguered Poland retrieved her independence, 4,500 hectares of forest were protected as a national park, along with four bison scrounged from various zoos. Poland's precarious position made hunting in Białowieża the great gift in Polish diplomacy. Which is how, from 1934, Hermann Goering, Prime Minister of Prussia and Grand Master of the German Hunt, creator of the Gestapo, future architect of death camps and Hitler's deputy, became a regular visitor.

King of the Beasts

Herr Goering can be watched hunting in snowy Białowieża serenaded (you've been warned) by a jolly German folk song and accordion, on

YouTube. Furry aviator hat, fur collar, necklace of soft small animal skins, big overcoat worn as a cape, his breath visible in the chilly air. He aims his rifle. *Bang.* Four men carry a dead wolf to their master, one at each leg, and lay him on the snow. Three lynx are arranged together like slices of cake. Goering gets into a sled, rugs up. Looks to camera. Two more dead wolves. Goering resting his wide rump on a shooting stick. Rifles being loaded. Man dressed as a goblin blowing a hunting horn. Beaters sweep through the forest. (Don't forget the jolly song still playing.) A herd of wild boar are flushed into the gun sights; almost airborne they leap across the forest road, their prehistoric silhouettes straight out of the Chauvet caves. *Bang, bang.* The vanquished beasts are jiggled to life for a bow to the camera. Cut. Goering inspects a fleet of antlers as if his was the hand of their creation, puffing on a long cigar, his bulk belted into an immense leather jerkin. With an absurd black plume in his fedora and his long hunting knife he looks like a colossal pantomime Robin Hood. Next up, *crackle, crackle*, a hooded falcon is placed on Goering's gloved hand. He looks uncomfortable. 'No, no, no,' he tells the bird. He purses his lips in a flatulent kiss and strokes the breast feathers with a single finger. Cut. Live birds flying across the sky. Dead birds falling. Cut. A long picnic table. Goering helps himself to sandwiches from a silver tray. Beer down the hatch. He looks to camera. And he winks. He does. Then he winks again. Oh my God, the most chilling wink in history.[115]

Goering loved Białowieża. By 1941 the forest was his, with a swastika fluttering over the royal hunting lodge. Goering dreamt of restoring it to become the largest hunting reserve in the world replenished with primeval (German) game: bear, bison, re-creations of tarpans (wild horses hunted to extinction by 1919) and aurochs (extinct since 1627).* Nature was the touchstone where purity would justify purification. The forest must be cleansed. Of course, Jews and partisans were forest animals now. Mass executions in Białowieża were commonplace; 7,000 Poles were deported to camps; 34 villages were burnt to the ground.

The paradise hunting grounds were not to be. As Stalin's red pen drew a line through the forest, Goering's little white teeth bit down on a capsule of potassium cyanide.

It is seven in the evening. Jonathan and I are waiting in a forest hide in Transylvania, 5 kilometres from the nearest road at the foot of the Carpathian Mountains. This is my second visit to Romania. I was here in 1972 on a sixth-form school trip when Romania, then the darling Communist country of the West, opened its doors, briefly and very narrowly, to tour parties such as ours. My abiding memory is of never-ending forests as we were coached around from one Soviet-style 'hotel' to the next, each with the same indescribable chemical smell, which I had never come across before, or since, but would recognise immediately. Another indelible memory is how after each meagre meal a heavy munic-ipal teacup arrived in which a single gooseberry floated in water. We hooted with derision, then ate all the sweets we had brought for the children in the orphanages. Not that we had been allowed to visit any orphans. The allure of those 10 days was thundering past those great forests thinking wolf, bear. Ever since, I have wanted to see a bear in the wild, and now we are here. We wait.

One hour, one hour twenty. Silent as soot falling. There are 15 of us perfume-free, €50-a-head spectators, along with two forest guides, who pack revolvers on their hips. We stretch our gaze through the sentinel trees into the gloaming. A fox. A magpie. There is no guarantee bears will come. But all this way, after all this time . . . I check my watch as I make little deals with the god I don't believe in. Please come, bear.

---

* The Heck brothers studied cave paintings of aurochs and by selecting fighting bulls from Spain and Hungarian steppe cattle had, by the mid 1930s, back-bred a long-horned, aggressive, albeit smaller version of the enormous aurochs, called Heck cattle.

A sharp intake of breath to my left. 'There!'

Out from the shadows of the gloomy trunks comes a large brown bear. 'A female,' whispers the guide.

The bear is nervous. She hesitates, sniffs the air, looks around. We all gasp – she has two cubs with her. There are biscuits on the rock in the clearing, not many, we were assured, but enough to be worth paying a visit for (looks like an awful lot of biscuits to me). She makes her way towards them. Suddenly she is spooked and makes off with both cubs. Another bear, twice her size, appears directly in front of the hide. His raw presence sucks the blood heat from my solar plexus. He is huge. The guide whispers that he weighs nearly half a ton. I hardly dare blink. The guide tells us that last year he killed a female bear and both her cubs right in front of the hide, and there was nothing they could do but watch.

To hunt them or not to hunt them, that is the Romanian question. We are in a not-hunting-them stage, but this will change. Local mayors regularly call for 'problem' bear culls, citing damage to crops and livestock, but the main pressure is the money to be made from trophy hunters, $6,000–$10,000 a pop. That's just the fee for the Forest Department; far more than our non-perfumed group. Since a surprise hunting ban in October 2016, there has been a media spike in scare stories to whip up local hostility against bears. Hunter/politician Csaba Borboly's blog claimed bear attacks had more than doubled, reasoning that the bears' newly protected status caused 'genetic and behavioural deterioration'. His suggestion: 'vigilante justice', meaning bread soaked in antifreeze or rat poison dipped in honey; but his real aim was to reinstate the hunting quota as 'the only way to keep people and their children safe'.[116] The quota proposed by hunters is 10% of the population. That's 600 brown bears a year based on their estimate of 6,050–6,640 bears. The numbers are counted by the hunters, who are now the game managers who manage the forests. Economic incentives drive high estimates. But here's the thing: take out the dominant males and matriarchs – which is what trophy hunters do – and it disrupts the natural order of sexual selection. This in turn opens the door to 'sexually selected infanticide', which is when a new male moves in and kills the dead male's cubs. Killing off the matriarch, similarly, not only takes out a breeding female but puts her cubs into danger. Hunting breaks familial groups, bear learning, hierarchies and territories, generating pressures far exceeding what would naturally occur, and *this* is what causes the trouble. Because *they* are supposed to be the top pred-

ator. Killing 'trophy' bears has detrimental side effects and mortality losses
for the whole population, and we could run that scenario through all
the charismatic predators.

I watch the bears and I am transported. A forest is completely different
with wolves and wild bears. The forest knows it. A vital ingredient has
been returned. A re-enchantment. *Unheimlich* is a German word for which
there's no good English equivalent. The etymology of *Heim* is home;
sometimes wrongly translated as creepy, *unheimlich* describes the estranged
familiar, both real and unreal, something hidden, both known and
unknown. There is a shiver in the word. A sense of something deeply
ancient that has been alienated.

The big male bear leaves; another bear arrives, a smaller male with
a latte-brown face. He rubs his back against a tree, standing to show
how tall he is. A fox arrives, nimbly snatches some biscuits and scurries
off. There's a crack of a branch. The big male is back. His whole
demeanour is different. He rears up to put his front paws on the log,
his fur rucked up at the shoulder. The smaller bear keeps his distance,
then scarpers. The big male pads around the clearing, takes a few biscuits,
and is gone. The light is fading and we must take our chance to leave
the hide safely. The guide whispers urgently, we walk fast and tight
together, guns out at the head and rear. As we cross a stream, I photograph an enormous fresh paw-print.

A few days later friends take us to the ruins of a hunting lodge built
for President Nicolae Ceauşescu, who ruled Romania from 1965 to 1989.
Ceauşescu had been a shoemaker's apprentice at 11 years old, and although
big-game hunting was an aristocratic sport, once he had cemented his
position as leader of both the Party and the country, he became obsessed
with the idea of shooting the world's largest bear. In the Communist
regime, power over life and death had an aura, and Ceauşescu's hunting
parties became spectacles where he shot the biggest bears and forced his
top officials to watch. Like Goering before him, Ceauşescu made himself
Hunter-in-Chief of the Romanian forests as well as commander-in-chief
of the military. Gamekeepers were assigned to 2,226 hunting areas, all for
Ceauşescu's personal use. His cohorts soon realised the political capital to
be gained by putting a great bear before him, so each district competed
for his visits by offering up bigger and bigger bears. As Romanians starved,
wild bears grew fat on maize, fruit, chicken mash and the cadavers of
horses and cows.

After one hunt, on 15 October 1983 (500 beaters, three drives), 24 bears, weighing between 400 and 500 kilos each, were dragged back to the hunting lodge and laid out in two rows, garnished with freshly cut brush for Ceauşescu's wife, Elena, to admire. After the photo shoot Ceauşescu and Elena helicoptered away.

Because Nicolae Ceauşescu jealously kept bears for himself and no one else dared enter the forests, the bear population in Romania doubled, from 4,000 when he came to power in 1965 to nearly 8,000 well-fed bears by 1988. An enormous number, given their range, which is about the same size as the Greater Yellowstone Ecosystem, which supports about 700 grizzlies – figures quickly seized on by trophy hunters in Romania. In their natural state bears sparsely populate vast areas. They give prolonged parental care to their few offspring, but it is their social stability that provides the right conditions for a healthy population. To thrive, bears need secure habitats without human persecution in order to find food, cover, den sites and mates. Where hunting is rife, and where human activity invades their habitats, conflicts increase. So it's alarming to see the amount of building going on in the Piatra Craiului National Park. We are staying in Magura, until recently a village of simple wooden-shingled mountain dwellings, now big holiday homes with all the modern accoutrements creep where flowery meadows are dug up for driveways and garages. The infrastructure demands yet more infrastructure. Visitor centres, car parks, fast-food outlets. Less space for bears, more chance of antipathy towards them. It is easy to fear for Romania, such a heart-stompingly beautiful country, with its scythes and its silences, the percussion of bees in the lime trees, the quick generosity of the people. And no fences. No fences! Where you still see grazing communally shared among the flocks and self-regulated prudently. How priceless is that?

Ceauşescu ruled for a quarter of a century with increasing megalomania. He enforced his industrialisation programme by transplanting rural populations into urbanisation schemes. While he feasted, the people were rationed 200 grams of sausage per day. He prohibited contraception and banned abortion, a rule enforced by compulsory monthly medical examinations. Mortality rates for mothers and babies soared, as did the number of children consigned to the orphanages that so shocked the world in the 1990s. By 1989 the people had had enough. Crowds, once cowed, began to jeer. Then the army took against him. On 25 December, Nicolae and his wife Elena were taken to a military barracks and put before a kangaroo court that lasted 55 minutes. In the video recordings Nicolae wears his black bearskin coat; Elena, an overcoat with a dark fur collar. At 4 p.m. the death sentence was read out. Hands tied, the Ceauşescu were taken to the backyard and lined up against the wall. The firing squad sprayed more than 120 bullets. The execution bore an uncanny resemblance to one of his bear hunts.

On our last day in Romania, at the door of the gothic Church on the Hill in Sighişoara, we chatted to a friendly guide. He told us that Romanians would not stick a pig on Christmas Day, but they stuck Ceauşescu.

THE NECROPHILIAC'S EMBRACE

You have a struggling mammal in your mouth. You clamp your teeth into his wriggling furry warmness. Not good, oh no. All that squealing as the warm blood trickles down your chin. The terrified eye. You chew into silence. No, no, we don't like to eat like this.

Our predatory nature is supported by the structure of our gut, 20 feet of small intestine to break down a (cooked) meat feast, in comparison to an ape's intestines with a larger colon, which deals with a primarily fibrous vegetarian diet. Weapons gave us strength from club to spear to bow to gun. Pulling a trigger distances us, as does butchery, cooking, shopping and euphemism. In contrast, big, dangerous beasts have safety devices built into their showdowns where one animal knows the other is stronger and backs off; gorillas beat their chests to *avoid* combat; the alpha bear doesn't go on a bear hunt, though he might kill a few children to assert his reign; and you won't see a walrus go on the rampage, however brutally he fights over his harem. Snakes will wrestle with their fangs in check. Lions will walk away, hair standing on end, tail swaying; young males learn to wait their turn. Although E.O. Wilson suspected that 'if hamadryas baboons had nuclear weapons, they would destroy the world in a week',[117] killing is limited in the animal world. Chimps do it, but much is down to shrinking habitat pressure. *Homo sapiens* are the deadliest predators, without brakes. Animal fear of us, now ingrained, has come from an indiscriminate excess in killing them in ever more inventive ways. An ecological conscience has not had time to evolve.

For the modern human, with a few exceptions, the necessity, craft, respect, or indeed *art* of hunting for sustenance is history. It's not economically viable. American hunters claim they hunt for food, but know they could dine on caviar in suits woven from mussel silk for what hunting costs them in kit, ammo, licences, time and travel. So what is the pleasure, and where does it come from? The idea that human evolution has been influenced by our historical hunting behaviour is called the hunting hypothesis.[118] The successful hunter got the girl, so passing on the killer traits that enabled him. But as much as hunting might explain the source of human aggression, there is a stronger case for cooperation. Our hunting history has been the rationale for strong male coalitions and gals staying home with the babies, for toolmaking and invention; more emphasis on hunt than gather.

Bear hugs

Hunting that is not about sustenance is called sport. Once the pursuit of nobles, nowadays 'bloodsports' are available to anyone who can afford it. The Spanish philosopher José Ortega y Gasset's *Meditations on Hunting* sees hunting as a 'vacation from the human condition'; 'a hunter does not hunt to kill', he writes, but 'kills in order to have hunted'. The hunter is alert, as animals are alert. A primal nature is restored in a divine rite where the hierarchy of man finds his rightful place. Yet Ortega y Gasset concedes that 'reason' is 'the greatest threat to the existence of hunting'.[119] For pure gratification, an animal, who let's face it wants to keep their life, must lose it. Lives with their own values, their own society, their own relationships, in their own wild world must be snuffed out. 'Never stop and think or it will break your heart' is the advice from writer Vance Bourjaily,[120] for hunting is natural and real and morality just a human construct. Killing solely for pleasure is difficult for others to understand. It works both ways, hunters say, eyeing our leather shoes.

In a Romanian guesthouse I came across the memoir of August Roland von Spiess. Von Spiess, an Austrian officer born in 1864, became a Romanian citizen and was appointed Director of the Royal Hunts by King Ferdinand I in 1921. His memoir contained photographs that baffled the eye. Something contrary was going on. There was no apparent difference in the affection displayed towards the living or the dead. Alongside the predictable – Spiess with foot on dead boar, Spiess resting gun on roe deer, Spiess sitting on dead bear – were photographs of the orphaned offspring taken in as family pets. Live and dead creatures were posed within the same loving embrace; Spiess looks into the hazy distance, arm around a dead chamois who lays her head on his chest; his daughter, Silvia, at two and a half cuddling a dead upside-down capercaillie; Spiess with live eagle; dead eagle; dead bear; live bear cubs; piggy-backing a dead wolf; a live wolf called Peter; live badger cubs, dead lynx; everyone gets a hug. The most arresting and confusing photograph is when this is all happening at once: Spiess holding two bear cubs while standing over a large dead bear with a dead dog laid on top of her, and a live dog leashed to her back leg. A few pages on, Silvia, with the cubs, one under each arm. Next, the upper body of a snarling bear, arm raised mid-swipe, mounted next to the taxidermied head of the snarling dog. I was curious enough to visit the Spiess hunting museum in Sibiu.

On 25 January 1911, Spiess tracked a bear to a cave in the Laita Gorge, deep in the Carpathian Mountains. His favourite hunting dog, Hadubrand, was sent to investigate and all hell broke loose. A commotion of rabid snarling and roaring echoed from the cave until a sharp cry rang out. It hadn't come from the bear. Spiess sent his reluctant assistant to investigate. When he returned, pale-faced, it was with the news that Hadubrand had been killed by a she-bear defending her cubs. Spiess flew into a rage for the loss of his best dog. He made his way along the ledge with his other hunting dog, who tore snapping into the cave. With a terrible howl the bear rushed at the entrance. Spiess fired and the cave filled with smoke.

As the smoke cleared, he could make out the dark hulk lying before him. He was astonished, knowing full well an unaimed close shot like that would normally result in a misfire – and probably a dead hunter. The account in the museum does not explain how Spiess brought back the two orphaned cubs, or what happened to them. Past the endless trophies and thorny antlers, I find the cubs' mother. Her roar arrested for more than a century, her nemesis, the dog, beside her. The proximity seems vengeful. There were no photos of her middle-sized cubs frolicking on the Spiess lawn. Silvia became an ornithologist.

Funny kind of love

What is it that over the centuries maintained such passion for the kill? Reverend David Cashman, Bishop of Arundel and Brighton from 1965 to 1971, was 'mad about' shooting birds and mammals, he said it was the nearest thing to heaven he knew. Keen fox hunter Reverend Charles *Water Babies* Kingsley (1819–75), smitten by the sight of the swift grace of a fox (in spite of his 'great naughtiness'), wondered if his quarry was a messenger of the gods. 'Who knows? Not I, I am rising fast to Pistol's vein. Shall I ejaculate? Shall I notify? . . . Shall I break the grand silence by that scream which the vulgar call the view-halloo call? It is needless;

for louder every moment swells up a sound which makes my heart leap into my mouth and my mare into the air . . .'[121]

David Barrett is a British trophy hunter besotted with hunting. He has agreed to meet the South African-born cricketer Kevin Pietersen for the podcast *Beast of Man*. KP and radio presenter Sarah Brett interview him at his home, where African trophies stud the walls. KP is looking at a photograph of Barrett next to an elephant with a scarlet hole in his forehead.

'To shoot an elephant is an awesome thing to do. A stunningly, stunningly awesome thing to do, which is why I did it,' Barrett explains.

KP asks if he feels a kind of urge.

Barrett concedes there is an urge. 'The buzz you get is enormous . . . a tremendous buzz . . .' He tells Kevin he dropped the elephant in a full charge at 10 yards. 'I paid a lot to shoot that elephant.'

Barrett is a 'hunting person', he says. A couple of years ago he shot a sable. 'A stunningly beautiful animal . . . I mean it's a very stupid animal. I only had to walk about 20 yards, and he stood and watched me the whole way there, and I actually felt quite guilty. I thought, You are dim; or he wanted me to shoot him, I don't know which. So. Bang! Down he went.'

'A bit rubbish,' Sarah suggests, 'that you can only shoot stupid animals now.'

Barrett agrees. He wants to find the *right* animal to shoot. Plan it. Shoot it. Wonderful! 'I love hunting, deeply, deeply, deeply,' he tells KP. 'I used to get up in the morning and think, God, I've got to kill something today! . . . It's what turns me on.'

Barrett says he loves animals. That he feels an odd mixture of emotions. 'I am honestly sad for that animal . . . and I'll pat him on the head and say, I'm sorry, mate.'[122]

The trophy hunter's photo album is the confusion of the necrophiliac's embrace. To love and caress the victim, the false affection of possession (after the chase). To covet something so much we must kill it. The power over life and death – but only if you prove it. Remember, the victim is *chosen*. Not an individual, but spirit messenger of the gods. Snap: modern selfies of mastery, or delusion, depending on who's holding the gun. Philip Glass, poor trophy hunter, weeps into the mane of the first lion he's killed.[123] Glass comes from the 'I kill wild animals because I love them' school. He thanks God for the animals. It's reverence killing. His tears are pride and joy, he says. 'It was the most gorgeous lion,' he says. We, watching, who have heard him describe a bullying psychopathic father ordering him to shoot animals, suspect these tears are the psychological outpourings of a very messed-up boy who must prove himself to Daddy. The dead lion is now the soft toy he can take to bed.

★

I think I could kill an animal if I needed to eat. My father once ate a fly, with theatrical flourish, that someone had swatted on the table. His lesson to us was simple: if you kill it, you have to eat it.

★

In 2007 an astonishing spectacle of a white-tail deer attacking a hunter is witnessed by millions on YouTube: 'Deer attacks a hunter!' (3,350,335 at today's count). The deer, standing on his hind legs, is boxing with his front legs until the hunter, dressed in camouflage gear, falls to the ground. The deer pins the hunter down. The hunter raises his arm to protect himself, but the deer continues to pound him with his hooves, lowering his head to butt him on the ground. It is as if a straw snapped and flipped his brain from *Flee!* to *Attack!* The video goes viral across the world: 'Sweet revenge for Bambi.' 'Bambi pummels hunter.' The hunter manages to get up and the deer kicks him down again. This goes on. We don't know the outcome. Except that now we do. It turns out, Lynn Chestnut sprayed himself with deer urine. He did this so his wife could video what happened for his hunting buddies. He holds up a bottle labelled Elk Fire. So now we know who the dumb animal is.

MEMBERS & COD PIECES

I am watching undercover footage taken of horses being ridden into hunt protestors. It's quite gripping. Clattering hooves on tarmac. Shouting. A man in pink charges, mouth twisted into a grimace that would gladden Goya. His whip lashes out.

Then his tongue. A stream of obscenities accompanied by a white-gloved single finger.

'Turn your camera off, you dreadful old whore!'

Clunk.

Chasing foxes on horseback was an inalienable human right, until it wasn't. Guerrilla warfare broke out in the British countryside. Saboteurs in dormobiles laid aniseed trails to disrupt the fox's scent; their manifesto was to avoid harming the hounds or horses, and if it came to violence, run away. Sweet. They went after vicars. 'Nothing sicker than a hunting vicar,' they chanted from the back pew; draped fox carcasses across altars (not so sweet); filmed huntsmen *not* calling the hounds off the fox, and riders hitting cameras with crops. Hatred between the camps ran deep, dark and personal. People died. Fifteen-year-old Tom Worby was killed in a hit-and-run by the hound van on 3 April 1993, driven by huntsman Anthony Ball. A 'way of life' threatened; more accurately, a way of death.

Fox hunting is a minefield. Banned, sort of, on 19 November 2004, the British government criminalised hunting a fox with a pack of dogs, although two hounds were permitted to flush one out. What for? So he could be shot. The battle for the new Hunting Act was the Labour Party's Brexit; it went on and on, cost a fortune and it didn't help the

fox.★ The well-to-do lashed out at the less well-to-do to defend tradition. To preserve the headiest unconstrained arcane taste of ravening
they will ever know. The rufous red rust streak flying low over the
ploughed edge of the horizon. Run, run. The electrified frenzy; the
lathered horses in full spate; the reek of mud, leaf, bark, tree, moss,
wind. Red blood berries and hoar breath. The filthy language of the
grown-ups. Ice-cracking puddles. A baying chorus with the moon in
it. This is as close as it gets. But what about the fox? The Burns Report
– charged to settle whether hunting was cruel – considered a tidal
wave of statements in evidence from interested parties and concluded
that to be chased by a pack of hounds all day before being ripped
apart 'seriously compromises the welfare of the fox'.

The rules

> and if, on occasion, I never quite saw the point,
> I was always the first to arrive, with my father's gun,
> Bound to the old ways, lost in a hand-me-down greatcoat.
>
> John Burnside, 'The Fair Chase'

Killing animals for pleasure is governed by rules. Each group must appropriate a mythology and wrap it in protocol. The right clothes, the correct
lingo – which you will have to learn to become a member. Get it wrong
and they will know. To start, you need a live target, obviously. A stag,
boar, tiger, snark, etc; but not a horse, or a sheep, or a swan, or anything
just wandering by. The animals must be free to run away, not tethered or
fenced in (although the canned trophy hunters of South Africa have
dumped this etiquette), so not park deer or Exmoor ponies. And wild(ish),
so not tame and just standing there. Mind you, some American bear
hunters will shoot a bear wearing a radio collar and a bloody great pink
bow tied round their neck at a feeding station. No pets. Nor farm animals;
that's stealing. And you have to kill them; that is, they must die, or it's

---

★   Once farmers tolerated foxes for the benefit of the local hunt; now a new generation
    of gamekeepers have stepped up their eradication methods of snaring and shooting foxes
    to protect the increasing numbers of pheasants released to be shot. Thus the fox has
    become more effectively eradicated in the British countryside (and still chased across it)
    than he ever was. The fox, vilified and loved, has moved to town for his survival.

not hunting. To chase or stalk your quarry makes it seem skilful and therefore more highly regarded in the eyes of those doing it. So you can't poison them or catch them first with a snare. Well, of course you can, and many do, but that's called gamekeeping or stewarding the countryside. Hunting must also be premeditated; you can't just run them over by mistake. Hunting is an armed and unprovoked intentional attack to kill a wild free animal, so not self-defence either.

Hunting for pleasure requires the attitude that an animal's wild life is an insignificant, expendable, biodegradable product. To protect this product there is another mantra: predator control.

The pits

I am staying with friends in Edinburgh and have casually picked up a magazine (Scottish moorland on the cover) and opened it at random. Sharp intake of breath. I've never heard of a stink pit. Now I'm looking at one. A gamekeeper stew of coagulated heads, tails, wings, legs, ribs, sinew, fur, feathers, paws sticking out. I turn the page. A *mountain* of dead mountain hares in their winter whites. Jesus Christ, it's a hell hole. In 2017 the cull was estimated at 38,000 hares, themselves an important food source for golden eagles and their newly reintroduced white-tailed cousins. Hares are killed, it transpires, to control a tick-transmitted disease, louping ill virus (LIV), in red grouse. The magazine *Revive* is making the case for the reform of Scotland's driven grouse moors.[124] It's profoundly shocking. As the name 'stink pit' suggests, the smell is the lure to attract foxes into the traps. Spring-loaded iron jaws at the ready to snap their tiptoe feet, then hours of waiting to conclude only one way. Anything that threatens the number of grouse is the enemy of the estate. There are nasty secrets to unpack about grouse moors in the UK.

Wild grouse are driven by a line of beaters flapping flags and blowing whistles towards a waiting line of guns. Red grouse are wild birds, but intensive grouse moor management entails industrial levels of trapping native predator species — weasel, stoat, fox, etc; mass outdoor medication for the grouse in trays of grit coated with Flubendazole, a drug for the strongyle worm parasite; miles of unplanned roads; heather burning and peat drainage — to provide punters with huge numbers of birds. The only animals welcome on these vast estates are red grouse. The Game and Wildlife Conservation Trust estimated 700,000 wild grouse were shot in the 2012/13 season, a conservative estimate via voluntary bag returns. The annual figure probably exceeds a million. I flick through the magazine thinking what British taxpayers don't know. General licences to kill native wildlife in order to protect game (for the gun) are issued without a mechanism to control numbers of traps deployed, or what or how many animals they kill. Flick, flick. Break-back traps in cages set on logs crossing ditches; a stoat's neck flat as paper in a spring clamp; a dead hedgehog caught by his nose; a buzzard dangling in a pole trap upside down; a satellite-tagged golden eagle spreadeagled; a strangled badger in a wire noose snare. Anything that creepeth. It's really not okay.

Robbie Marsland, Director of the League Against Cruel Sports Scotland, has received a dossier. Seven estates, representing 4% of Scotland's grouse moors, have been surveyed over nine months by an investigator using Scotland's rights of responsible access. The findings were logged, timed and photographed, and the locations mapped with GPS coordinates. Here's one of them. Millden Estate: 78.34 square kilometres; 3,001 grit stations; 761 traps and snares; 646 shooting butts; 150 kilometres of road; game-keeper vehicle patrols driving around slowly; rail traps across streams at 70-metre intervals, on top of walls and across drainage ditches; box traps along roadsides, in rubble piles and even in drainage culverts; decoys, baiting stations, rabbit traps, bird scarers, 'falconry' birds and what looked like raptor shooting hides. 'Everything about this estate is extreme and none of it is desirable.' Each estate told a similar story of sick ecosystems prevented from healing themselves by the management. 'Nothing remotely wild or natural about Easter Clunes . . . in a very sick ecological state . . . an intense grouse farm, nothing more.' Lead ammunition, gas guns, electric fences, a heat map of raptor persecution. 'There is a war going on in our peaceful glens, every single second of every single day, it is relentless, intense and our animals are dying in their thousands, all so that more grouse can be killed.'

In these landscapes it's what you don't see that screams the loudest. No raptors, no trees, no scrub, rocks placed on peregrine nesting ledges to prevent breeding, whole ecological groups of animals exterminated. And a catastrophic lack of imagination by the owners of these vast estates. The tacit agreement between gamekeeper and his master is unspoken, a

nod, a euphemism. You know what the master wants. He doesn't want
the detail, just the results.*

Grouse shooting estates occupy 13% of Scottish land, yet their business
contributes less than 0.02% to the Scottish economy, and fewer than 3,000
jobs (including part-time and seasonal) that average out at a salary of less
than £11,500 a year.[125] On 18 June 2020, the Scottish Government banned
the unlicensed culling of mountain hares. The gamekeepers were up in
arms at the unleashed devastation that would unfold. But the operative
word was unlicensed . . .

INTERMISSION

> I have had many teachers — from the smallest dung beetle to the
> ponderous elephant. But by far my cruellest — and yet most healing
> — lesson came from ingwenya: the crocodile.[126]

Sicelo Mbatha grew up in a rural village on the edge of the Hluhluwe-
Imfolozi nature reserve in South Africa. To get to school the children
had to walk 10 miles and cross three wide, but shallow rivers. The six-
year-old Sicelo always walked with his best friend, eight-year-old Sanele,
who taught him about the birds and animals. One day in a summer storm,
the teachers sent the children home early before the rivers rose. The last
river to cross was up to their waists and flowing fast. The children checked
carefully for logs that might be crocodiles, held hands in a long line and
began to cross. Halfway across a log swept past and broke the line, then
suddenly Sanele's hand was wrenched away from Sicelo's. Sanele had gone
under, but his hand was flailing above the surface. Sicelo grabbed it and
tried to pull him back, but in the thrashing foam a great cloudburst of

---

* Alan Wilson, former gamekeeper at Longformacus, kept lists: 42 foxes, 32 cats, 75 rats,
103 stoats, 37 weasels, 90 hedgehogs, 5 mink, 622 rooks, 81 jackdaws . . . In 2019 he
was charged with shooting two goshawks, a peregrine falcon, four buzzards; despatching
an otter and three badgers using snares likely to cause partial suspension of an animal
or drowning, and being in possession of two bottles of the EU-banned, highly toxic
pesticide Carbofuran. His sentence (second conviction) was to carry out 225 hours of
community service.
   In his 1843 *Highland Note-Book*, Robert Carruthers celebrates over 4,000 'vermin'
destroyed by gamekeepers on the Glengarry Estate (including 198 wildcats, 48 otters,
15 golden eagles, 98 peregrine falcons, 78 merlin, 71 nightjars, 63 goshawks, 27 sea
eagles, 18 osprey, 462 kestrels, etc., promising readers 'fond of natural history' what
may be achieved with 'steady and combined efforts, for the protection of game'.

blood turned the muddy river red. Sicelo felt Sanele's hand go limp as the lifeless fingers were dragged from his grasp. By the time the elders arrived with their spears, the crocodile and Sanele had gone. Sicelo kept the pain of his friend's violent death locked in feelings of grief and anger. Unable to afford the fees to realise his childhood dream of studying nature conservation at university, Sicelo volunteered as a trails officer in the wildlife reserve. One hot day in the dry season, Sicelo was patrolling the banks of the great Imfolozi River with Induna, a wise Zulu elder, when they heard a great commotion of vervet monkeys and the warning calls of birds. As they approached the noise, a mongoose fled in front of them. They could hear the laboured breathing of a large animal. They peered through the dry reeds and saw a large male buffalo up to his haunches in a mud pool as a visitation of crocodiles tore at his flesh. Sicelo was catapulted back to holding the lifeless hand of his best friend in the churning river. They watched as the buffalo's horns sank into the mud. Induna turned to Sicelo and told him the buffalo was now at peace and they must consider what his death could mean. For the first time Sicelo understood something important about the cycle of life and death. That the crocodile who took his friend was simply hunting for food to sustain himself. 'Perhaps only the crocodiles could bring this message to me,' Sicelo wrote. Sicelo let his grief go, forgiving both the crocodile and himself for not saving his friend.

Twenty years on, Sicelo Mbatha devotes his life and work to teaching humans the Zulu concept of *ubuntu*, the human virtue of compassion, by guiding people from all over the world into the wilderness of South Africa to tap into a spiritual connection with the wisdom of nature. This is his gift to his childhood friend, Sanele, and a homage to his teacher, Induna.

VIPs

'All predators must be culled and the otter is no exception', so a member of the Eastern Counties Otter Hounds told Philip Wayre (founder of the Otter Trust) in 1976.[127] In the first half of the twentieth century otters had played under London's Westminster Bridge. By the close of the 1970s they had gone (DDT, polluted rivers, roads, draining, water bailiffs, hunting). The Otter Trust, founded in 1971, together with Angela King of Friends of the Earth, campaigned tirelessly for the otter to be legally protected. Otter surveys were based on presence and absence, indicated from spraints and footprints (nowadays we have camera traps). It took an almost total absence before the government was moved to help. Otter-hunting was finally banned and the otter protected under the Wildlife and Countryside Act in 1981. By 1996 the Otter Trust had bred and released 130 otters into the wild, but it took three decades – otters do not reproduce quickly – and hardworking river restoration volunteers for this magnificent native animal to come back. So it is galling to see, in 2021, the toxic waste of industrial agriculture – from the proliferation of intensive poultry units – being allowed to pollute and suffocate the life out of British rivers once again.

The energy-flow through ecosystems dictates there are always fewer predators than prey. Predators suffer high mortality in early life; prey are fleeter, safer in numbers, camouflaged; or, like a gazelle, spring into the air on all four feet (called pronking) to say, *Forget it, I've seen you.* Vilified as vermin, as if they are doing something bad, the belief is that native predators are to blame for the loss of bird life, game or livestock, and that killing them is the solution. But the presence of apex predators like goshawks keep the middle-order predators in check, by both catching them as prey and by creating a landscape of fear. Farmland bird decline is the result of changing farming practices, lost habitat, lost food and nesting sites, hedge flailing, tsunamis of insecticides, etc. A fox population is self-regulating; it remains more or less static if let be. Continue to kill foxes, as gamekeepers do, and the survivors will breed earlier and more prolifically; young, dispersing foxes will fill a vacancy and by inexperience can exacerbate conflict with humans, where an established fox will have learnt to avoid danger areas, like the hen coop. Until of course there is a localised extinction, a good thing in the view of the shooting community, but not for an ecosystem where rats and rabbits have no predator, and poisoned animals poison other animals, where functioning communities have broken, where pressure for food is exacerbated by huge numbers of introduced game species – astonishingly 60 *million* intensively reared, non-native pheasants are released without licence in Britain every autumn for sport, who will help themselves to the hard-pressed wildlife: young snakes, slowworms, amphibians. Grain hoppers put out to supplement their food support the explosion of rats. And so it continues. These days restaurants turn down game birds because of the antibiotics and lead shot they contain.[128] There are plenty of problems when you suddenly double the biomass of birds in the environment.

But if we are really talking food for the pot, could not skilled hunters provide venison (organic, free range) and help moderate the unchecked populations of deer in the absence of their apex predators?

What other sport with weapons might satisfy the appetites without spilling innocent blood and damaging the environment? I think fencing. It ticks a lot of the boxes. A display of manhood and virility with a long weapon and sexy get-ups. Frilly shirts, codpieces and lunging. Flashy swords engraved with family crests. Etiquette, rules, clubs, membership. That it is a skilled sport might count against it, but it could take place outside in flowery meadows with plenty of fresh air, followed by the same slap-up lunches . . .

We inherit our social behaviours, passed down from forefathers. Parents are the key, whether Mum rescues the spider from the bath, screams or drowns him. We are all members of different clubs with membership rules, little feudal systems with family allegiances. The growing sciences of ecology and animal intelligence has not transferred into a general understanding. We are taught geography, human history, presidents, kings and queens, but not the importance of predators, nor who lives in the tree outside our bedroom window, or in the soil under our feet. We reside in echo chambers of an increasingly partisan world where we think what others think and it is increasingly hard to think (or voice) what we might think for ourselves.

★

A female hen harrier raised in England's Forest of Bowland stares indignantly to camera as she is fitted with a satellite tag. The following year she is found dead on a grouse moor in the Yorkshire Dales.

'Bowland Beth'

That she made shapes in air

That she saw the world as pattern and light
moorland to bare mountain drawn by instinct

That she'd arrive at the corner of your eye
the ghost of herself going silent into the wind

That the music of her slipstream
was a whisper-drone tagged to wingtips

That weather was a kind of rapture

That her only dream was of flight forgotten
moment by moment as she dreamed it
That her low drift over heather quartering home ground
might bring anyone to tears

That she would open her prey in all innocence
there being nothing of anger or sorrow in it

That her beauty was prefigured

That her skydance went for nothing
hanging fire on plain air

That her name is meaningless
your mouth empty of it mind empty of it

That the gunshot was another sound amid birdcall
a judder if you had seen it her line of flight broken

That she went miles before she bled out

David Harsent

LOVE, ACTUALLY

Some years ago our friend Dan gave us a book, *The Gypsies: Waggon-time and After*, written by his father, Denis Harvey. The cover shows a large piebald carthorse pulling a decorated gypsy wagon. Walking alongside, holding the reins, is Denis, a striking-looking man – dark, handsome, with sideburns, a moustache and a porkpie hat, wearing a three-quarter-length coat with big lapels. Horse, home, man. In the 1950s and '60s Denis lived the wandering life, he understood south-country Romany speech and had a deep knowledge of gypsy customs and ways. My enthusiasm encouraged Dan to lend me 'Plaiting the Magic', the unpublished story Denis wrote about his unusual life. In it, he tells a three-way love story. A young man at the end of the Second World War in 1945, Denis was working as a forester in Dorset at the Springhead estate, close to where Jonathan and I have our small nature reserve. Early one morning, in a nest of grass on the banks of a stream, he found an abandoned otter cub, about six weeks old, weak and starving. The mother, he concluded, must have fallen victim to the otter hounds that hunted the area. He took the cub to Joanmary, the young wife of the estate's administrator, an uncommonly arresting young woman of gypsy descent, who preoccupied his feelings. What better excuse for Denis to visit?

Denis knew Joanmary had a great compassion for creatures from a story she'd told him: the cause of her departure from her home in Devon, away from the malicious gossip. The trouble began when a young game-keeper took a rabbit with broken legs out of his gin trap and tauntingly broke the animal's neck in front of her. In her anguish and fury, which she described as 'a physical pain in her head', she cursed the young man, telling him that the same fate would befall him, and she hoped it would be soon. Three days later the man's legs were crushed when the trailer he was sitting on the back of reversed into a brick wall. In the eyes of his friends this confirmed Joanmary as a witch.

As Joanmary bottle-fed the otter cub back to health, her fondness for this enchanting yet powerfully wild creature grew. They called her Kelpie, after the shape-shifting spirit of the lakes in Celtic legend. To help Joanmary wean her off the bottle, Denis began to spear fish and then learnt the gypsy method of catching trout by tickling them with his hands. Soon Kelpie was part of the household with Joanmary's cats and the family collie. Kelpie learnt to use the cat door and came and went as she pleased, answering to her foster family's whistle, and keeping to her natural nocturnal cycle. Though utterly wild in her essential being, there was one difference – she had no fear of humans. This caused Denis concern, not least when the otter hounds were in the area. When a meet was advertised Denis laid a false trail of scent – a concoction of Kelpie's bedding and faeces – along the upper reaches of the stream, and rejoiced as the baying hounds tore off in the wrong direction. As Kelpie grew and became harder to control, so did Denis's feelings for Joanmary. To prevent himself from causing trouble by declaring himself, Denis decided to return to his Surrey home near Box Hill outside Dorking, and take Kelpie with him.

There, Denis lived in a converted 1920s stable block where Kelpie made her holt in the garden. Each evening she emerged with a fluting whistle of excitement, bounding over to him, somersaulting onto his feet, then straddling all four legs around his shin in a tight hold, where she liked to ride as Denis walked about. This game lasted until he picked her up to stroke her thick brown fur, but cuddling doesn't last long with an otter, and soon she'd be wriggling to get free to jump into the sink where she would try to turn the taps on herself until Denis obliged. There she would lie under the running cold water, 'her hands raised beside her face, the little black palms outwards in a gesture of surprise and wonder'. Soon bored, she was off into the garden where everything was turned into a game and Denis was constantly amazed at how quickly and cleverly she worked things out. Though Denis provided fishmonger fare, Kelpie resorted to her own diet of goldfish and ornamental carp on her night thieveries. Curiously, the local policeman took an inordinate liking to the new burglar on his beat. He was fascinated by Kelpie, and always laid the blame on herons to complaining fishpond owners. He was delighted to discover the law couldn't prosecute Denis for damage because otters were indigenous. But Denis knew it would not be wise to let Kelpie run amok. No one understood more than he the problems an animal with acute intelligence and 'incurable wildness' would face living in a human environment. The issue was not that Kelpie would have been unable to fend for herself in the wild, for this seemed to come naturally to her. The danger was her lack of shyness. He tried to restrict her in a homebuilt run, but otters are not creatures easily or happily contained. Kelpie took over the neighbourhood, cavorting not only in garden ponds but in and out of people's houses. On one occasion she flashed into a kitchen by the back door and completely disappeared, only to reappear suddenly when the fridge door was opened, snatching from the hand of the stunned homeowner the filleted plaice he was about to cook for supper. By the time Denis arrived to retrieve her she was guarding her prize and tucking into the fish in the corner. Denis paid for the fish, grabbed her by the tail and marched her home objecting. On another occasion, she entertained a cocktail party at a large house with a lake, while thinning out their population of ancient carp. Her most infamous night was spent in the grand residence of one Colonel Whyte, where she was shut in by mistake when the unsuspecting colonel and his wife retired to bed. Denis knew something was wrong when Kelpie returned late in the morning, submissive and strange. Then the telephone went. Over to Denis:

I will tell the Colonel's side of the story as he told it to me, and no doubt to countless others. Twice in the night he and his wife had been disturbed by strange sounds somewhere downstairs and he had gone down (doubtless with drawn sabre or a brace of pistols) to look for marauders. He had looked in all the rooms except the dining room, the door to that had been closed he told me. In the morning he opened the dining room door, to prepare the toast for breakfast, and found the room wrecked almost beyond recognition. At first he thought that someone must have been searching systematically for something that was believed to be hidden there because the rest of the house was in perfect order. But the destruction was so comprehensive that his next thought was that it must be some supernatural or psychic manifestation. Until, that is, he spied what he termed 'a little furry face' peaking at him from under the sideboard.

Kelpie for her part, finding herself shut in after everyone had gone to bed, had gone berserk. She moved all the furniture away from the walls to see if it was blocking the way out. She swept all the books from the shelves, pulled out the drawers turning most of them upside-down, and leapt up to knock all the pictures off the walls. She rolled up the carpets, tore down the curtains, and finally scratched the wallpaper away from the skirting boards and ripped it up to the picture rails in long strips which hung in festoons and tatters around the room.

The Colonel and his wife only needed to show Kelpie the open front door to be rid of her. To me they were absolutely charming. The Colonel told me that nothing so exciting had happened there for years and his wife said the room had needed redecorating anyway. The only repair they allowed me to pay for was to the heavy mains radio that had been pushed off the sideboard face down onto the floor.[129]

Bless the Whytes. But Denis knew it couldn't go on. Kelpie was a wild animal with sharp teeth, and one day someone might get hurt. After much soul-searching Denis caught the train to London to meet Professor Julian Huxley of the Royal Zoological Society, who had offered to rehome her. When Denis met Kelpie's would-be keeper at London Zoo and saw the concrete enclosure with its sterile stream, he backed away; Denis had a deep understanding of the physical and emotional needs of his fostered charge and was not impressed with the zoo's insistence that animals, such as an incorrigible wild otter, were happier safe and sound. He demanded better conditions for Kelpie that reflected her physical, emotional and instinctual needs. But after being treated like a well-meaning sentimentalist who knew little of the scientific, Denis went home. What had come into his mind was a neglected hidden valley he had discovered as a boy, where the descending levels were connected by streams and waterfalls, constructed, it turned out, by the seventeenth-century gardener and diarist John Evelyn.

> I resolved to take Kelpie to this valley and abandon her to her own destiny whatever that might be – to find a mate if she survived, and if she did not, a good place to die – one I would have chosen for myself. How much better to die there than at a hospital in Hammersmith, or in a concrete tank beneath a square of smoke-polluted London sky.[130]

Denis put her in a tin trunk with a fish, carried her halfway up the long valley, and released her half an hour before sunset. He watched her sniff the ground, then race across the grass to dive into the nearest pond. She surfaced and looked back, then swam to the far bank, climbed out, shook herself, and without looking back disappeared into the beech woods. Denis wished her well and watched the place where she entered the wood for a long while.

GREEN FIRE & NAUGHTY FUN

The Faroe Islands lie northwest of Denmark in the rich cold sub-Arctic seas, on whale migration routes and feeding grounds. Faroe Islanders hunt whales for meat. That the hunt, called the *grindadráp* or grind,* is depicted in paintings, sung about in ballads and mentioned in a law from 1298** supports the principle that something has 'rights' if it has been done for a long time. The Faroese consider whale meat an integral part of their culture. Blubber for food and lighting fuel, skin for ropes, stomachs for fishing floats. The flesh of the whale became the flesh of the people. These days whale meat is highly toxic with an accumulation of pollutants: methylmercury, polychlorinated biphenyls (PCBs) and DDT derivatives – the persistent poisons we dribbled into our rivers – remain in the bodies of animals, becoming more and more concentrated as they work their way up the food chain. A noxious demonstration of the interconnectedness of life.

Høgni Debes Joensen, the Chief Medical Officer of the Faroe Islands, warned in 2012 that these food contaminants can adversely affect foetal development of the nervous system and the immune system, cause hypertension, arteriosclerosis of the carotid arteries, Type 2 diabetes, and high blood pressure in children.[131] Not good. So the Faroese government recommended one whale dinner a month, with a special warning for pregnant or breastfeeding women, and children. Nevertheless, the Faroese eat 200 tons of whale meat and 100 tons of blubber a year, which divides up into 13 pounds of whale per person. To provide this they kill 1,000 long-finned pilot whales and Atlantic white-sided dolphins every year. The pilot whales are about 12 feet long, weigh around 2 tons and travel in family pods.

The tradition may be old, but the speed boats, jet skis, mobile phones

---

* Rhymes with binned.
** The Sheep Letter, which mainly deals with sheep.

and killing equipment are all spanking new. There is no hunting season, and no quota. Passing whales are considered *their* birth right – not the whales'. As soon as whales are spotted near land, the Grind Master is alerted and out they go. A testosterone high carried on the hot wind of nationhood. Men on the shore wait with hooks, rope, lances and knives. The boats drive the pod into one of the killing beaches with a wall of noise. The men on shore wade out and thrust their hooks into the animal's blowhole, then pull them ashore. On the beach the whales are stabbed with a lance to sever their spinal cord to paralyse them. I can't help thinking there is a lot of whale, from epidermis to spinal cord, for a lance to get through.

The international marine conservation charity Sea Shepherd has been fighting the grind for 40 years. Their volunteers are the scruffy types that nice society looks down on. Job description: 'No pay, long hours, hard work, dangerous conditions, extreme weather.' They go out on boats for weeks at a time to try and steer the whales and dolphins away. Committed, courageous and informed, they are willing to be arrested, go to prison, whatever it takes: their vested interest is the life of the whale. In 2015 Sea Shepherd filmed graphic video footage of the grind for those who had the stomach to look at it. Not me. But I did see a scarlet sea busy with wading people and thick with dying whales. Among the bodies an unborn calf slithers across the guts of his dead mother. Sea Shepherd logged 490 slaughtered pilot whales before their boats were confiscated by the Danish Navy and police.* The whales take more than 20 minutes to die. The Faroese government admit the spinal lance is not very effective and they are going to invent a better tool. Kate Sanderson, an advisor to the Faroese government on 'responsible hunting', says an uncontrolled environment is never clinical: 'Sure, it's a tradition but it's a form of food production. It's a local resource. It's part of the diet. It's not a sport. It's a way of getting food for the family.' A grind foreman says the hunt is 'environmentally friendly' and that it is their *duty* to use what they can get free from nature.[132] Any pilot whale passing is swimming straight to the butchers. Every member of the pod is killed, pregnant mothers, babies, juveniles, in front of each other. Two hundred whales on a beach. People

---

* The Faroe Islands (ranked one of the highest GDPs per capita) receive a block grant from Denmark, their protectorate, and benefit from a free trade agreement with the EU, but are conveniently not a member; under EU law it would be illegal for them to kill and harass cetaceans.

shouting. Whales panicking. Blood. Thrashing. Because they breathe in water they can't scream. A child dances on top of the stilled, shiny torpedo of a dolphin. Young men cheering and laughing pose for photographs. Men with dolphin blood all over their big bald heads.

The grind is a parable. Looking to the past instead of the future. The wrecking ball before our shut eyes. The circular narrative of connectedness from distant poisonings to intimate poisoning, from extinctions to ecological downfall. We are the grind. Deadly, weaponised, caught in the thrall, baying for our freedoms and entitlements.

★

Deep male voice-over, strong Australian accent:

'I'm Mick Fanning. After being on the professional surfing tour for 14 years I had the encounter with the great white.'

Tracking shot to surfing competition. A surfer paddling on his board. Blue sea.

Commentator: 'I can see a little splash. Holy shit!'

Twenty-seven million people have watched Australian surfer Mick Fanning's encounter with the great white shark, filmed during the final of the 2015 J-Bay Open surf competition in Jeffreys Bay, South Africa, on 19 July 2015. There is a splash, a large fin, Fanning turns his head, then spins round on his board; there is some kind of commotion. He is knocked off the board into the water. We see a glimpse of the board, then white spray, then a rising wave blocks the view. Nine long seconds as the wave builds, and builds, and slowly curls over.

We see Fanning swim away from his board. He is picked up by a jet-ski, severely shaken. Later Fanning said that when the shark circled and came in again he punched him on the back, and the shark went for the board, severing his leg rope. 'I was petrified,' he said.

Fanning's way of putting the fear and flashbacks behind him was to find out all he could about his near-nemesis. What he discovered prompted him to team up with National Geographic Australia and four shark experts to make a documentary about saving them.

Female shark expert voice-over: 'You had an encounter with *the* apex predator.'

Male shark expert voice-over: 'They're not there for a holiday, they're there to eat.'

Chisel-jawed Fanning walking towards the camera: 'So I'm gonna go out there on this quest and understand what is really going on with sharks.'[133]

Something good has come out of a close shave. Fanning describes sharks as magical, majestic and misunderstood creatures. Of the 1,200 species of shark, over a quarter are threatened with extinction. They are a barometer of the ocean's health. Sharks kill about 10–15 humans a year; humans kill *100 million sharks a year*.[134] Fishermen slice off the fins and throw the animal back to drown. The point and hope of the film is that with understanding comes the desire to protect them. Jonathan, who likes to think he is still a surfer, tells me Fanning would have won the championship if it hadn't been for the shark. I could have told him that Fanning won it the following year, but I'll save that one-upwomanship for when he reads this.

Stag Nation

In April 1944 Aldo Leopold, the forester we met canoeing with his brother in the Colorado Delta, sat at his desk to write a confession. Thirty-five years earlier, fresh out of forestry school, he had travelled to Arizona to work for the US Forest Service, surveying vast tracts of land and counting trees. One day he took a break to eat lunch on the rim of a canyon and was watching the turbulent river below. He noticed a doe crossing, but when she shook herself dry on the far bank, he realised his mistake. It was not a deer. A family of wolf pups sprang out of the willows to welcome their mother. The 'pile of wolves writhed and tumbled' at the very foot of Leopold's canyon perch. Within seconds the reunion was blown apart by Aldo's gun, 'pumping lead into the pack, but with more excitement than accuracy'. He scrambled down, thrill accelerating his heels, just in time to reach the mother wolf and see 'a fierce green fire dying in her eyes'. This was Aldo's epiphany. In that dying fire he saw something known only to the wolf and the mountain. He was young and his fingers 'full of trigger-itch'; he'd thought that fewer wolves meant more deer and that no wolves would mean a hunter's paradise. But after seeing the green fire die, he sensed 'that neither the wolf nor the mountain agreed with such a view'. In the years that followed Leopold saw state after state extirpate its wolves, and the mountains slowly die, wrinkled with more deer trails, browsed to baldness until the bones of the deer 'dead of its own too-much' bleached with the bones of the sage brush.[135]

In 1935 Leopold moved his family into a run-down cabin called 'The Shack' on a sandy, worn-out farm in Wisconsin in an area known as the Sand Counties. There, he planted trees and began to develop his concept of a land ethic in what became *A Sand County Almanac*. The mature Leopold saw around him a sense of entitlement to a stocked natural pantry, without obligation, without any notion of its complex workings. He had come to understand the importance of predators like wolves as the apex of a towering biotic pyramid, the vital regulatory control of an ecosystem, and to believe that man could control something so unfathomably complex was hubris. In that he was ahead of his time. 'In short, a land ethic changes the role of *Homo sapiens* from conqueror of the land-community to plain member and citizen of it. It implies respect for his fellow-members, and also respect for the community as such.'[136] Leopold called his short requiem for the she-wolf and her pups *Thinking Like a Mountain*. That humans have not learnt to think like a mountain is why 'we have dustbowls, and rivers washing the future into the sea'. The year Leopold's seminal book was published, in 1949, he died fighting a grass fire on his neighbour's farm.

A veil

There's a hoo-ha because Prince William has said there's a place for trophy-hunting. The prince's friend, the engaging TV presenter Ben Fogle, thinks the money from trophy-hunting is important. 'I don't like it, but what is the alternative?' Fogle asks, as if there isn't one.[137]

Here's the trophy-hunting argument: the huge fees paid by hunters to kill charismatic megafauna pays for their conservation and gives local communities employment; demand ensures their protection; trophy hunters are not the problem, poachers are. Will Travers has heard every in and out of this argument for 35 years. It's true, trophy hunters do pay *a lot* of money to kill threatened species. The rarer, the more they pay: $50,000 a rhino.* It's tosh that this helps conservation or the local community. The money stays within the industry and a small number of corrupt officials. Of course it does. Trophy-hunting feeds the industry, not conservation. To keep up with demand, lions and other animals are intensively

---

* And remember the detrimental effects of taking out the dominant males and matriarchs.

bred to be trophies. Legal activity opens loopholes to make illegal activity easier. Remove the stigma, ivory becomes acceptable. Ivory is bling, a sign of affluence, demand goes up and outstrips supply. Poaching increases. That is what happens. Trophy-hunting brings in $200 million to the African continent, and around 50,000 jobs full and part time. In Kenya, where wildlife-hunting is banned, wildlife tourism generates *two billion dollars a year* and sustains half a million jobs. Ten times the trophy-hunting business of the entire African continent in wildlife-based tourism. Suck that up, 'hunting conservationists'.

Tourism belongs to Africans, ivory belongs to elephants. That is where the future lies. The poverty for humanity is that we are forced to measure elephants or lions by jobs and money for ourselves. Morality doesn't come into it. The British restaurant critic A.A. Gill wanted to feel what it was like to kill a primate so he went to Tanzania and shot a baboon. 'I know perfectly well there is absolutely no excuse for this,' he said. 'Baboon isn't good to eat, unless you're a leopard. The feeble argument for cull and control is much the same as for foxes: a veil of naughty fun.' Trophy-hunting is a throwback to a colonial past, killing wild beasts for *their* big tusks, magnificent manes, beautiful skin. For what the punters lack. Reason *is* the enemy of hunting. We don't have sustainable slavery, we have education, and what slavery remains carries shame. That is more effective. The 'place' for trophy-hunting is in the bin.

30 April 2016. President Uhuru Kenyatta sends a smoke signal to the world. Eleven pyres of crescent tusks tower into the sky: 105 tons of poached ivory, from some 8,000 elephants. Each confiscated tusk, inscribed with identification: origin, age, sex, herd, weight. Like a war memorial. The flames leap hot and high into a deluge of rain. The hiss of cracking tusk, the acrid smell of spitting kerosene and burning ivory. The principled message: This is not a commodity that will be traced back to our country. 'For us, ivory is worthless unless it is on our elephants.'[138]

★

In the 1970s, the Indian naturalist Arjan Singh was hand-rearing leopard cubs to re-establish a wild population where he lived, miles from civilisation, at the edge of Dudhwa Forest, on a river on the border between India and Nepal. In 1976, a female leopard who had successfully been released was raising two cubs of her own in the forest when she was threatened by rising monsoon floodwater. To Arjan's amazement, she brought them back to his house, one after the other, and deposited them in a bedroom, where she stayed with them. When the rains were over a week later, she picked up the first cub in her mouth, and began to evacuate the house to return to the jungle. On reaching the river crossing she found the brown turbulent current too fast and deep to carry her cub across. She turned, under Arjan's watchful eye, and walked to his canoe. With her cub still in her mouth, she jumped in. What could Arjan do, but paddle her across? When they reached the other side, the leopard leapt out and disappeared into the jungle. Arjan waited for her, knowing the other cub was still in the house. Eventually he gave up and rowed back, but no sooner had he reached the other bank than she appeared, calling him to return. Back he rowed, in she jumped and Arjan ferried her back to the house, where she collected her second cub and returned to the canoe to be ferried across. Arjan obeyed with wonder, marvelling at her trust and her ability to plan a safe strategy for her cubs.

From princely Sikh stock, Arjan had been an avid big-game hunter in his youth. His epiphany came the night he shot a leopard in the lights of his car. In sudden revulsion at what he'd done, he vowed to devote his life to saving the big cats of India. He built his HQ and home, Tiger Haven, raised his orphaned leopards, and waged war on poachers and bureaucrats.[139] His irascible tenacity gave him the title 'Honorary Tiger'. He secured 200 square miles of Dudhwa Forest and grassland as a national park and one of India's tiger reserves. Arjan died on New Year's morning in 2010, aged 92; he was cremated on a pyre at Tiger Haven and his ashes scattered among the graves of his beloved animals, according to his wishes.

VII: IT'S THE ENVIRONMENT, STUPID

. . . I wanna kill educated ignorance.
Dis is me naked, revolting in front of you, I'm
not much but I give a damn.

Benjamin Zephaniah, 'Naked'

HERON BLOOD TULIPS

I am a river monitor. A few years ago the Wiltshire Wildlife Trust advertised for volunteers to attend a training course and I applied. This would
be my perfect vocation. Aside from the joy of poking around in rivers
in waders, I would be able to say I was a river monitor, removing the
dilemma I faced when people asked me what I do. I could tell them
about taking three-minute kick samples of the river and counting specific
species of invertebrates to give an indication of the health of the water
quality and habitat. Far easier than explaining how I sat in my shed trying
to write. Even better, I could avoid the dreaded question: what's your
book about?

So I went on the river monitoring course and learnt to identify the
larvae of the cased caddisfly and the caseless caddis, the mayfly, the blue-
winged olive, the olive, the flat-bodied mayfly, the gammarus (freshwater
shrimp) and stoneflies. These, apart from the gammarus, were the aquatic
larvae forms who would later hatch out into flying river flies, the food
of many freshwater fish. They were also sensitive to pollution, so if their
numbers fell below a certain level an investigation by the Environment
Agency would be triggered;★ they were like canaries of the river. The
Welshman who taught our group gave the example of a stretch of a river
in Wales where samples were showing no gammarus present after heavy
rain. They set up gammarus traps (a netted cage from which the shrimps

---

★ Although these days limited funds means limited investigations.

can't escape), and after every heavy rainfall the gammarus would be dead. This baffled them. Until they traced the cause to a nearby woodyard with uncovered piles of tanalised timber. When the rain washed through the stacks, the toxic preservative chemicals leached out straight into the river. Milk was another problem. One teaspoon of milk in your garden pond, he said, can kill everything. If dairy farms had a spill or cleaned equipment, the natural bacteria that break down milk use up the oxygen in the water more quickly than it can be replaced – so aquatic life suffocates. Fishermen of course have a particular interest in a healthy river, because lots of river flies mean lots of fish.

So it transpired the next time someone asked what I did, I was able to say I was a river monitor. We talked about chalk streams, which interested him a great deal because he had one. Let's call him Roger Swift. A few days later an email came in with the subject line: River A— AGM. Roger Swift was inviting me to attend his river association's AGM to 'give us an overview of your interest and work on river entomology'. A 10-minute talk was all he required, and there would be other guest speakers talking about chalk streams, fishing, ecology, hydrology and wild trout. The meeting would be held at a great house with its own river, lake and cricket ground, even its own chapel, and there would be lunch afterwards. Chalk streams, ecology, lunch . . . I was in.

A few days later the association's secretary sent me the agenda, the list of invitees, the address and directions. I checked out the invitees. Two lords and a sir. A doubt crept in . . . The meeting was to be held in the Shoot Room.

The day before the meeting I decided to take a sample of the River A—. I cleaned my kit and drove to a bridge I knew where a footpath crossed an old water meadow. I scouted for a good entry spot, filled a bucket of river water, set up my inspection tray, grabbed my net and stopwatch and clambered in. I held my net on the bed of the river and gently roughed the gravel upstream with the toe of my boot for a minute, then emptied my catch into the inspection tray. Another minute sweeping my net under the river weed, then a minute along the bank and under some stones. I crouched over the inspection tray waiting for it to settle and for its occupants to reveal themselves. I never tire of it. The rocking-horse motion of the blue-winged olive nymph; the fluttery leaf-like gills along the body of the large dark olive; the swaying cobra motion of the flatworms and leeches; the comical shrimps buzzing about, swimming on their sides, their transparent amber bodies and whizzing legs. The cased caddis were my favourites. These soft-bodied creatures make mobile homes from tiny stones and twig debris which they stick together with silk from their salivary glands. The minuscule tubes have a studded, jewel-like quality; the caddis lives inside and lugs the case about, like a hermit crab. There are enterprising human jewellers who provide caddisfly larvae with precious gem chips, pearls and crumbs of gold to make their cases, which they collect after metamorphosis and sell for a fortune. Which has nothing to do with this story which needs to get me into the Shoot Room.

I drove through the big iron gates, down the grand drive flanked by battalions of blood-red tulips, and *swish*, onto the sea of gravel outside the stable block where I parked among a fleet of Range Rovers. A man in a flat cap directed me up some wooden stairs to a long, elegant room. I glanced around. My shoulders sank, for everyone was in the uniform: Bracken tweed shooting coats, green waistcoats, gingham, Barbours, brushed cotton Viyella checked shirts, or smart blazers and club ties. Good God, one guy was wearing breeks, those short tweed trousers with woollen stockings and garters. There was a silver tray loaded with glasses of elder-flower cordial. The long table was set with water carafes and 25 places laid with a single-page agenda by our name tags. Mine, embarrassingly, read: 'Keggie Carew Riparian Entomologist'. I made a circuit of the table. There was a 'Game & River Keeper', another 'River Keeper', a hydrologist, someone from the water company, a fish farm manager, an estate manager, the head gardener, someone from the Angling Trust, the Wildlife Trust, the Wild Trout Trust. One label read, 'Retiring Cormorant Licensee'. All men, bar me and the secretary taking the minutes.

By now of course I had realised I was in a country 'sports' room. I scanned the agenda: Apologies; Chairman's Introduction; Matters arising from last year's meeting; Accounts; Subscriptions; River Reports from the various estates in the area (we're not talking council); Reports from Guest Attendees. No. 9 on the list was: Cormorants & Predation. I could feel my skin cool with foreboding. I was in The Wrong Place.

We began. This river was divided into beats. Of course it was. The first anglers had been through and caught 'reasonable numbers'. One river keeper reported that one fisherman caught nothing and complained, but then he discovered the man was using a dry fly! Ho, ho. (I have no idea, but can imagine.) A stoat had been spotted. I was quickly getting the picture. A stoat in this company was most probably bad. A cormorant did a fly-by. Also bad. A pair of swans were resident but not causing any damage, so the keeper was happy for them to stay. There were three 'persistently active' herons. Uh oh. And a kayaker. I piped up. What was the problem with a kayaker? He might carry the spores of the crayfish plague. Otters, it was noted, carry them too. Double-bad. And there were a few incidents of poachers . . .

I began my spiel, telling them for the purposes of the meeting I had done a kick sample of the River A— the day before. The red-faced river keeper three seats down shuffled loudly in his seat.

'If I may . . .' he coughed.

I stopped and turned. I suspected what was coming.

'If I may ask *where* you took this sample and *how* you accessed the river?' he said, his cheeks flaring brighter.

I told him where.

'Did you have permission?' he asked.

'From whom?' I replied.

'From the landowner,' he snarled politely.

'I did not,' I said. 'I accessed the river from a path by a public bridge at a place where some schoolboys were fishing for minnows.' (I knew that information would annoy him.)

'I don't know if you understand,' he proceeded, 'that the river is easily contaminated by dirty nets, your waders and the like.'

'I was very careful, as I was taught, to make sure *everything* was scrupulously clean,' I replied, thinking of all the flying ducks and other creatures that travelled to and from the rivers, along the banks.

Roger Swift motioned me to continue.

I gave my results, explained what invertebrates I was looking for and why, then sat down, a little red-faced myself.

Ten minutes later we got to Number 9. Cormorants & Predation. This began with mink traps. Minks have wiped out the water vole population, so to be expected. And fair enough to the man from the Angling Trust who donated a 180-metre stretch to offending poachers – to be managed and stocked by them – and has not suffered any poaching for three years. The thing that was exercising them most was 'cormorant trouble'. There was a simple solution, of course. Shoot them. A licence to shoot 10 cormorants had been granted, but this was a drop in the ocean to the 120 reported.

My jaw was beginning to lock. I understood cormorants were 'supposed to be' coastal birds, but they were only up the river because these days there was so little for them to eat on the coast. I felt for these fine birds, trying to survive, trying to feed their one brood a year, trying to cut out some niche in their impoverished world. But the next thing on the troublesome list was herons. And herons bloody live here. They, as far as I was concerned, were entitled to be on their river. Herons *were* the river, for it was in their bones, and river light filled their eyes. Monitoring the river and keeping it healthy was, I thought, for *all* the indigenous wildlife. For the well-being of the ecosystem and the complex web of life that we were trying to put back together again. Because now we understood that to mess with bits of it messed with the whole thing. And things got out of kilter, for these were fine-tuned, beautiful, elaborate systems where everything was interlinked and had impacts on everything else.

It was noted, eyes cast down, that the authorities required 'proof of evidence' to grant a licence to shoot herons. The red-faced river keeper tittered. I saw a knowing look flicker around the room. Lips folded firmly over smirks. The estate manager opposite me smiled a smile with thunder playing around the edges. And I knew. I shrank into my expensive wooden chair scanning the men in the circle. Proof of what evidence? That they were herons? That herons ate fish? I looked across at the guy from the Wildlife Trust, but he wasn't saying anything.

In Margaret Atwood's novel *Surfacing*, a dead heron hangs upside down in a tree. The abiding image is of two grey wings wide open as if flight had fallen out of them. Trappers had snared the bird and tied his feet to a branch with a nylon rope. But why had they strung him up like a lynching victim? Crucified upside down, until his life had fallen out of his wings.

I loved herons. Their bigness, their patience, their creaking pterodactyl flight. Old Franky is his Sussex name from the sound of his cry, *Frarnk, Fraarnk!* We have a small pond in our garden. Last winter, each morning for a fortnight in the coldest, shortest days, a heron stood hunched over it. She was there at 7 a.m. and still there an hour later. Waiting. With her own spear. Even when the pond was frozen, as it often was after a bitterly cold night. Every day the heron flew away empty-beaked, for the frogs were staying below, comatose in the mud. I watched from our bedroom window and my heart ached for her. Her sharp round yellow eye with its fathomless black pupil, fixed on the bare larder of our pond. I imagined her stomach twisting in tighter knots. She waited, motionless, except when the breeze picked up her long soft-grey over-feathers.

I left dog food out in vain hope. But herons need heron food and it is hard to conjure it up. I watched heron videos on YouTube to see if there was anything I could do. The babies have mohawk hairdos. In one of the films a young heron, fallen out of its nest, climbed all the way back up the tree, pulling itself up with its unformed wings through the branches. A feat of wonder, and the indomitable instinct for survival. This is what always amazes me: the lion-hearts around us, that anything survives at all on the leftover crumbs of our dominion. Atwood's heron was valueless to those who took his life. Here, the value was in what the heron was perceived to be taking from these river keepers and landowners. That the heron's magnificence was not seen, or valued as part of the experience, that eradication was the default, never cohabitation, or that we couldn't factor in the loss, from — let's face it — very deep pockets indeed. At least stop playing 'guardians of the countryside', please. For there is little guarded that is not to be murdered later. I could understand the attraction of fishing. I really could. What can be more lovely in this world than a river? Its rush flowing around you, the quiet contemplation, the long braids of emerald crowsfoot with its trapped buttercup flowers, the flash of a kingfisher, the azure blue of a damsel, or a sudden cloud of emerging mayfly — once so common, now such a rare sight. (Us again — phosphate and silt.) There was skill in catching a fish, and it didn't involve gunpowder. The fascination, the knowledge, learning to think like a trout. I got it. Dry fly tying is a mysterious art. And frequently the fish got away with their lives — I appreciated it wasn't great being dragged into a suffocating atmosphere by a hook in your mouth, but at least you lived to get dragged out another day. I got fishing. And it has a vested interest in keeping the rivers clean. But I didn't get killing a four-and-half-foot-tall feather boa-ed heron for it. I didn't get that at all.

Three months after the heron's winter visits I came across a skull in the meadow. I held the long stabbing bill, imagining the missile speed of its dagger strike. Heron cell upon heron cell, built up in the nest by both parents' ministrations. The bone had a shell-like lustre. In the eye cavity was the stretched meniscus of dry meat. Tiny feathers were stuck above the eye socket where her long black extended eyebrow once flicked over the back of her head. I contemplated the heron's skull, remembering her patience and cold hunger. She didn't touch the lifeless dog food, of course. The place I feel this is in the centre of my chest. It rises up and clenches its fist tightly, then stays like that in silence. In the harsh winter of 1963, the heron population fell to just 2,000 pairs.

I sat, quiet as a mole, slunk into myself. Eyes down, fingers fingering my pen. Herons are protected under the Wildlife and Countryside Act 1981, but that doesn't mean much on private land where herons and otters mysteriously disappear for no one to know and no one to miss them. I toyed with the notion of speaking out, but I knew my voice would only climb to an alarmingly high register. My tongue would clog in my throat, my brain would forget words, and what, anyway, would be the use of it? I was outflanked and outnumbered. I took the English path and kept my counsel.

The chairman wrapped the meeting up. If anyone saw a silver Subaru Forester and four large guys, the estate manager would like the registration number. And what to do with all the money in the river association's coffers? *Feed the herons?* He thanked the lord (human, I think) for having us. To a scraping of chairs we funnelled out towards a table groaning with sandwiches. I love sandwiches. I've always loved sandwiches. And there were cakes and vol-au-vents. A longing look was all they got from me.

'Not staying for lunch?' Roger Swift had caught me speed-walking for the door.

I fumbled an apology, looking at my watch. My legs sped me out into the fresh air and into my little car which sped me up the long drive, flanked by the battalions of blood-red tulips.

HURLING STARFISH

To save every cog and wheel is the first precaution of the intelligent tinkerer.

Aldo Leopold, *A Sand County Almanac*

1963, Makah Bay, Washington. Robert Paine hurls a purple starfish, as far as he can, into the Pacific Ocean. He throws another, and another. He has begun one of the most important experiments in the history of ecology. What regulates the population sizes of living communities? How does nature work?

At the time, we believed a stable ecosystem was regulated by food availability at the bottom of the chain. The limit of plant food limited the number of herbivores, which limited the number of predators, in a kind of pyramid. Each level limited the next level up. This, however, did not explain why the terrestrial world remained green. What stopped herbivores eating *all* the plants? While we understood prey numbers regulated predator numbers, the idea that this mechanism functioned the other way round was oddly quite radical. And hardly easy to put an ecosystem into the laboratory to test out. Robert Paine's brilliance lay in selecting the one place, particular and small enough, that he could both manipulate and observe: an 8-metre stretch of rockpools in the intertidal zone. Paine wanted to quantify the influence of a species in a community. What happens, for instance, if you remove the top predator?

In Paine's rockpools of starfish, barnacles, mussels, limpets, sponges, sea urchins, seaweeds, anemones and molluscs, he could see that the large predatory mussel-eating starfish, *Pisaster ochreceus*, was at the top of the food chain. Paine prised the starfish from the rocks in his study area and hurled them out into the deeper water far beyond the rockpools. Every fortnight he returned to repeat the expulsion. Starfish out. The reprieved mussels began to spread. After a year and a half of ridding the pools of starfish the number of different species in his patch had dropped from

15 to 8. After seven years the mussels had monopolised the rocks, pushing all the other species out. The starfish, by feeding on the mussels, had created space for a community of species. To describe the phenomenon of a species that has a disproportionately large impact on the ecosystem, Paine used the term *keystone*, analogous with the apex stone of an arch; dislodge it and the whole thing falls apart. This was the first field experiment that clearly demonstrated top-down predator control.

While evicting starfish at Mukah Bay, Paine had noticed some rockpools full of sea urchins with very little kelp. He began removing the sea urchins from some pools, leaving others untouched. Where the urchins were banished the algae and kelp returned, while the untouched pools became urchin barrens (nothing but urchins) – which pointed to something awry in that ecosystem. His answer came on a trip to Amchitka Island in the Aleutians in 1971.

The Aleutian Arc of volcanic islands runs for 12,000 miles in the North Pacific Ocean – from Alaska to the Kamchatka Peninsula in Russia, where East meets West. The arc's 57 volcanoes form the northern part of the Pacific Ring of Fire. Three of the six largest earthquakes ever recorded occurred here between 1957 and 1965. Nonetheless, it was on the remote and seismically active Amchitka Island, a wildlife refuge since 1913 (sounding familiar?), and 'one of the least stable tectonic environments in the United States',[140] that the United States Atomic Energy Commission chose to detonate the largest ever nuclear explosion a mile below ground in 1971, codenamed Cannikin: 5 megatons; 400 times the power of the Hiroshima bomb. At the behest of the commission, PhD student James Estes was in Amchitka to monitor what happened to the island's wildlife – the bald eagles, peregrine falcons and migratory birds; and the largest concentration of sea otters in the world in the offshore kelp beds. Otters, decimated by the fur trade, were absent from most of their former range, but here a remnant population had survived and rebounded to fabulous numbers. As the bomb test galvanised the environmental movement to launch Greenpeace, Estes was galvanised by the otters. Estes was convinced the thriving kelp forests supported the thriving otter population; when Robert Paine arrived on the island he told Estes to look at it the other way around. What do otters eat? Sea urchins. What do sea urchins eat? Ah . . .

Estes went to compare an island without otters 200 miles away. The most dramatic lesson of his career came the moment he stuck his head under water and saw an ocean floor covered with urchins, and no kelp:

'Any fool would have been able to figure out what was going on.'[141] The abundant species in Amchitka – fish, seals and birds – was absent.* Sea otters, he realised, were the keepers of the kelp, and the kelp was keeper of a diverse community of marine life, who were keepers of a rich community of coastal bird life. Where otters returned, kelp beds returned, providing food for urchins, who provided food for otters, who kept them in check, which allowed the kelp forests to grow. This wonderful world. Sea otters were the keystone species.

November 1741. The Russian ship *Saint Peter*, on an expedition to chart the Arctic coast, flounders on rocks off a frozen island in the frigid oceans between Russia and Alaska. The crew manage to get ashore, but 28 men and the Danish captain, Vitus Bering, die on the island that will bear his name. The survivors build a makeshift shelter and remain for 10 months while they build a new boat from the wreckage. Among them is a naturalist, Georg Steller. It is thanks to Steller that we have a scientific eyewitness account of one of the most extraordinary relics of Pleistocene megafauna that survived in this one far-flung outpost, living quietly, away from human eyes: Steller's sea cow, *Hydrodamalis gigas*, a colossal 10-ton creature measuring up to 33 feet long and heavier than a *Tyrannosaurus rex*. This was their last refuge. Herbivorous, gentle, monogamous, living in family groups, the sea cows spent their days grazing in the kelp beds, blowing and snorting, moving gracefully, one foot, then the other, in a half-swim half-walk buoyant ballet of the shallows. So placid were they, Steller found he could wade out and touch their hides, tough and rough as bark.

In the spring they come together in the human fashion, and especially about evening in a smooth sea. But before they come together they practice many amorous preludes. The female swims gently to and fro in the water, the male following her. The female eludes him with many twists and turns until she herself, impatient of longer delay, as if tired and under compulsion, throws herself upon her back, when the male, rushing upon her, pays the tribute of his passion, and they rush into each other's embrace.[142]

---

* Sea urchins can survive without food for a long time. Kelp that regrows is quickly eaten, and some urchin barrens remain for decades without change.

The sea cows sustained the marooned sailors until their boat was seaworthy to set sail for the mainland. Being so buoyant, they floated at the surface and were easy targets to harpoon. The meat tasted like corned beef scented with almond oil and with its high salt and fat content was slow to spoil. The oil was used for cooking and lamps, the hides for clothing and sails. Steller gathered as much anatomical detail as he could. In place of teeth he records horny plates and long bristles on their upper lips (for mashing kelp); the stomach measured 6 by 5 foot; and their blubber, built for the sub-Arctic waters, was 4 inches thick. It is Steller's descriptions of the sea cows' 'uncommon love for one another' that is most affecting. The only noise they made were sighs and snorts.

Within 27 years of their discovery, these gigantic gentle creatures were gone. News of the fur-rich seas with plentiful otters was carried home by the shipwrecked survivors and brought Russian trappers to the islands. To feed themselves while decimating the otters, the hunters helped themselves to these convenient larders. But that is not what did for them. In just 12 years the initial tally of 8,000 otter pelts was down to just 25. Without check, urchins ripped through the kelp beds leaving the cupboard bare for everything else. The domino effect of Paine's trophic cascade*

---

* Trophic from the Greek, meaning nutritional. A trophic cascade is a succession of changes that *cascade* through the trophic levels of the food web.

took all the breath out of the coastal waters. The mighty sea cow annihilated by the lowly urchin. For ever.

There was another twist to this watery tale. Jim Estes returned to the
Aleutians in the 1990s to find otter populations had collapsed by 95%
throughout the archipelago. A layer to the food chain had been unwit-
tingly added. Or rather taken away. The consequence of a sequential
megafauna collapse, begun by Japanese and Russian whaling in the North
Pacific after the Second World War, was that starving orcas had shifted
their diet from dwindling supplies of large mammals – whales, porpoises
and sea lions – to the smaller mammals available. Sea otters have no
defence against orcas, and orcas need to eat 10 times as many.* Sea otters
were not the only predator of the kelp-eating sea urchins in trouble. Two
thousand miles south, after a record-breaking marine heatwave in 2013,
Californian starfish were hit by a mysterious wasting disease that allowed
urchins to proliferate to 60 times their previous density. They chomped
their way north to Oregon, where on a single reef 350 million urchins
were counted in 2014. The 10,000% increase in their population closed
a recreational abalone fishery worth some $44 million.[143]

What Darwin called the 'entangled bank', a dynamic community of living
organisms interconnected through energy flows, nutrient cycles and the
physical elements of their habitat, was termed an *ecosystem* by the British
botanist Arthur Tansley in 1935. We mess with these interactions at our
peril, for they are so immensely complex we do not understand them.
Conservationists have learnt single species cannot be protected in isolation.
Top-down and bottom-up, the focus needs to be towards entire ecosystems. Our banners must shout more expansively: Save the whale! Save
the krill! Save the phytoplankton! Save everything in between!

Great kelp forests once marched across the cool, nutrient-rich ocean
from northeast Asia to the American Pacific coast. As tall as 175 feet, with
enormous floating canopies, kelp forests are one of the most dynamic
ecosystems on the planet. Home of shrimp, lobsters, rockfish, perch, starfish,
abalone, bristle worm, marine snail, urchin and otter; provisioning gulls,
terns, egrets, herons, cormorants, eagles. And humans. Fishing communities

---

* The knock-on effect – fewer otters, more urchins, less kelp, fewer fishes – oddly benefited the bald eagle, who switched prey to seabirds. This in turn relaxed pressure on
  sand eels, which benefited mackerel, who were scooped up by . . . bald eagles. There
  are a multiplicity of top, bottom and side-on effects in these complex trophic webs.

who once followed the kelp highway for food *and* safety, for the kelp forests buffered the swell of the big seas, were probably the first colonisers of the Americas. To trigger the collapse of a such an ecosystem, to them, would have seemed unimaginable.

Aldo Leopold's wolf epiphany manifests in George Monbiot's film, *How Wolves Change Rivers*, celebrating how Yellowstone National Park came back to life with the reintroduction of wolves after their 70-year absence. This is the textbook example of a terrestrial top-down trophic cascade. With their return, the wolves reinstated a landscape of fear: the wolves hunted the elk and kept them on the move by their wolf presence, allowing willow and aspen to regenerate. In a few years beavers reappeared, building dams to create pools which slowed the river, diminishing erosion, and thus a host of other animals returned to the ecosystem. Fear was an ecological driver. Time-lapse photography shows the change in landscape, affected not just by wolves but also by the return of mountain lions and bears, rippling all the way through the food chain to the scavengers and the decomposers who recycle the nutrients for the vegetation. But the Yellowstone story was to take another turn, as we shall see . . .

NICE TO HAVE, BUT SO WHAT?

11 September 1826

Between Somerford and Ocksey, I saw, on the side of the road, more *goldfinches* than I had ever seen together; I think fifty times as many as I had ever seen at one time in my life. The favourite food of the goldfinch is the seed of the *thistle*. This seed is just now dead ripe. The thistles are all cut and carried away from the *fields* by the harvest; but, they grow alongside the roads; and, in this place, in great quantities. So that the goldfinches were got here in flocks, and, as they continued to fly along before me, for nearly half a mile, and still sticking to the road and the banks, I do believe I had, at last, a flock of ten thousand flying before me.

William Cobbett, *Rural Rides*

Each generation redefines normality. The expression 'shifting baseline syndrome' was coined by fisheries scientist Daniel Pauly in 1995, when he noticed how each generation of scientists judge the normality of fish stocks according to what existed in their youth. In a TED Talk, Pauly remembers a visit to Ghana in 1971 when there was a thriving blackchin tilapia fishery, with the average fish measuring about 20 centimetres. Twenty-seven years later the fish had shrunk to 5 centimetres, the new normal. Pauly had been involved in disastrous development projects that introduced industrial trawler fishing into countries like Indonesia, previously an artisanal fishery. He went out in a new trawler in the Java Sea and they hauled out the whole seabed onto the deck. Within a year the area had transformed into a muddy mess: 'We adjust our baseline to the new level and we don't recall what was there.'[144] Shifting baseline is a kind of generational amnesia. To read about the flocks of passenger pigeons, 300 miles long, a mile wide and so gargantuan they darkened the sky for four days, the feathered roar of their migration arriving before them, becomes mythology. William Cobbett's goldfinches almost sound fanciful: 10,000! By the time the damage is done an animal has become 'rare', so we don't lose 'abundant' animals, we lose rare animals, by which time they are not noticed as such a big loss. Our baseline drops. Daniel Pauly thinks the acceptance of depletion is because people don't know it was once very different. Scientists don't like anecdotal evidence like *Grandad caught hundreds of big fish here*, because it's not scientific, so it doesn't count.

Thrillingly, marine protected areas *can* almost recreate the past. Goat Island, New Zealand's first marine reserve, was established in 1975, not far from Jonathan's childhood home, at the northern edge of the Hauraki Gulf. He took me snorkelling there and I was not entirely in my comfort zone. It was literally *shoving* with fish, humungous snappers, huge blue cod, crayfish walking about, parore, trevally, absolute whoppers. Species unfamiliar to me. A snapshot of our ancient seas. As I stood in the shallows fish fins ruffled the water around me. They were nibbling my knees! Not only had Goat Island recovered from overfishing, but it was seeding the surrounding seas like a nursery. This was dynamic, living proof that no-take marine reserves work, yet astonishingly they have not really caught on. What a hoo-ha every time one is mooted. How reduced the vision and result becomes – especially in the UK – with lily-livered concessions and lack of 'no-take' enforcement. *We've always fished this stretch, blah, blah,* even though there is nothing left to fish. So the baseline drops and drops.

My dream baseline for a healthy sparrow population is loads of them everywhere. If I hear a bush of sparrows I now stop to savour it. The thing about *sparrers* is the party, when after teatime everyone comes home and bowls into the bush with the day's news. In his book *The Moth Snowstorm* avian linguist (and naturalist), Michael McCarthy translated their gloriously noisy single-double noted cacophony:

> Hey!
> What?
> You!
> What?
> You!
> Eh?
> Who?
> Him.
> *Him?*
> Nah.
> *Her?*

Nah.
Me?
Nope.
Him?
Yup.
Really?
Yup.
Me?
Yeah.
Oh.
Yeah.
Why?
What?
Me.
Cos.
What?
You.
Eh?[145]

When we moved into our rural cottage nearly 20 years ago, we had more than 30 sparrows in our garden careering about the place. Our neighbour, Ruby, had 30 more. Every spring we watched pairs of squabbling boys fall to the ground in balls of combat. Alarmed, I tried to break them up. In *The Birds of the British Isles*, published in 1920, T.A. Coward recalls how he retrieved a pair of fighting sparrows that fell between the panes of an open window and that when he pulled them out they continued to fight in his hand. Mostly we took them for granted. We thought they were safe at the end of our quiet lane. You don't notice it at first. And then you do. In 2014 a male sparrow stationed himself on the roof of a nesting box in view of our bedroom window. Week after week he held his lookout, and chirped and chirped. To human ears it became deranged. On and on it went, it was obvious what he wanted, but no love came.

I missed them. I really missed them. In 2018 we were down to our last pair. If I conjured their merry chatter and the rustling commotion in the bush we called the sparrow bush, something crushed inside me. That sparrows are highly social birds is their charm and maybe their downfall. They whoosh from bush to bush in flurries of feathery cloud bursts. They dustbath together, stretching their wings out as they rumba in dry puddles. They are choristers rather than soloists. They nest close together like a village. The young from neighbouring colonies get together and go off (but not very far) in foraging flocks. And *not very far* is another problem. Sparrows rarely journey more than a mile, so the colony can only be replenished locally, which it relies on to keep the gene pool healthy. And something else goes on. Something psychological. Sparrows *need* their social group to be themselves. I noticed it. If numbers are too low their normal gregarious behaviours are unable to play out. Where in bands they are opportunist and bold, singletons are timid, they skulk about, nervous and flighty. A dunnock pecks unconcerned at the bird table, while Ms Sparrow on the branch behind cranes her neck this way and that. Will she come? Yes, no. She lands and then immediately flees back into the hedge without a seed. Nervous behaviour takes up energy with no reward. It's not impossible to starve in the midst of plenty. As the flock shrinks, they shrink into themselves. Sparrows have lived in bustling colonies for thousands of years and have not been able to adapt in rapid time to a solitary way of life. From the outside, it appears as if a kind of depression sets in. There is a term, the Allee effect, for the phenomenon of a reduced population affecting the individual's fitness. It has been described as a suicidal tendency, which is misleading, as if they choose to wipe themselves out. But for some species when numbers decline for whatever reason – lack of food and habitat, pollution – there is a tipping point when they can't scrape back, they lose their chutzpah, and the consequences are far-reaching. It's hard finding mates if there aren't any and it's against your inherent nature to pack your bags and go seafaring. Ultimately it's the loss of this congregational behaviour (herds, swarms, shoals, flocks) which affects their ability to survive (confusing predators, group vigilance, lookouts, warmth, knowledge, mating partners, genetic variability, cooperative hunting etc.) and can trigger an abrupt collapse. This plays out further, as does everything, to destabilise the predator/prey balance and competitive dynamics.

The mystery of nose-diving house sparrow populations baffled scientists. Once they were an agricultural pest, giving rise to Sparrow Clubs whose vocation it was to kill them. In November 1925 a young man called Max Nicholson counted 2,603 sparrows in Kensington Gardens. He returned 75 years later with Michael McCarthy and the Royal Parks Wildlife Group; they counted eight. They were vanishing from Edinburgh, Bristol and Dublin, in Prague, Moscow and Hamburg. On 16 May 2000 the *Independent* newspaper offered a reward of £5,000 for a convincing theory. The contending suppositions included: loss of places to nest (as a result of modern housing with plastic fascia boards, roofs without eaves, barn conversions); tidier gardens; the fashion for patios and decks; intensive agriculture; lack of seeds in winter; lack of insects in summer; insecticides; disease; predation from cats; climate change; air pollution from the chemicals in lead-free petrol and diesel particulates. Yet these conditions remained the same for other garden birds who did not appear to be suffering such catastrophic losses. New York still had sparrows. Paris seemed to have them. The biggest population Jonathan and I saw was thriving in a French motorway station. Fifteen years later the prize money remained uncollected for no complete explanation had satisfied the awarding panel.

What *was* discovered was that a high proportion of hatched chicks were dying in the nest of starvation at two to six days old. Sparrow hatchlings need insects to flourish – only later does their diet switch to grain. What faecal and feather analysis revealed was that the parent birds were unable to provide enough insects in June and July. Only nestlings fed on aphids, beetles, weevils, caterpillars and spiders had a high enough body mass to fledge and survive. Weaker fledglings were vulnerable to ticks, mites, fleas, E. coli and predators. To keep populations even, sparrows need to raise three broods a year, and at least five chicks.

So where had all their insects gone? Insect declines were so acute it *was* noticeable.* Nothing spattered on the windscreen; no craneflies spreadeagled in the glass of water beside the bed; no moths bashing against the windowpane at night. No glow-worms on an evening wander down our lane. Things were not okay. Yet few people seemed bothered. Insects were buzzy and bitey, and not nice to swallow, and ate the lettuces. But, as we should know, insects run the place. They are the soil makers, waste disposal units, cleaners, pollinators and food source – the base of the pyramid. Intensive agriculture does not tolerate attrition and supports virtually no insect life. An ordinary arable field can be sprayed with noxious chemistry up to 22 times a single growing season; in the

---

* In 2004 the RSPB devised 'a splatometer' to count the insect splats on vehicle number plates; 40,000 drivers recorded an average if one splat every 5 miles. By 2021 there had been a further decline of 58.5%. If it takes 200,000 insects to raise a swallow chick, hardly surprising there is a parallel decline in the numbers of swallows, swifts and martins.

Netherlands a study in 2022 found 129 different pesticides in the soil and manure from 23 cattle farms (69 were found in certified organic farms).[146] Killer fact: there is typically 10 to 20 times more insecticide present in human breast milk, than in cow's milk.[147] The war against nature is not over.* In August 2002, the house sparrow was added to the RSPB's Red Data list of the UK's bird species of conservation concern. Our grand-parents would find that impossible, just as our nineteenth-century American ancestors would never have believed the rolling thunder of wings that announced the migration of passenger pigeons – that inex-haustible fresh meat fly-by – could be silenced.

*

November 2018. I am at Gladstone's Library in the sitting room describing this book to an American academic. She tells me that she and her husband recently watched a documentary on butterflies and their decline. Her husband criticised the programme afterwards, saying it did not make the case as to why it mattered. She said, 'I mean nice to have and all that, but so what?' I was a bit taken aback. *So what?* The blurred velvet diffrac-tion of an iridescent butterfly wing crashed across my visual cortex like a psychedelic cartoon. This was not enough? I was speechless. Brain zapped by the butterfly wing, blinded by beauty, the urgent need for utility that would impress my interlocutor's husband had me incoherent with a scrum of examples I couldn't quite think of. Even the sparrows abandoned me. How could they? In China, when Mao decreed his 'Smash Sparrow' campaign of 1958, he triggered an ecological catastrophe. Though the sparrow ate grain, he also ate locust nymphs. After the Chinese population set about sparrow bashing, unchecked locusts devastated harvests and somewhere between 20–45 million people starved in the worst famine in history.** Or the vultures in India, a disaster overflowing with facts and figures. Vultures were *nice to have*, with their serrated tongues and powerful

---

* The EU's Common Agricultural Policy (CAP) fast-tracked environmental degradation through untargeted subsidies – farming wetlands, ploughing grasslands, digging peatlands – and we paid for it (half the total EU budget). By 2021 in the Biodiversity Intactness Index, Britain, Ireland and Malta were the most wildlife-depleted countries in the EU.
** Officials, scared to report crop failure, commandeered grain 'surpluses' that did not exist, leaving the population without, while China exported grain so Mao could save face.

stomach acids, because they gave India excellent disposal services – until tens of millions were wiped out by the anti-inflammatory diclofenac (Voltaren), the cheap drug of choice given to ailing cattle in India, but toxic to vultures, whose kidneys had not evolved to assimilate human chemistry. As more than 40 million vultures across India disappeared, 12 million tons of dead cattle and water buffalo were left to rot. Feral dogs moved in; their population increased by 5.5 million, with 47,300 additional deaths from rabies. If only those figures had been on the tip of my tongue. From 1993 to 2006 the loss of the critical ecological role of vultures cost India an estimated $34 billion, and the costs continue to rise. That's a big *so what*.[148]

But of course what I *should* have said was that while bees pollinate fruit crops, butterflies are primary pollinators for a multitude of vegetables and herbs: the carrot family (dill, fennel, celery, parsnip), the sunflower family (artichokes, lettuce, chicory), the legume family (peas, beans), the mint family (lavender, basil, sage, rosemary, thyme, oregano) and the brassica family (cabbage, kale, broccoli, cauliflower, Brussels sprouts).[149] So there.

There are plenty of utilitarian reasons why animal lives are important. But a bit like pick-up sticks, where it's hard to extract one stick from the interlocking pile without disturbing the others, we don't know which is the crucial one until we remove it. Natural capital is the term economists use for the value of the free stuff that nature can provide, like clean water, air, food, healthy soil and carbon sequestration, which were taken for granted. These 'services' are now costed into 'so what' monetary values and measured in 'so what' contributions. Sometimes the best 'so what' is pure wonder.

Here's a little story about butterflies

Every winter in a mountain forest of pine and oak in the Mexican state of Michoacán, the trees are dressed in orange butterflies. The monarch butterflies begin to arrive in early November around the Day of the Dead, as if they were spirits of the forest or souls returning. Branches sag with the weight of butterflies as they cluster together in their millions. They have flown nearly 3,000 miles from Canada and the US to overwinter at this spot, favoured for hundreds of years for the cool temperature, the clear streams and the silence. This Monarch Butterfly

Biosphere Reserve is a UNESCO World Heritage Site, yet is threatened by illegal logging and pollution. In the sun, when the butterflies take flight, the beating of their wings sounds like rain. In 2014 there was an 81% decline in the American Midwest monarch population, attributed to a 58% decline in milkweeds, the plant on which their caterpillars feed. Milkweed was a casualty from increased use of weedkiller on genetically modified 'RoundUp Ready' maize and soybean crops. Monsanto has produced seeds which grow into glyphosate-tolerant plants known as RoundUp Ready crops, so you can slosh Roundup all over the place – if you use their seed (double profit). When the monarchs arrived in Mexico in 2019, the extent of butterfly cover was less than half the winter before.

In January 2020, 50-year-old Homero Gómez González began to post daily videos of himself covered in clouds of butterflies. Hands on hips, white shirt laden with orange wings, on his head, his nose, his bushy Mexican eyebrows, his moustache. A former logger, González managed the El Rosario butterfly reserve to protect the monarch wonder for the world. 'Come and see this marvel of nature!' he grinned. From a distance the millions of butterflies looked like swirling autumn leaves. His hands reached up in the air among them. The butterflies are lovers of the sun and the souls of the dead, he told us. On 13 January there was no post. Two weeks later González was discovered floating in a well. Two weeks after that, Raúl Hernández Romero, a part-time tour guide at the Monarch Butterfly Biosphere Reserve, was also found murdered. The pressure on the forest is from clearcutting to allow for avocado orchards, one of Mexico's most lucrative crops. In spring the monarch needs calm, sheltered trees and a food source for their caterpillars. In California, the introduced Australian eucalyptus has become the monarch butterfly's preferred roost. But the thing about eucalyptus is that they are highly flammable . . .

In 1985 Bayer patented their first commercial neonicotinoid (meaning new nicotine-like) miracle systemic group of poisons which attack the nervous system of insects. If you sow coated seeds the plants that grow from them are impregnated – roots, stem, leaves, pollen, nectar, even the seeds will be toxic. It is effective on sucking insects, soil insects, butterflies, bees, and has a long residual life. If the insects don't die, they become discombobulated, unable to navigate properly, and then they die. Oh, and

it's water soluble so you can spray it, or just irrigate your crop with it.
Just a short dash to the watercourses, where we now know it alters the
food web dynamics with huge losses of zooplankton biomass that in turn
affect fish and other species through trophic cascades. In the absence of
light in the soil, neonics can take four years to degrade. In 2000, more
neonics came onto the market, named Gaucho, Admire, Calypso, Cruiser
and even Gandalf. By 2008, neonics had a billion-dollar market and made
up 80% of all seed treatment sales. In 2011 more than 30 products in the
UK were on the market and being used in farms and gardens. By 2013
every stalk of industrially grown corn in the United States was treated
with one of these insecticides. Corn, canola, cotton, sorghum, sugar beet,
soybeans, maize, oilseed rape, apples, cherries, peaches, oranges, berries,
leafy greens, tomatoes, potatoes, cereal grains, rice, nuts, grapes. Shouldn't
someone somewhere have been alarmed?

Apparently not, according to the young, good-looking lawyer from
Brighton I was sitting next to at a fancy birthday dinner in a fancy
Moroccan restaurant. God knows how we got onto the subject of neonics.
I think it began harmlessly enough, talking about moving to the country,
where he'd like to live. I said I wouldn't want to live bang next to arable
land because of the spray drift. He snorted in derision.

'You *can't* be serious!'

'What, breathing in all that soapy toxic chemistry? Have you *smelt* it?'

He leaned back in disbelief. 'That's not going to hurt you! If it were
dangerous it wouldn't be licensed for use.'

'Who do you think tests it? Scientists employed by Bayer.'

'Oh, come on. This sounds like a conspiracy theory.'

'Forget about what it might do to us, what about the insects, the soil,
the watercourses?'

'I would stick to the scientific facts,' he advised.

'Actually there is *a lot* of scientific evidence,' I said.

'What papers? What papers? Name me the scientific papers. Where's
your empirical evidence?'

I was in the dock. 'Well, the Environmental Protection Agency was
sued recently by a group of beekeepers for allowing the registration of
neonicotinoids based on inadequate studies and tests.'

'What case was that?'

Of course I hadn't a clue.

'What was the outcome?'

Of course I didn't know that either.* He sensed blood. 'You need evidence before you make rash statements.'

'How about the crash in bee populations, the decline of farmland birds, our sparrows for instance? River insects, mayfly; all in freefall. Moths on the windscreen. Do you remember them?'

'All this is circumstantial.'

Not a good birthday party conversation. People were beginning to look at us.

'How will crops be pollinated if we wipe out the pollinators? By flicking feather dusters? Who's going to do that?'

Everything was rigorously tested as far as he was concerned, and I was hysterical. A woolly lefty nimby-namby emotional girly. He wanted numbers, names, data, journals, titles, authors, credentials, and they were not at my fingertips. Then I said that for all we knew, neonics could be our generation's DDT. Well, that did it for him. His face contorted in ridicule.

'Poppycock!'

He was the lawyer and I was the unreliable witness. If only I'd remembered the leak of a confidential memo from the US Environmental Protection Agency warning bees were at risk from other Bayer neon-icotinoid products; or Buglife's review of scientific papers on neonic effects on non-target species; or the findings reported in the journal *Science* of unsafe levels of neonicotinoid pesticides in 75% of honey samples from around the world. Or the study that showed neonics washed off seeds and contaminated wildflowers. Or the study in *Nature*, 'Be Concerned', which concluded that the level of neonics in envi-ronmental samples correlated strongly with the decline in insect-eating birds.[150] Or the Japanese paper about the collapse in the eel and smelt harvest caused by neonicitinoids washing off the land into lakes.[151] Or at least reminded him of the precautionary principle which is supposed to underlie environmental regulation and protection. I wish I could have dropped my voice a few octaves. I wish I'd been able to curl my lip back at him and, if I am honest, push his bullying mush into the tagine.

---

* The case was 'stayed' as of 2013. An attempt to suspend the use of four neonicotinoids by a 'Save American Pollinators Act' was assigned to a Congressional committee in 2013. Which is as far as it got.

Butterflies are nice to have, but another *so what* is that if they are not there, there will be other insects not there for probably the same reason. Pollinators are estimated to have a global economic benefit of $500 billion a year. In Germany, a supermarket emptied its shelves of all products that were pollinated by bees to bring bee plight to the attention to the consumer. The shelves were stripped almost bare, most of the bakery was empty, fruit, nuts, veg; 65% of crops needed pollinators. In 2021, a giant Woolworths store in Sydney did the same. No avocados, apples, cucumbers, pumpkins, melons, blueberries, nuts, cocoa, coffee, chocolate, tea, tomato, sugarcane, tequila, vanilla, nutmeg, spuds, beans. In 2018, two years after my set-to with the lawyer, the EU banned neonicotinoid outdoor use after they were found to be harmful to bees (the UK voted against it). Almost the day after Brexit, the British government granted 'emergency' use of neonicotinoid-treated sugar beet seed. Sugar beet crops, which lift a lot of topsoil with them, are grown for the production of refined sugar for Mars bars and Coca-Cola. Priorities. Let's spare ourselves joining up those miserable dots and get back to our sparrows.

Our sparrows. Ownership is a weird habit of humans when it comes to wildlife. Ownership is part of the problem, *our* trees, *our* hedges, *our* grass, without a thought for who else might live there. Over the years the landscape had changed around our neighbourhood with a shifting of the guards. Trees down. Hedges out. Ponds filled in. Then three silent padding killing experts moved in next door, tightrope-walking along the fence, swinging from the bird feeders, snoozing in the sun on soft feather beds of wren wings. Then one quiet day the cats left with their owners. And it was spring, and the surviving male and female sparrow brought straw and feathers into the cranny under the eaves, back and forth. The male did his business in front of our eyes in the conifer outside our bedroom window. I put out trays of mealworms and suet pellets on the windowsill. We watched their first brood fledge early one morning, a chaotic bowling of wonky flights into the lime tree. We heard the second brood chippering from the nest. I stocked up the mealworm tray. Then the third brood. Could I dare to hope? By summer a small flock of sparrows was back in the sparrow bush. In the autumn I counted nine. The following year three nest sites under the eaves were occupied. Each with successful consecutive broods. The sparrow bush began to chatter again. They were back home. In 2021 there must be forty. I cannot describe my happiness. *Will they ever shut up?* They whizz about. Busy, busy. Alive, alive-o. I can see their pale chests puffed out sunning themselves in the hazel. They lift me. Glory be to sparrows. We forget abundance at our peril. And theirs.

WHALE PLUMES & HIPPOPOTAMUSES

Ecology, *noun*, from *oekologie*, from *oikos*, the Greek for home, dwelling.

The niche where a species lives with a myriad of other species is a fine-tuned web. The consequences of interference are impossible to predict. You cannot do just one thing, as ecologist Barry Commoner's First Law of Ecology observes: 'Everything is Connected to Everything Else'.[152] The late biologist E. O. Wilson used the acronym, HIPPO, to spell out the most serious threats to natural ecosystems and wildlife through human agency. H: Habitat destruction; I: Invasive species; P: Pollution; P: Population growth; O: Overhunting.[153] In homage to HIPPO, I offer three more initials: I, I, I. Ignorance, Inconvenience and Indifference. For all unnecessary human acts of environmental destruction come down to one or all of these.

H

Habitat destruction includes climate change (all three I's). A vivid metaphor comes from David Quammen, who asks us to picture a fine Persian carpet, and to imagine cutting it up into 36 pieces. What do we have? Thirty-six rugs? No. We have 36 fragments fit for nothing.[154] Ecosystems function at the scale of their entire landscape. Take too much, or remove a vital component, and they collapse.

Consider the humble hedge. Before the Second World War a million miles of hedgerow criss-crossed the British countryside, contiguous and thick enough to act as stock fencing and boundary markers. These were dense pyramids of tangled blackthorn, hawthorn, holly, honeysuckle, dog rose, dogwood, elder, brambles, hazel, frothing with blossom, berries and song. Not only did they provide food, shelter and nesting sites for native birds, mammals and insects, but they mitigated floods; kept soil in the fields; prevented mineral leach; stored carbon; provided safe corridors and highways for hedgehogs; way-markers for bats, moths and migratory birds;

new stems for moths to lay their eggs, and crevices in which bugs could safely hibernate. Hedges stood in for lost mosaic scrubland. A decent hedge was like a nature reserve, churring with turtle doves. Now, half our hedges have gone. What remains is annually abused by great flailing blades on massive (soil compacting) tractors, just before redwings and fieldfares arrive to feed off the berries. And we wonder where our *hedge-hogs* have gone. In March, when birds are looking for nest sites, the modern hedge is a wind-whistling geometric waist-high scaffold of cracked branches hacked to the knuckles. The blackthorn blossom that once foamed along our lanes now appears only in spits. We wonder where the moths have gone. Into the mulcher, every one. Much of our wildlife loss is due to hedge loss. Little brother to the hedge flailer is the whining strimmer beheading every stem over 2 inches long, and any tiny hiding creature. We wonder where the glow-worms have gone. Every verge, hedge and garden in the land is under threat of the Tidy Disease. We have been recruited too. More sun, more view, more machinery, more noise, chop, slash, burn.

I

Invasive species arrive by the hand of ignorance (ours), mostly. They are newcomers who wreak havoc (predate; parasitise; compete for food, homes; bring disease; change the landscape) on residents who have no time to evolve defence. Cane toads taken to Australia from Hawaii to eat the grey-backed cane beetle that was destroying sugar cane plantations was an abysmal idea, as it turned out. Loaded with poison, they are deadly prey for snakes, goannas and crocodiles. Prolific breeders, the cane toad's estimated population is now 1.5 billion. Pest control, exotic importations or lowly hitchhikers, species invasions damage ecosystems and are one of the biggest threats to wildlife diversity. Grey squirrels, Zebra mussels, Asian hornets, Asian carp, American mink, Japanese beetle, Australian possums, and many more – all total disasters.

Yellowstone *is* America's best 'idea', a protected national park as a haven for native species to interact in their natural ecosystems.* It is hard to

---

* The operative word is 'idea'; we should not forget the hunter-gatherer native people who lived in the region for 11,000 years as part of the ecosystem, and that it is a land rich in Native American history, yet these inhabitants were dispossessed and forcibly removed, with their treaty rights erased.

think of a terrestrial environment that has taught us more about the web of life that shapes landscapes. Yellowstone was the poster-child for a fully functioning ecosystem with its native wildlife back home, as it was before Europeans arrived 500 years ago. However . . . From 1890 to the 1950s Yellowstone's Lewis and Shoshone lakes were stocked with non-native lake trout (from the Great Lakes) for sports fishermen. The lake trout have migrated into the rivers and done very well predating on the last stronghold of native cutthroat trout. In 1998 it was estimated 125,000 lake trout consumed 4 million cutthroat trout. The cutthroat's absence in the spawning streams (lake trout spawn in the deep lake) depletes the normal sustenance for grizzly bears with cubs – who have consequently shifted their prey to elk calves, otters and fish-eating birds. In 2017 just three osprey nests produced one fledgling, down from 67 fledglings in 1994. In the absence of cutthroat trout, bald eagles also struggled so began to prey on trumpeter swans, wiping out all the park's fledglings. Ducks, American white pelicans, double-crested cormorants, loons and terns were all in their sights. What to do, when the very success of Yellowstone had been to do nothing? Electrocution? Even importing harp seals to eat the lake trout was suggested! Gill-netting dented the population of larger fish, but what to do about the eggs? They tried vacuuming, electrocuting, even covering the spawning beds with the carcasses of the fish they caught, to remove the oxygen from the water and suffocate the eggs. This just brought in grizzlies. So they pulped the fish – so that the bears couldn't eat them – and used the pulp to smother the eggs. And it worked! Now they are making pellets of soy and wheat gluten to use when the fish pulp runs out. The good news is the cutthroat population is bouncing back. Let's hope Yellowstone is not another masterclass in a negative trophic cascade.

P

Pollution is big business (all three I's). The most endangered animal in North America is the modest freshwater mussel. Though a filter feeder with an inordinate ability to filter the toxins, silt, algae and bacteria out of huge volumes of water (for free), many North American rivers nowadays are beyond even their vast capacity. The rivers were a free drain for paper mills, steel production, textiles factories, and very expensive for mussels and everyone else. In June 1918 the Passaic River which runs through New Jersey was so choked with oil, creosote and acid that it literally caught fire, with flames racing across the surface. William Carlos Williams described it as 'the vilest swill hole in all of Christendom'. All it took was 100 years. Some mussels live for more than a hundred years; you can count their annual growth rings in their shell, like the rings of a tree. Mussels, like moths, have brilliant names. There's the Fuzzy Pigtoe, the Spectaclecase, the Purple Wartyback, the White Heelsplitter, the Snuffbox, the Elephant-ear, the Rough Rabbitsfoot, the Hickorynut, the Fat Mucket and – the mind boggles – the Giant Floater! Tierra Curry, a scientist with the Center for Biological Diversity, celebrated her birthday with her colleagues dressed up as different species of mussel. Thirty-eight species are already extinct. Adios Alabama Pigtoe, who curled up her ancient toe in 2006. Before we had plastic, mussel shells were collected to make buttons. Of course they were overharvested, but it is over-damming and over-pollution that really does for them. The Rayed Bean mussel wafts out her mantle as a fish-like lure to attract a rainbow darter fish to have a chomp so she can expel her fertilised eggs onto his gills, where they will develop and drop off someplace new – for goodness' sake.[155]

With tragic speed North America's great rivers have been desecrated with the malignancy of heavy metals, industrial chemicals, toxins, arsenic, acid, dye, sewage, slurry, antibiotics, dioxins, DDT and the virtually indestructible bioaccumulate PCBs in hydraulic fluid, coolants, lubricants and flame retardants. Dioxins are deadly, and TCDD (tetracholorodibenzodioxin) the deadliest of all, 60,000 times more toxic than potassium cyanide, but guess what, that went into the river too. In 2004, estradiol, a chemical in birth-control pills, was found in the three-spined stickleback. The females were noted to be more frisky and extrovert, and consequently were predated on in higher numbers than sticklebacks without estradiol. (Barry Commoner's Second Law of Ecology: Everything Must Go Somewhere.)

Animal advocate John Oberg posts a video on Twitter of two swans shoring up their nest on a canal in Amsterdam, traffic noise in the background. The incubating swan stretches her neck out and drags a plastic carton into her plastic nest; she stands to reveal a clutch of seven grey-green eggs, then sits down again; her mate is helping: a plastic cup, a bin liner, a can. Our pitiful reaction is a cross-face emoji. From a swan's nest to a 100-kilo 'litter ball' in a sperm whale's stomach, and the millions of unwitting stomachs in between, plastic fantastic is the water world's ruination.

In 1958 the American biologist Rachel Carson received a long and furious letter from her friend Olga Huckins, describing her 2-acre bird sanctuary in Massachusetts 'made lifeless' after the area was sprayed by plane with DDT to control mosquitoes. 'The mosquito control plane flew over our small town last summer. Since we live close to the marshes we were treated to several lethal doses,' Huckins wrote. When Carson visited Huckins, a spray plane flew over while she was there. They took a boat out into the estuary. The sight of dead and dying fish, washed-up crayfish and crabs staggering on the shore with damaged nervous systems, galvanised Carson's determination to expose the full horror.[156]

DDT was the wonder chemical originally developed in the 1940s as an insecticide to combat malaria, typhus and other insect-borne diseases. In the US it was employed to eradicate mosquitoes, gypsy moths, fire ants, the white-fringed beetle and other 'pests'. It was sprayed over crops, livestock, people, homes and beaches indiscriminately, in the blind trust that if it were dangerous it would not be in use. Yet across the globe, millions of birds, insects and other animals were being poisoned in the 'amazing rain of death' of organochlorine pesticides.[157] Carson began to catalogue the widespread destruction. The more she studied, the more aware she became of the dangers of scientific and technological progress removed from ethical concerns. The more angry she became at the reckless drenching of persistent chemical pesticide without any idea of the long-term impact. Her 1962 book, *Silent Spring*, exposed her findings, and in clearly showing how DDT remained in the food chain illustrated the central ecological tenet that everything is connected to everything else. Her title not only alludes to the loss of birdsong, but the silenced truth. The pushback came with the strong-arm of the chemical industry and the unprincipled politicians who served it. Litigious intimidation of Carson's publisher and a propaganda offensive tried to denigrate her as a Communist spinster without children (she raised her orphaned nieces, and later adopted her orphaned grandnephew, but so what). Her mental strength, supported by her scrupulous science, was not matched by her body; Carson had been diagnosed with breast cancer, but this did not stop her testifying at the Congressional hearing President Kennedy had summoned to investigate pesticide use in 1963.

So came about the first federal policies intended to protect the environment. Rachel Carson died in 1964. *Silent Spring* lived on to influence the course of history; it changed policy, ignited the modern environmental movement and led to the formation of the US Environmental Protection Agency in 1970.★

In his book *Silent Earth*, Dave Goulson wonders if the success of *Silent Spring*'s alarm bell paradoxically gave us a false sense of security that this could never happen again, and might be why we dropped the ball and neonicotinoids slipped through the net.

The bleeding obvious

Sixty years later, politicians and corporate interests are driven by the same forces that downgrade damage to the environment. We remain tranquillised by short-term convenience. Insect numbers are impossibly difficult to count and quantify. Historic data and long-term studies are thin on the ground. It doesn't matter that *everyone* around in the 1960s remembers car windscreens splatted with insects. That's not evidence. And in April 2019 there was a big spat over an academic paper that hit the headlines: *Insect Armageddon*. A 27-year study in 63 German nature reserves showed that insect populations were down by 76%. The study used indiscriminate traps that caught anything that flew – wasps, bees, flies, beetles – then weighed the biomass. The biggest decline was in the summer – at 82% – when numbers should peak. The likely reason was the industrial scale of dosing farmland with pesticides, and loss of habitat. Although the reserves were protected, the thing about *flying* insects is that they fly, into sprayed crops, which are everywhere. The shit hit the fan when two researchers went one step further to extrapolate (from this study, another from Puerto Rico and one from the US) their conclusion of a global decline that *may* lead to the extinction of 40% of the world's insects in the next few decades.[158] That drew a lot of attention. Academics don't like emotive language and Francisco Sánchez-Bayo and Kris Wyckhuys were jumped on for using words such as drastic, dreadful and devastating; 'such strong intensifiers should not be acceptable in research articles' their critics charged; "Insectageddon"

★ DDT was banned in the US in 1972; in Europe in 1978 (1984 in the UK); and globally in 2004. DDT is still used in Africa, South America and Asia for malarial control.

is alarmist by bad design'. A BBC investigative radio programme repeated '*INSECT ARRR-MA-GEDD-ON!*' in an over-the-top sci-fi voice as their reporters shone their spotlight on the paucity of long-term widespread hard data. *Only three studies!* Yet one of those *three studies* (silly voice) was far from a paltry dip in the sparkle jar; plus there were 304 scientific papers referenced and consulted. It depressed me. By debunking the study, they debunked the issue. The scientists were right to ring the alarm bells. Because scientists cannot say something *unequivocally*, we must wait until they can? We study another 30 years for confirmation? We *wait* to mitigate the unproven trend? Even when the signs are in plain sight: clean windscreens, decline of insect-eating birds, no insects in the bedroom at night; and the drivers are all there: unregulated pesticide use on an industrial scale, climate change, habitat loss (chop, strim, flail, spray). What data exists certainly tells the story of a decline in diversity, and less variety means fewer specialist pollinators, which creates species vulnerability in fauna and flora. In China, where they ran amok with pesticides, the fruit tree orchards require hand pollinators, flower by flower. I don't need hard data to tell me there has been a drop in insect numbers in the last 50 years, because I can see it. What are the dangers of dismissing what is in front of one's very eyes? How many years were animal behaviour scientists unable to uphold emotions in animals because the animals couldn't *confirm* their pain, suffering, loneliness, whatever? I monitor aquatic organisms in streams; the numbers keep going down, down, down. Any fisherman over 50 will tell you the mayfly blooms are not as they used to be. When I walk a 10-acre field yellow in dandelions on a sunny day and I do not see a single bug I know there is something wrong. It's obvious. Which is why the silly voice mocking *Insect Armageddon* depressed me. The Australian insect ecologist Dr Manu Saunders says she can see no real value in starting to collect long-term data on insect populations now because 'the point is we need to act'. She says we *know* the threats to insect populations, and we have for nearly a century.

> Saunders: We don't need to prove that every single insect on Earth is declining, to do something about it . . . Why are we focusing on whether there is or is not an Armageddon happening? Why can't we just do something about what we already know?

Interviewer: You want to say insects are in trouble?

Sanders: Everything is in trouble. The Earth is in trouble. I don't
understand the argument . . . Biodiversity, the whole entire
biosphere is in trouble and we have known this for quite a while.
It is deflecting attention from the bigger issue, that we go: Oh,
insects might be in trouble, let's argue about this for a while and
then we don't have to do anything about everything else.

That is what she said.[159]

Scottish salmon farms have the go-ahead to trial the insecticide Ectosan
to control fish lice. Ectosan contains the neonicotinoid imidacloprid,
which is highly toxic to all insects and aquatic life. Four billionths of a
gram is a lethal dose for a bee. A teaspoon, 5 grams, is enough to kill a
quarter of a billion bees. Imidacloprid is 7,000 times more toxic than
DDT.[160] We shouldn't be discussing this. Good people have to spend
valuable time fighting this nonsense.

P

Population is the *inconvenient* elephant in the room. When my father was
born there were 1.8 billion people on the planet. When he died in 2009,
aged 90, there were nearly 7 billion. A decade later we are nearly 8 billion.
In my lifetime alone the human population has increased by 5 billion.
Our species' footprints radiate across land, sea and sky. Growth is our
mantra. The more we consume the less viable space there will be for other
species, which means they must die there and then, or die out slowly. E.O.
Wilson described the decline of wildlife as 'the gradual dimming of light'.
Not so gradual the searchlight image of the orangutan (an extremely shy
canopy-dwelling primate) attacking a bulldozer as her rainforest home is
annihilated for palm oil plantations. She is beside herself with all the
emotions we can recognise. She beats and screams. She climbs up the last
limbless standing tree, all around devastation. Vanquished, she cowers in a
muddy ditch, crying.

Some years before lockdown, the British poet laureate, Simon Armitage,
wrote a poem inspired by the road closures in his Yorkshire village to

allow the Tour de France cyclists to pass through in 2014. 'It was like we'd gone back 200 years or gone forward 200 years, to a world where there wasn't any traffic,' he said.

. . .

an asphalt fairway
  vaulted by sycamore, rowan, beech,
    woods to both sides

reinstated as woods |
  when a thought approached
    in the form of a child

tightrope-walking the white lines
  between cats' eyes | she said:
    If I breathed the word

that disappeared all people
  in the world,
    leaving the world

to the world, would you
  say it? Would you
    sing it out loud?[161]

★

Even moth parasites control their populations to protect the habitats they rely on. Dicrocheles mites live in the ears of noctuid moths. Although their occupation means breaking through the tympanic membrane, which results in deafening the moth's ear, they will never migrate to the other ear. That would not do for a moth who needs to detect the echolocation calls of hunting bats, and it would not do for the mites if the moth became prey. So, by the never-ending wonder of the world, the mites send scouts across to the clear ear to lead any stray wayfarers back home.

★

O

Overhunting invokes all three I's. We can only dream of the abundant seas of our ancestors, yet we know exactly what to do about it. Here is one stunning fix, if only we could *sing it out loud.* Whale defecation. Whale defecation fertilises ocean ecosystems. A Southern Ocean sperm whale defecates 50 tons of iron each year in the upper layer of water where the sunlight reaches and plants grow. An enormous floating plume of faeces that brings nitrogen consumed in the feeding depths of the ocean to the surface, acting as a biological pump – the whale pump. These plumes provide key nutrients that run the marine food web. The iron and nitrogen fertilise the microscopic plants (phytoplankton) that feed the plankton who feed the krill who feed the little fish who feed the bigger fish. Whales are the keystone species who maintain a healthy ocean. When whales migrate they shift the nutrients around the globe. The presence of whales increases productivity; the lack of whales decreases productivity of marine life *and* the ocean's capacity as a reservoir for carbon. Phytoplankton captures carbon through photosynthesis and is estimated to take up as much CO_2 in the atmosphere as terrestrial plants.[162] The more phytoplankton, the more carbon in the atmosphere is removed. Save the whale to save the planet. A great big cycle of nutrients and carbon is regulated by the sphincter of the whale.

Before nineteenth-century commercial whaling devastated populations, the effect would have been phenomenal. By 1900 the whalers' original targets, the grey, bowhead, right and northern humpback whales, were nearly extinct. Whalers moved to the next species, blue, sperm and fin whales, with a new invention, the exploding harpoon. In the ocean's upwelling around South Georgia the abundance of whales was so great, whalers could hear their blows resounding across the bays. Scientists estimate about 3 million whales were killed in the twentieth century. In the 1930s worried whaling nations convened to regulate the hunts, reduce catch and agree shorter seasons, with some species out of bounds. So ships killed as many as possible, faster, in the shorter time allowed. A new body, the International Whaling Commission, was set up in 1946 to regulate whaling, with prohibitions on hunting right, grey and hump-back whales, but its motivation was its view of the whale as a resource and the population numbers as stocks. The IWC set the limit at 16,000 blue whale units a year (one blue whale unit = one blue whale, or two fin whales, or two and a half humpback whales, or six sei whales), but without enforcement. By 1964 the scientific guidance was a limit of 2,800 blue whale units, but the IWC adopted a quota of 8,000. By 1970, just 1% of the pre-whaling blue whale population remained. In the 1970s Save the Whale consumer boycotts targeted Japan and Russia. In 1978 the IWC called for an end to international trade in whale products and adopted a whaling ban to commence in 1986. Japan, Norway and Russia refused to play. These days whaling products feed pets, fur farms, sled dogs and humans who don't mind the toxic load. Tusks, teeth and vertebrae make carvings. That's pretty much it. Yet the value in environmental benefits of a healthy single great whale over his or her lifetime, calculated by humans who do those kind of sums, is around $2 million – aside, obviously, from the intrinsic value of the whale's life to herself, her calf and her family.[163] Today's population adds up to about a trillion dollars' worth of 'natural capital', shabby as it is to reduce such a creature so. As if one might go down the Natural Capital Shop and buy some more. (If only.) With some whale populations in recovery, there are still only about 5% of the numbers that once cruised the world's oceans, tending the sea.

The more we understand of the marine ecosystem the more incalculable the consequences of the loss of this gigantic biomass. Not only in life, but also in death, their bodies sinking to the ocean floor, sequestering carbon (the reverse of when we process their bodies), providing habitat and food for the community of creatures living there.* The ocean's gardeners, who fertilise phytoplankton, provide oxygen, sink carbon, keep oceans cool and sustain fish stocks are the almighty reasons to do our utmost to help *all* whale species get back to pre-whaling days.** And then leave them alone. In the bright light of the ecological role of whales, Japan – you don't say – has withdrawn from the IWC.*** Oh, and now we harvest the krill (on which *they* feed) for glucosamine, a health food supplement for our aching joints, and our pets' joints. (Clinical trials show that 1,500 mg/day

---

* The body of a great whale takes 33 tons of CO_2 to the ocean bed, captured over a 60-year lifespan and removing the sequestered carbon from the atmosphere for centuries.
** The Blue Boat Initiative is a complex whale early warning system for vessels that includes a network of smart acoustic buoys to monitor the locations of whales and provide shipping with alternative routes. This project is sponsored by the Chilean Ministry of the Environment in conjunction with the charity Fundacion MERI. A world first to protect whales for the role they play in carbon capture. <fundacionmeri.cl>
*** Japan, Iceland, Norway, Canada, the United States, Russia, South Korea, Greenland and the Faroe Islands continue, one way or another, to hunt whales.

does bugger all.) The blue whale, who relies on krill, needs to eat more than 3 tons a day. There is no ecosystem immune to us.

In 2015 Australia came up with a 35-year plan to manage pollution and other threats to their mind-bogglingly beautiful *wonder* of our world: the Reef 2050 Plan. But in 2013 Greg Hunt, Australia's environment minister, had approved three new shipping terminals for a coalport, including the dumping of 3 million cubic metres of dredged seabed in the Great Barrier Reef marine park. That's right: sunny, solarised, burnt-to-a-crisp Australia, intent on mining an *extra* 70 million tons of coal a year. So the World Heritage Committee stepped in and the Australians agreed to dump the seabed on land. What is it with environment ministers? The world over, their appointments are perverse. What's more, there is a rabid Queensland 'shark control' programme going back 50 years: shark nets and long unmanned drum lines of baited hooks kill sharks, dugongs, rays, whales, green sea turtles and dolphins indiscriminately, to make tourists *feel* secure. Thousands upon thousands of marine animals taken out of the ecosystem. Meanwhile the sea warms, coral bleaches, fishing continues,★ cyclones rip through, agriculture intensifies, pollution rises, water quality declines, ships run aground, oil spills, seabirds die, tourist infrastructure grows, coastal wetland disappears, and the crown-of-thorns starfish runs amok (predator deficit) busily digesting the largest living structure in the world.★★ So Australia, with your plans, targets, reports, goals, summits, zoning, reviews, assessments, criteria, presentations and due process, you'd better bloody hurry up.

★

★ Only a third of the Great Barrier Reef is protected from species removal. But let's hope by the time you are reading this Australia's new government of 2022, with a greener agenda, has stepped up protection of this unique ecosystem.

★★ The crown-of-thorns starfish, once rare and quite helpful in preventing fast-growing corals from swamping the slower-growing species, became invasive when the suite of predators that kept their numbers in check – who fed on the eggs, larvae, juveniles and adult forms – was in deficit. The giant Triton's trumpet, for instance, an enormous marine snail, was over-harvested for the misfortune of its beautiful shell. You might well have one. The adult crown-of-thorns, up to 2 feet across with as many as 23 arms covered in venomous spines, feeds on coral by embracing it with its stomach, which it extrudes out of its mouth. With the missing components the ecosystem unravelled and reefs came under siege. Outbreaks were reported in Japan, Australia, Guam, Samoa, New Guinea, Malaysia, Thailand, Hawaii and Fiji. In Okinawa, Japan, 13 million COTS were destroyed from 1970–1983 costing over 600 million yen.

Some friends asked Jonathan to give them advice on an overgrown pond. Inspired by our nature reserve at Underhill, they wanted to do their bit to restore habitats. The following night, Jonathan dreamt they were peering through a tangled thicket looking at a hidden pond, and lo, there were three hippopotamuses wallowing in it! They stared in wonder. The hippopotamuses waggled their little ears. The revealing thing was, they all knew *not* to tell anybody.

VIII: IF YOU KILL IT, YOU HAVE TO EAT IT – 2

This gone-already feeling
here in no place with our heads on upside down.

Les Murray, 'Pigs'

A HAM SANDWICH

A small pony is tethered without food or water outside a supermarket. His front legs are tied together. There is an outcry on Twitter, a torrent of comments, all shocked and indignant. Would the incensed citizens apply the same ethical criteria inside the supermarket? To, say, a ham sandwich? A nameless pig not tied *outside* for a few hours but confined *inside* for his whole short life? Bad luck to be born a pig, our most intelligent farm animal. Is there anything about this pig on the ham sandwich label? *This pig could undo a latch with his snout and liked his head scratched.* Not likely. No head scratching in pig units these days. Nothing about his life, or his death. Just swallow it.

Pig. *Sus scrofa domesticus.* Life 'expectancy': 15–20 years. Domesticated from *Sus scrofa*, Eurasian wild boar. Omnivorous, roaming, sociable, highly intelligent, woodland dwelling, lives in small matriarchal groups of related females and their offspring. Likes to wallow, forage and play. Communicates in squeals, grunts, snuffles and chirrups. Smarter than most pet dogs – you don't have to housetrain a pig, they naturally dung away from their nest in latrines. If allowed. But *Sus scrofa domesticus* are not allowed.

Once upon a time there was a deal; a one-way deal, admittedly. Nevertheless we gave you shelter, food, safety, and you could root around, wallow, have your babies. Then we ate you. Being eaten is not an uncommon fate of pigs in the wild. But oh my, have we been squeezing the deal. The Israeli historian Yuval Noah Harari has said that one of 'the worst crimes in history' is taking place on industrial farms.[164] He means here, now, in our midst. That is a strong statement. So strong we might think we should want to know what, and why.

Five days before this sandwich's mother gave birth to him, she was put in a 'farrowing crate'. Note the term 'crate' for a barred metal cage no bigger than the actual pig. Unable to turn around, Mama sow stood on concrete slats so her waste could be conveniently flushed away to a 'lagoon' (note the word lagoon). The moment Ham and his siblings were born, oh joy, they were whisked away and 'fastened' in the 'creep area' to the side of the farrowing crate, so Mama's whopping modern bulk could not topple over and squish her little sausages. After 20 minutes Baby Ham was 'introduced' to his mother's udder. But note, not his mother. She was too fat and unsteady in such a small space to be trusted, so she was confined to her cage for the next *28 days* while her piglets suckled. When Ham was not suckling, he was put back in the 'creep area'. No nest-making, no mother–young interactions, no licking, no nuzzling and 'no normal postural adjustments'.[165] And *that* is diabolical. *More* diabolical is life for a pregnant sow in the US who spends *all* her pregnancy in a 'gestation crate' (200 by 60 centimetres): three months, three weeks and three days, before she is moved into the farrowing crate. Some sows are so big they cannot lie down and have to sleep on their chests. Hardly surprising they exhibit signs of chronic desperation: bar biting, tongue rolling, head weaving, and a behaviour called 'learned helplessness'. 'Learned helplessness' means they remain passive when prodded or when a bucket of water is thrown over them. Learned help-lessness. How's that on our list of achievements? Sores, urinary infections, muscle wastage, cardiovascular problems, bone density problems, foot problems, joint problems, and of course, demented problems. But it's okay, because the US National Pork Board Vice-President *and* veterinarian, Paul Sundberg, has assured us the sow 'doesn't even seem to know she can't turn. She can eat and feel safe, and she can do that very well in individual stalls.'[166] In 2012 a National Pork Producers Council spokesman said, 'So our animals can't turn around for the 2.5 years that they are in the stalls producing piglets. I don't know who asked the sow if she wanted to turn around . . .'[167]

I think about this on the four-hour flight back from Romania, strapped between Jonathan and a man with his arm on the *shared* arm rest and a leg angled into *my* area. After two hours my teeth are on edge. I'm hot, stiff and I'm getting quite grumpy. Once we flew to New Zealand on a bargain deal (four stopovers) which took 48 hours. We could move around the plane and walk through four transit lounges, but it was unmitigated hell.

★

Jonathan, who was raised on his family's dairy farm in New Zealand in the 1960s, remembers Oink. Oink was a piglet runt whom Jonathan's mother, Dorothy, fed from a bottle and kept in the laundry cupboard. In those days, New Zealand dairy farmers kept pigs to eat the whey, a by-product from the milk. They lived in pens with straw and access to a paddock where they rooted and wallowed. Oink lived in the house and played with Jonathan and ran behind the farm dogs to fetch in the cows. Everyone loved Oink. One day Oink, now a big girl, bashed down the back door trying to get in the house. Which was when Jonathan's father, Eric, said she had to move in with the other pigs, where she wallowed and rooted and socialised and ate the apples Jonathan brought her, for all her natural life.

In Herefordshire, Gloucester Old Spot pigs were often put into the orchards after the harvest to clean up the fermenting windfalls. Sometimes they were discovered snoozing it off under an apple tree and could only be stirred by a bucket of cold water.

These days piglet runts drain resources, so tend to be dispatched by whacking their heads on the concrete floor. Other problems are easily fixed. Tail biting? Chop off their tails. Teat biting: clip their teeth. Any bits that give us any trouble we usually chop off. Tails, tusks, toes, testicles. The Royal College of Veterinary Surgeons says *mutilation* is too emotive a term, so instead we brand, dock, debeak, debark, dewattle, decomb, declaw and despur. We clip wings, notch ears, ring noses. Without pain-killers. Desnooding is removing a turkey's snood. Mulesing is to remove strips of skin around a sheep's buttocks. The scar tissue that forms does not grow wool so is less likely to suffer flystrike.

Cash cows and cash brides

Cattle: from the Anglo-Norman *catel*, from the Latin *capitale*, principal sum of money, capital. Cattle, *catel*, chattel, moveable cows, moveable wealth. Cow world and human world entwine. The word 'fee' comes from Old English *feoh* meaning cattle or livestock, moveable property. Money.

One cannot overemphasise the importance of cattle to the Dinka of South Sudan. These legendary pastoralists graze their herds during the dry season in the swamplands of the Nile. They live together, animal and human, day and night, milking, walking, grooming. At night the animals' dung feeds the campfires, from which the white ash decorates human skin and protects it against mosquitoes. Noted for their height, Dinka men, some over 7 feet tall, walk in rhythm beside their bulls, as their ancestors – bull and man – had walked, millennia before them. The Dinka hold their slender black arms up in the air mimicking the shape of their white beasts' long horns, as if carrying the world-sky. Cattle are the beating heart of their language, culture, ritual, marriage, sustenance, and their power.

Years of civil war have shattered the way of life of the Dinka people. Cattle raids – a long tradition of 'wealth exchange' between the feuding Nuer and Dinka tribes, largely checked by ritual, the limitations of the spear and authority of the elders – became mercilessly exploited by the country's opposing political factions.* Young herders on both sides were tooled up to fight, with promises of more arms and raided cattle. Kalashnikovs replaced spears. The raids became uninhibited. Counter-raids fanned the spiralling violence. Traditional systems collapsed. Hundreds of men, women and children were massacred. At $500 a head, cattle were everything. Without a gun to guard them you would be shot. Something to fight for, something to steal. The language of all currency. Status. Wealth. Personal power. The Nuer name for bullet, *dei mac*, literally means 'a gun's calves'.

The truce of 2018 has not stopped the violence. Drought, failed harvests and inflation of the bride price – up from 20 head of cattle to 100 – not only perpetuate the raids, but push families to marry off their daughters in return for precious cows.

Saving their bacon

'Gestation crates' are now banned in the UK, Sweden and some American states, but 'farrowing crates' (in which a sow can be incarcerated for up to five weeks) are legal.** Compassion in World Farming, calls to ban them too. The UK's National Pig Association (NPA) says this would be 'draconian', and 'an insult'. They defend these 'piglet welfare' and 'stockperson safety' systems. The touching concern for 'piglet welfare' is, we know, concern for the welfare of the pig factory purse. Ham, once plucked from his teat-mother, lived on concrete slats (1 square metre of floor area per 100-kilo pig) where he was fed and fed and fed until he was six months old, and then he went on a little journey. Or indeed a very long, terrifying journey. Moreover, the 'stockperson' might be safer and pigs less inclined towards 'savaging behaviour' if their incarceration were not diabolical in the first place. Thing is, if we had not bred Mama sow to be 'hyper-prolific' and unnaturally top-heavy (300 kilos plus) for her weak (unexercised) legs, and if we gave her enough space to make a nest and something decent to make it with, she would not crush her piglets in the first place. It is incredible to think that instead of

---

* The Dinka, loyal to President Salva Kiir; and the Nuer tribe, supporting former Vice President Riek Machar.

** Prohibited only in Sweden, Norway and Switzerland.

someone thinking, *Hang on a minute, these huge sows we're breeding are falling over and piglets are getting squashed, they're aborting and acting aggressive* [to the point of savaging their own piglets to death], *maybe we need to give them more space to be pigs and cool it with the genetic engineering.* Instead, someone thought, *Let's put them in cages so small they can't fall over.* Take a bow, T. Euel Liner and Roy Poage, the Texan swine breeders who, in 1964, invented these torture chambers. In 1969 tie stalls were introduced, in which a sow was fixed in position by a collar around her neck, or a belt around her girth. 'As sows are always isolated, social problems are eliminated and complete control of feed intake is possible.'[168]

January 2021. In Brexit Britain's shrunken export market 100,000 pigs grew too fat to be slaughtered. Genetically programmed to grow, they couldn't *stop* growing. Compounded by abattoir lockdown holdups, pigs grew too heavy to meet supermarket specifications. 'Once a pig hits a certain weight it no longer fits in that packet that the supermarket wants so they penalise us massively.'[169]

To think, 100,000 fat pigs in limbo, and just 20,000 lions left in the wild.

Eyes wide shut

The NPA condemns secret footage taken inside pig units as irresponsible. Photographs of these cages of despair, metal bars pressing into huge rumps, little trotters sticking out, sunken dull eyes, row upon row, drill into your soul. It is hard to fathom how an industry, even from within, has normalised this. The NPA say there is 'no market pull' to finance change. That's us blamed, the consumer. A common trick. The convenience of low-information eaters. How pig factories keep the status quo is by the invisibility of the suffering, and that bacon is very tasty indeed. If every consumer could see a snapshot of their lunch's life, I suspect the factory purse would have to unzip fast. And surely our purses too. Pain, loneliness, confusion, biological needs frustrated, ill health, self-harm, for Chrissake they can't even turn to scratch. Every hormone screaming, *Make a nest!* A being's very being-ness denied. And BORED OUT OF THEIR MINDS. In sum, a life worse than their death. Yes. Industrial pig factories are legal concentration camps of shame on us. Monumental in scale, medieval in imagination. What happened to us? How did we get here?

INTERMISSION

Tiger shark eggs hatch inside the uterus to be born fully developed. Sand tiger shark females have two uteruses and produce two pups in a breeding season, although they start out with a dozen or more embryos. The first to hatch, or fittest, will begin to eat his siblings when he is around 10 centimetres long, about five months into a year-long gestation. When they have polished off their brothers and sisters the tiny shark will eat any unfertilised eggs. In Japan researchers have witnessed the shark embryos actively on the hunt not just in their uterus but between uteruses. This form of sibling murder is called intrauterine cannibalism.★ By the time they are flushed into their marine world, they are fully fledged killers, as big as baseball bats.

Lobster sat opposite me and Buzz was on my right

One summer's day in 1935, on the remote island of Skokholm off the coast of Wales, where five-year-old Ann Lockley lived with her goats, her mother Doris, her father, the naturalist Ronald Lockley, and tens of thousands of nesting seabirds, a fishing boat came into the harbour. Ann was sitting near the landing steps listening to the puffins chatter as they prepared to leave for the winter. The fishermen had brought a gift. A large lobster. The first of the season, they told her, '. . . and by his claws, a gentleman'.[170]

Ann carefully carried him home for tea, as he clicked his yellow-brown whiskers. But before the grown-up tea, which she didn't like to mention to him, she invited him to tea with herself and Buzz, a young buzzard who had been rescued after falling out of his nest. Buzz turned his head to the right, eyed the goings-on intently, then flapped his wings. Lobster waggled his whiskers. Then he cleaned his mouth with 'his other pair of whiskers, the very small pair', his maxillae.

'What time is high tide?' Lobster asked, hoping it would come in and fetch him home soon. Then Ann heard her mother call out from the house that the water was nearly boiling. Lobster looked at Ann anxiously.

So off they set, across the meadow to the harbour and down the landing steps. Lobster curled his tail and splashed backwards into the sea. Ann watched his long whiskers vanishing into a dark hole surrounded

---

★ One story goes that a shark embryo bit the hand of a researcher during a dissection of the shark's mother.

by ribbony seaweed, which led to a secret cave called Skylight Cave, where a seal calf waited for his mother and *his* tea.

How much does it cost, does it really, really cost?

Our brains buzz white noise when we pick up a ham sandwich from a supermarket. Producers pay camouflage people to conceal the cruelty with words and pictures and other smokescreens. Thus we have a global billion-dollar business blinding us to the violence required to get an egg or three on our plate, or a ham sandwich in our lunchbox, for a pittance. Reduce costs, increase production. Squeeze space, time, food, labour. More science, more pigs, larger litters, bigger pigs, who get even fatter, quicker, on less food. Protein has never been cheaper, as long as you externalise (ignore) the cost of the seas of slurry seeping into our watercourses; antibiotic and drug overuse; disease; deforestation and ill health. The poorest workers do the dirtiest work, ex-cons and migrants in the slaughterhouses and processing plants on minimum wage. One dead cow can make 2,267 Big Mac burgers at $3.99. Multiply that by 68 million customers a day. Here's the point. Food and agribusiness is a $5 trillion global industry[171] and counting, and *that's* political. Which is why you could be taken to court if you locked up your dog, but Saucy Sausages, Inc. can put 10 million pigs in metal cages so they can't even turn around, let alone put their nose in a drift of oak leaves, snuff the breeze, or ever feel the sun on their back. That is the reality for farmed animals. Me, humanising them? No. Animalising them. Mammalising them.

There is no such thing as a factory 'pig farmer'. It's a factory. The owner is behind a desk looking at numbers, spreadsheets, calculating inputs and outputs. The stockperson has a job to do and a family to feed. The migrant is desperate for work. The consumer likes sausages. Government will not protect pigs because they don't give donations to political parties. The profit is in the volume, volume. Don't let's kid ourselves we are talking anything more than cheap sausages to gratify our greedy guts. This is not about feeding the world – the defence dredged up every time there is a question about animal welfare. Forget that a third of our food is wasted (1.3 billion tons or 3 trillion meals a year); over a third of global arable land grows crops for livestock (driving deforestation); that 7 kilos of plant protein produces 1 kilo of beef*; and that there are now more obese humans than starving ones (who have *never* eaten a sausage). People living in poverty are used as pawns to defend rich men's profit or assuage the average person's guilt. What twentieth- and twenty-first-century agricultural scientists and factory-farming tycoons have achieved is to turn the farm animal, once again, into Descartes' machine. And somehow not enough people have noticed. We are complicit. We keep buying. We keep looking the other way. There is no outcry. Apart from the *lunatic* fringe. We hate talking about it, we *loathe* reading about it, so we don't. Who wants to read a book called *Eating Animals*?[172] Certainly not me. Then I did. (It's a page-turner.) And life was never the same.

In the UK if you write to your MP about farrowing crates they will reply: 'The UK has some of the highest animal welfare and environmental standards in the world.' The bar is *so* low that's not hard to achieve. Here's what the DEFRA 'Code of Recommendations for the Welfare of Livestock: Pigs' *recommends* for floors. Floors must, 'where no litter is provided, form a rigid, even and stable surface'. The EU Directive for minimum standards

---

* Recent studies consider calorie conversion efficiency. For 100 calories fed to animals, we receive 17-30 calories in meat and dairy products. Like feeding a hundred bucks into the slot machine and winning back twenty. Every time. And we are supposed to be the rational creature. What a staggering waste of cereals, land, water, energy and medication. Not to mention the required agrochemicals and resulting pollution. This is what threatens food security and drives poverty.

for the protection of pigs makes depressing reading unless you consider sufficient feed and water, light (40 lux for eight hours), 'a clean lying area', and to be 'in sight of conspecifics' (other pigs) is enough. A week before farrowing sows 'must be given suitable nesting material, *unless it is not technically feasible for the slurry system used in the establishment'*. No directive on *how much* nesting material – a handful of straw can tick that box. The USA's libertarian approach to regulation allows a very low baseline for animal welfare standards; South America, Oceania and Asia (India being the notable exception) are poor in protective legislation with Codes of Practice here and there that make unenforceable recommendations; Africa, well. One can hope or imagine.

Some people are trying. Efforts to raise the bar double-up as marketing tools; they are called farm assurance schemes. Their purpose is to comfort us and help us make ethical choices. These schemes are modelled on the 'Five Freedoms' drawn up by the Farm Animal Welfare Committee, FAWC.* In 1965 these freedoms for animals were: to stand up; lie down; turn around; groom themselves; and stretch their limbs. Scary. In 1979 FAWC updated and codified them:

1. Freedom from hunger and thirst: ready access to fresh water and a diet to maintain full health and vigour. [*Little need for vigour if you are barely allowed to move.*]

2. Freedom from discomfort: by providing an appropriate environment including shelter and a comfortable resting area. [*A confinement crate. Concrete slatted floors.*]

3. Freedom from pain, injury or disease: by prevention or rapid diagnosis and treatment. [*Antibiotics and more antibiotics.*]

4. Freedom to express most normal behaviour: by providing sufficient space, proper facilities and company of the animals' own kind. [*The RSPCA welfare suggestion was to give them a football. True.*]

5. Freedom from Fear and Distress: by ensuring conditions and treatment which avoid mental suffering. [*Sigh.*]

---

* In 2019 FAWC changed their unfortunate-sounding name to just AWC, Animal Welfare Committee.

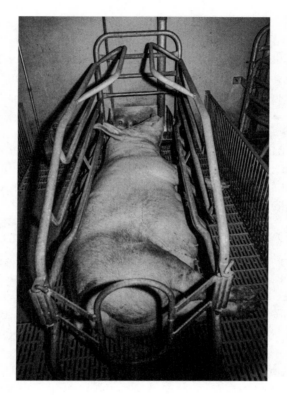

Let's pretend

It's all open to interpretation. Even if these 'freedoms' are fulfilled it does not imply a positive experience of life – aside from the factory farm football teams. In 2009, the FAWC went further, outlining a position that farm animals should at least, 'Have a life worth living, from their point of view, and that an increasing proportion should have a "good" life.' Heartbreaking if you think about it. Who are these code-makers kidding? Us ham sandwich eaters, that's who. The British 'Red Tractor' label (which is supposed to assure us high standards of British food) allows farrowing crates, concrete slatted floors, tooth clipping, tail docking, nose ringing for outdoor pigs (prevents pig rooting by causing pain), and some don't even get straw. A 2012 survey found Red Tractor barely complied with minimum legislation let alone any freedoms. Photographs of some Red Tractor-approved farms in 2015 are deeply unsettling (tightly caged pigs gnawing bars, a terrified piglet trapped *under* the slatted floor in the slurry). But we can relax because the product is 'sourced from UK farms and checked from farm to pack'. The NPA will tell you UK consumers have unprecedented choice, with 40% of sows farrowing outdoors. Well, that is because they can then shove 'Outdoor Bred' on the label and charge us double. 'Outdoor bred' or 'Bred free range' means *born* outside, but then it's inside Piggy on the fast track to the sausage factory. Space allowance: 0.47 square metres total area per 50-kilo pig. What a nightmare for a curious animal who wants to root around and be a pig.★ The logos and labelling are a minefield. 'Outdoor reared' means: born outside, stay outside for half their life (3 months/30 kilos), but 'finish' inside. 'Free range' should mean: born outside, remain outside, until 'sent for processing'. Just 1% of the pig industry operates organic standards. Of course, most animals wouldn't qualify for any of those labels. If you're talking chickens, 'free range' is often claptrap. There are so many chickens in the enormous shed they can't get to the door. Salmon farming? These days we eat salmon like we used to eat baked beans.

If meat is the cheapest thing on the shelf, then there is something suspicious about how it got there. In America it's not the chlorine in the chicken (which isn't great), it's the sores the chlorine is covering up. Chlorine baths remove slime, odour and bacteria. Bird bodies are injected with broths and salty solutions to give them the chicken look, smell and taste that their raddled, diseased bodies lack. Chicken and turkey *natural* products can balloon by up to 10–30%. Ever heard of ractopamine? It's a chemical growth promoter fed to most American, Canadian and Mexican pigs, one of a class of drugs known as beta-agonists. Banned in the EU because it causes lameness, stiffness, trembling and shortness of breath. Even Russia and China ban ractopamine. In 2009 the European Food Safety Authority concluded there was not enough data to show ractopamine was safe for human consumption at any level.[173] When the US Food and Drug Administration conducted a human health study with six healthy men, each man reported his heart pounding, then one dropped out.

LOVE, ACTUALLY

Joanna Kossak remembers how the first visits she made to her aunt, Simona, at six years old, were imprinted with a deep enmity for Lech Wilczek, Simona's neighbour. How he and Simona fought! Joanna recalls

---

★ John Oberg tweets, 'If we cannot bear to watch a minute of film of industrial farm conditions – How do we think the animal feels her whole life?' I am sure the irony that this is a footnote will not escape you.

Simona complaining about Lech's polluting motorbike, and how he
trampled her garden declaring her plants alien species and a danger to
the genetic stability of the ecosystem. But the following year Joanna found
that things were very different.

The impasse had been broken with the arrival of Zabka, a one-day-old
wild boar piglet Lech had been given to look after by Warsaw Zoo. Zabka
needed feeding every two hours, so Lech was compelled to ask Simona
to help. Lech's photographs track Zabka growing up: tiny Zabka asleep
with the dogs; little Zabka with the chicks sharing the mash; a bigger
Zabka following Simona in the snow; an enormous Zabka licking Lech's
face and snuffling up the crumbs on the oak table.

This clever Carpathian boar (and versatile omnivore) quickly learnt
that if she left a few crumbs on the ground, the chickens would come
pecking. Her trick was to lie in wait, quietly snoozing . . . peck, peck,
chicken or turkey, *bwaak, bwaaak, pukaaak*, beaks like bell hammers, peck,
peck . . . when Zabka would explode from her slumber and grab the
nearest unsuspecting fowl. Feathers flying, survivors squawking, the hapless
victim down in one. Simona and Lech had no idea where the chickens
were disappearing to until the day Zabka got caught.

Every Christmas Eve, Lech and Simona decorated the spruce tree in the clearing in front of the house with a feast. The branches were festooned with rowanberries, lard, apples and dried fruit – the wild birds' favourite food. They left a stack of hay for the deer who might pass by Dziedzinka on their way through the forest. Zabka would get a loaf of bread, bowls of acorns and a basket of apples garnished with fir branches.

Zabka lived with Simona and Lech at Dziedzinka for 21 years. Going for walks in the woods, snoozing in the sun, being terrorised by Korasek, following the donkey, luring chickens to their doom. When the time came for her, she went in search of Lech and Simona. She led them to her place, looking behind to make sure they were following. She lay down in her nest of hay bedding in the garden where she had lived all her life. They crouched down beside her, stroking her, talking gently. They all knew. And then Zabka died.

JUST DESSERTS

A cow is a cow, wants to be a cow not a bird

Gabe Brown was a city boy from North Dakota whose closest contact with nature was mowing lawns. In the ninth grade he took a class in agriculture and learnt about fertilisers, pesticides, balancing rations, feed lots and artificial insemination. After school he landed a job clearing rocks from the fields of a local farm; then married the farmer's daughter. In 1991 Gabe and Shelly bought out her parents' ranch. From the start he questioned the logic of the conventional model which held that the more you work the soil the better it is. Because every July they were praying for rain. This was once prairie country, drilled by wind, sun, hail and snow, 16 inches of rainfall a year, with temperatures that could soar to +45°C and plummet to −45°C.

So Gabe went to the library. He read the descriptions in Lewis and Clark's 1804–6 journals of fertile prairie country teeming with life, of immense herds of buffalo, elk, antelope 'in every direction feeding on the hills and plains' with 'multitudes' of prairie dogs, pronghorn, mule deer, coyote, and seas of grass swaying 4 feet high.[174] The richness of life sustained the explorers as they made their way through the Missouri River landscape, upstream across the Great Plains, crossing clear water creeks glutted with shrimp, catfish, salmon and pike; great flocks of geese, 'hundreds of pelicans'; groves of plums, wild cherry and 'great quantities of grapes on the banks'. They recorded more than a hundred previously unknown animal species, including the trumpeter swan and Lewis's woodpecker. This was America's Eden. Gabe came to see his land as a sick, degraded substrate that had ceased to function without animals. Four years of crop failure set him – to raised eyebrows – on an unexpected change of direction, from industrial agriculture to biological, regenerative farming. Moving away from conventional grazing, he began to integrate the animals back onto the farm, allowing their biological processes and natural behaviours to regenerate the numbers of pollinators, predator insects, earthworms and the rest of the microbiology that drove the ecosystem. Gabe had bought all-purpose

Gelbvieh cattle (milk, beef, draught), a red, toffee-coloured breed origi-
nating from northern Bavaria that were strong yet docile with good
mothering ability but, wanting to maximise growth and milk production,
he tracked the genetic markers for those qualities. This came back to bite
him in 1997 when a white-out blizzard and 50-mile-an-hour winds
swamped his open barn with 10-foot drifts, cutting off his herd for four
days. Although the cows crowded into a corner, many of the calves died.
The traits he'd bred for size needed supplementary food and were not the
ones to survive extreme conditions. He asked himself what on earth he
was doing, feeding grain to ruminants. They were not birds with gizzards.
Cows' stomachs were designed to digest grass. Gabe began selecting the
smaller animals to breed from. He decided to get his cattle back, as far as
he could, in sync with nature to restore their *natural* resilience. He calved
when the deer gave birth in spring, allowing the calves to stay with their
mothers to learn the herd dynamics, where to shelter, to eat snow for
hydration; he weaned them gradually, without stress, mother and calf in
view of each other, able to touch noses, separated by an electric fence to
prevent the calves nursing. Gabe studied Allan Savory's methods to restore
grasslands. Savory, born on a 40,000-acre ranch in Zimbabwe, had noticed
how large herds of animals grazing intensively were kept on the move in
a prey–predator environment, and how well the land recovered, despite
the pressure of such large numbers. For Savory, this realisation had come
after 'the saddest and greatest blunder of his life', having instigated a
government programme to cull 40,000 elephants in the mistaken belief
their numbers were destroying the habitat. Savory advocated grazing cattle
in bunched herds, moving quickly, fertilising as they went, and crucially
not returning too soon, to allow the grass to recover. This was the same
'landscape of fear' dynamic that had moved the great herds of buffalo
across the plains 200 years earlier, until the land was broken under the
plough, and in a few decades had become a dustbowl. Savory's bunched
herds hark back to how the rich grasslands of the mammoth steppe recy-
cled nutrients in the last Ice Age. Another biological pump. The fertility
of the land had evolved with grazing, browsing, rootling animals, and died
with them too. Jonathan says his father grazed their dairy herd in New
Zealand the same way; they called it break feeding.

It worked for Gabe. In a short time he was enjoying healthy pasture, a
healthy grass-fed herd, 17 species of dung beetle, and free-range hens. The
free-ranging hogs were given a few large round straw bales to burrow

into and make a winter den where they could all pile in together. Gabe asked, why harrow uneaten hay and manure when the hogs can do it for us? By 2010 he had eliminated the need for synthetic fertiliser. No fertiliser, no pesticides, no wormers, no antibiotics, no vaccines, no artificial insemination, no pollution, no tilling. No need to take more habitat to grow grain to feed to ruminants. No more worries about blizzards, sick calves, frozen ears, or 'bragging about how hard we work'.[175] Massive savings, bigger profits and a deep aggregated soil to show off about – which he does in spades. In the drought of 2017 the Browns' 5,000-acre ranch had 5.6 inches of rain, but they ran the same number of cattle as normal.★

Meanwhile, on a Cotswold escarpment, at the other end of the continuum, Rosamund Young has opened all the gates on Kite's Nest Farm's 390 acres for her grass-fed cows to wander wherever they want, eat whatever, hang out with their calves, or not. Watch the bees, find some herbs, taste the clover, chew a branch. Go down to the shed. Or stand under the apple tree. Make friends with whoever. Born outside, they live outside, although they can come indoors if they want to. Rosamund knows them all as different characters. Each cow is so individual that she can distinguish them by the timbre of their moos. She watches them. They have plans. Friendships. They work things out. They experience grief and joy. She loves them and they all have a life.[176] And they frolic.

★

Now we are told we should buy white eggs. We assume brown eggs are healthier, but they come from Rhode Island Red chicken stock who are more aggressive and peck each other, so the chicks are 'debeaked' when they are a day old. They peck each other because *they are overcrowded*, without natural light, unable to behave like chickens. No wonder they are stressed, frustrated and confused. A Mississippi vet has suggested poultry producers should stop calling it 'debeaking', and call it 'beak conditioning' instead.[177] A beak has an extensive nerve supply that senses stimuli. That the chickens feel pain after 'debeaking' is disputed of course, however, there is documented weight loss from reduced food intake. We might assume this is because it bloody hurt and is now sore, which our lips might be if they were

---

★ In 1991 Gabe's soils could store 40,000 gallons of water per acre; by 2017 they could store over 100,000 gallons per acre.

hot-wired and then fell off. Here I go, sentimentally anthropomorphising again. Alternatives to beak trimming? One of the industry's suggestions: dim the light so the birds cannot see each other. Blind chickens were found to experience less stress in factory farms. Isn't that the pits, that you have to be blind to deal better with your situation? There are layers and there are broilers, the ones we eat. These days, a broiler takes 28–35 days from egg to slaughter. That is the lifespan of a housefly. The 'normal' life expectancy is 12 to 15 years ('normal' would be winning Chicken Lotto). Broilers grow so fast that if you want them to reach sexual maturity (15–18 weeks) you have to starve them, otherwise they get too big to mate. 'Enriched cage environment' is another preposterous term that rarely gets called out. Enriched cages are the improvement on battery cages, banned in the UK in 2012. They are communal cages decreed by European legislation to allow 600 square centimetres of 'usable' space per hen (that's less than an A4 page), a perch space (width of a hen), a nest area (no minimum size), a scratching area (a piece of Astroturf) and a cage height of 45 centimetres (that's below my knee).[178] The extra space, compared to battery cages, works out at less than the size of a beer mat. They cannot flap their wings. They cannot dustbath. They have to compete for the nest area and often lay eggs on the floor. They cannot escape bullying or pecking. That is enriched for you. There is more room for them in the oven with the spuds. The poultry industry is not a good place to be.

INTERMISSION
The longest-living vertebrate we know of is the Greenland shark. A large grey shark, up to 23 feet long and the largest fish in the Arctic Ocean. We thought they could live for 200 years, but as soft-bodied sharks without calcified tissue deposited in layers, their exact age was hard to determine. So it was with some surprise and great wonder that a study in 2016 – using radiocarbon dating on the specialised proteins in the lens of the shark's eye – revealed that they can be twice as old as that, maybe more. Of 28 sharks caught in by-catch (there we go), one 5-metre female was estimated to be 400 years old, and possibly as much as 500.* This shark, in all likelihood, was swimming in the North Atlantic when the *Mayflower* set sail from Plymouth for the New World. Certainly before Cook mapped

---

* The age range was from 272 to 512, with, curiously, the most likely age estimated to be 392.

the coastline of New Zealand. And possibly when Magellan sailed around the world to prove it was round. Greenland sharks swim slowly in cold, deep water, growing just 1 centimetre a year. They don't reach sexual maturity until they are about 150 years old. Being once killed in large numbers for their livers (used for machine oil) has meant that sexually mature sharks are now rare, and rarer still are pups or juveniles. Most surviving Greenland sharks are teenagers who will take another 100 years to reproduce, so their recovery is profoundly slow and fragile. Another urgent reason, if we needed one, to change how we fish to prevent by-catch. I imagine there is something chilling in discovering that the animal you are examining, killed incidentally by humans, was 400 years old.

The Greenland shark was almost as old as the oldest individual animal ever discovered: the 507-year-old clam named Ming (after the Ming dynasty), born in 1499 and dredged up off Iceland in 2006. A long life cut short by Bangor University researchers who opened the shell to find out how old she/he was. The clam's reproductive state was recorded as 'spent' so we don't know Ming's gender. Ming initially was declared to be 405 after a count of the growth rings on the clam's hinge; however, the rings had converged so tightly in the small space over such a long time that in 2013 the more evenly spaced bands on the outer shell were counted. (Alas for Ming they did not think of that first.) The revised age, supported by radiocarbon dating and comparing patterns with other shells alive at the same time, is thought to be accurate within one to two years.[179] Ming, born in 1499, was filtering the ocean before the Inca city of Machu Picchu was built, before Michelangelo had carved David, before Henry VIII was crowned King of England.

A gigantic glowing, floating, marine super-organism in the shape of a hollow tube wide enough for a human to swim into wafts slowly along emitting waves of blue-green light. It is a giant pyrosome. (Pyro – fire; soma – body.) It is made of thousands of clones, called zooids, in a colony that can measure up to 60 feet long. Each clone is a complete animal that filters water continuously for food and flushes out waste. This combined activity propels the cylinder along.* Because the super-organism is made up of clones they can regenerate damaged parts, and theoretically

---

* They belong to a group of marine invertebrates called tunicates, a class with a noto-chord (a cartilaginous rod to support the body) and our closest invertebrate relative. The name comes from their outer protective covering, called a tunic.

live for ever. The colonies reproduce either asexually, budding little starter colonies, or when two giant pyrosomes meet and their individual clones engage in sexual reproduction.

Crates, creeps, sops

Seventy-two billion animals join a queue every year. Each one has a unique character that he or she was never allowed to express. Few had 'a life worth living' and none of this was what I was intending to write about. But this is our story, and theirs. We are being misled if we don't know about it. Most of the 72 billion have had an unspeakably miserable existence of anxiety and inhibited instincts, and a terrifying and inescapable death. Birds are hung upside down by their ankles in metal shackles on a conveyor belt (their wings flapping for the first time) as their heads are dunked into a low-voltage electrical bath. In Europe chickens should be unconscious or dead before being bled and scalded; in America chickens are exempt from humane methods of slaughter. Next stop the automated throat slitter. Cows get a retractable bullet fired into their brain. Sheep are stunned with electric tongs on their heads. Then knifed. Pigs are most likely to be gassed in large groups for cost efficiency. 'Gas is very aversive, it's not pleasant at all,' says an RSPCA spokesman. It can take five minutes. They suffer. Jonathan had the misfortune to see a video of it. He wept. 'Their faces,' he said. One proposed solution is genetic selection for pigs who do not find CO_2 aversive. I KID YOU NOT. The renowned animal behaviourist Dr Temple Grandin has dedicated her life to make farm animal deaths less terrifying. Her famous clamp method, known as a 'hug box' which gives even, deep pressure along the body, serves to calm the animal down. Grandin is not going to stop the slaughterhouses so she decided to design them as best she could, and to work in that area is the gift of her great courage.

Farmers need to make a profit, so maybe governments should subsidise healthy, affordable food, like organic fruit and veg, instead of spending our billions propping up unhealthy, polluting factory farms. Things have not improved for industrial farmed animals. They have simply become more systematic, more hidden away behind locked gates signed No Entry or BioSecurity. There are things going on that should not be okay by anyone's standards. Gestation crate, veal crate, enriched cage. Animal welfare is a game of words, of enriched terms, of get-outs and loose interpretations. Guess what they call slightly bigger crates where a sow can turn round? Freedom Pens. The National Pig Association says that raising welfare standards would be uncompetitive and result in importing pork from countries with lower standards. A pig who never goes outside doesn't know he's in Yangxiang's 12-storey pig building on a mountain in southern China. These days 'welfare' is conflated with safety and disease transmission. I want no part in this. The methods, designs, and deaths we administer and allow in factory farms are legitimised EVIL.

There's a nipple on it

We all do kind of know. We know that farm animals are not *really* any different from our pets at home, that pigs are as clever as collies, they grunt with pleasure and can do tricks, they want to protect their piglets. Yet every year we need more factory-farmed animals to feed our beloved pets with. All animals are equal but some are *stratospherically* more equal than others. Sometimes I'm fool enough to bring up factory farm conditions in conversation. Invariably a hand shoots up to the face, fingers splayed to form as wide a barrier as possible against the information. I drop it; a win for the hamburger king. In 2013, shock-horror, British and Irish beef burgers were discovered to have been 'adulterated' with 29% horsemeat. Horses! The trail went back to a Dundalk beef baron, multi-millionaire Larry Goodman. He blamed rogue managers, a Dutch supplier, a Spanish abattoir, a Cheshire slaughterhouse. Not me, guv. Half a million pounds' worth of cannabis was found concealed in a horse transporter; some of the horsemeat tested positive for the banned analgesic Bute (phenylbutazone); Polish workers claimed they were paid cash to mix horsemeat from British and German abattoirs with defrosted beef that was several years old. Tesco, Burger King, Aldi and the Co-op apologised to their customers. Other customers on social media were reeling in disgust after someone tweeted a photograph of a piece of bacon with a nipple on it. It went viral. Some said they would never eat bacon again. What did they think bacon was? Far, far better the life of the anonymous tethered pony outside Aldi without a drink, and probably the anonymous horse in your burger, than any factory-reared pig in your sandwich or on your BBQ. When I searched in vain to see what happened to the Aldi pony, I found a jumble of tethered ponies and distressed citizens. Pony tethered to electricity pylon in Suffolk, 'How can this be right?'; pony tethered to a roundabout, 'Cruel beyond words'; pony tethered to

a sofa; a purple pony tethered to a beer fountain; and best of all, a tethered pony stretches his rope in front of a toilet block, fastening the door
and trapping a person inside for 45 minutes!

Once upon a time vegans were malnourished, mad, shadowy humans
only found in America. A person who didn't eat meat was unloved at
dinner parties. Who's the fool now: the vegan who faced the facts, or the
most of us who avoid them? I am not a vegan, nor a vegetarian, and I
am not convinced by my 'ethical' meat-eating position, but here it is: we
eat meat *sometimes*. No Cruel Meat.* No pigs, indeed nothing of unknown
life and death. We are lucky to be able to source (and afford) wild venison,
beef from a friend's farm allowing cows a good cow life, and free-range
eggs from running-around garden chickens. That's where we are at. I am
sure I am full of hypocrisies. The ecological case for roaming, grazing
and nutrient-cycling fauna to restore soil fertility and the health of our
ecosystems gives us our apology – for now. It is expensive, but much
cheaper because we eat so little of it, once a week max, sometimes once
a month. But overall, the numbers tell us 9 billion humans will need to
move to a plant-based diet. Or screw the planet.

In the margin of my notes for this section I have written, *I now need
counselling.* So we'll leave it there. From the deep recess of my mind come
the jumbled words of Ted Hughes in the leggy leaping form of a March
calf, freed to the pasture, scattering his joy, finding himself to be free, to
be a calf, to be himself, to stand and moo.

---

* And the difficulty of finding this information out is why we need proper labelling.

IX: DAMNED

I know already who wants to buy it, sell it, make it disappear.

　　　Adrienne Riche, 'What Kinds of Times Are These'

And I had done a hellish thing,
　　And it would work 'em woe:
For all averred, I had killed the bird
　　That made the breeze to blow.
Ah wretch! said they, the bird to slay,
　　That made the breeze to blow!

　　Samuel Taylor Coleridge, *The Rime of the Ancient Mariner*

NO SWEET BIRD

So, there we have it, man slays the bird that makes the vital breeze to blow and, as we know, man has been slaying the bird ever since. The albatross, following a ship driven by storms into the frigid seas of the South Pole, had become an omen of life to the sailors, a living thing in the fog-frozen landscape. After the bird's death, the wind drops, and the ship languishes day after day, drifting into the doldrums of the tropics. Without water the crew begin to perish and curse the Mariner for shooting the albatross, thus the harmless bird who watched over them is hung around his neck.

As a weird teenager, I murmured Coleridge's *Ancient Mariner* aloud, over and over, hardly daring to imagine this hell could ever be on Earth. The elemental metre beat its drum inside me. Gustav Doré's accompanying illustrations added another layer of awful wonder. The tiny speck-ness of the ship alone in the immensity of an endless ocean, the looming mists, the icicles dripping from the rigging. The crime against nature hangs around the Mariner's neck. There was no mistaking man was the culprit here; the bird's blazing white archangel wings pointed to his feet. While I would have understood the parable, would I have understood its reach? Atonement would come in seeing the beauty of the water-snakes as they coil and swim (previously among the *thousand thousand slimy things*), so utterly other, so far away from the human world – their glossy green-blue and *velvet black* in tracks of flashing gold, *O happy living things!* The albatross fell from his neck as he *blessed them unaware.*★

The albatross stayed with me. In 1988, when we lived in New Zealand, I became re-obsessed. The only mainland site in the world where albatrosses nest is on the Otago Peninsula of the South Island. One of these northern royal albatrosses had been returning to Taiaroa Head since she was banded in 1937. Her name was Grandma. It was hard to compute: 1937 was before the outbreak of the Second World War. Grandma was still breeding at over 60 years old. In October, I joined the nation's nervousness for her expected return. Her mate of the last 17 years, 34-year-old Blue-green, named after the colour of his band, had already arrived. He held lookout over the horizon. To the country's relief Grandma touched down in November. The pair hadn't seen each other for over a year; they reunited in a bill-clacking wing-waltz and settled down to make their nest. Albatross parents talk to their egg before it hatches, clacking and checking it, then tucking it under their brood patch, sky pointing, *clackclackclack*.

One chick will take both parents eight months to rear. Two kilos of regurgitated squid a day. A single fishing trip might be 10,000 miles, cruising in 90-mile-an-hour gales. When a young albatross takes off for the first time he will not touch land for five years. Alone he will range the southern latitudes, round Cape Horn, his 3-metre wingspan locked onto the great winds of the Southern Ocean, bright-white pinions tipping ice-cold waves. Distance is nothing: 6 million square miles of ocean; 100,000 satellite-tracked miles in a year. He is the living wind. His heartbeat, existing in this layer of air so wild and unimaginable, expends barely nothing at 65 beats a minute – same as on the nest. The facts of a twenty-first-century albatross have no need of parable. Not for this mighty gliding bird, nor sadly for the human agency of his fate. The Mariner's crossbow has been replaced by the tuna fishing industry's longlines, each one unreeling 60 miles of thousands upon thousands of squid-baited hooks. Or the 4,500 kilos of plastic swallowed by albatrosses every year; plastic debris not only looks like squid but smells like it. Their highly sensitive tubenoses pick up on the dimethyl sulphide released as the plastic breaks

---

★   Although the albatross around our neck as a curse remains.

down, which being a petrochemical product is similar to the oily smell of squid. The stomach contents of a Laysan albatross chick found on Midway Atoll contained 52 items of plastic.

In August 1990, the last chick that Grandma and Blue-green reared together, a male, poetically named R159, fledged. Grandma never returned. Her family tree had produced 51 direct descendants spread over six generations, four great-great-grandchildren, and one great-great-great-grandchild. In her 60- to 70-year life she could have flown more than 5 million miles. I watched the footage of Grandma's last flight over and over. We could stare for ever at the flight of the albatross and get nowhere near.

It's September. I'm watching the live stream of the northern royal albatross at Taiaroa Head in New Zealand on Royal Cam. Grown chicks are getting ready to go. Their globe-voyaging clock ticks inside them. They hold the armature of their wings out horizontally, as unbalancing and unwieldy as carrying long ladders in a gale. They face the wind for some hover practice. Wings bowing, wobble, oops backwards, hover a few feet, legs paddling, up, down, tail feathers ruddering side to side, up again, wobble, all over the place, more like treading water, ah − there, lift, straddle the eddy, gust-snagged again, everything akimbo, hover a moment, ah, getting the hang of it. Each day, another sets off. Until a lone last chick is standing on the windswept cliff. The Otago Peninsula laid out before him, all his world till now. Wings out, he lifts. Heart in my mouth.

MAD

Albert Einstein worked out how *a lot* of energy can come from a minuscule amount of matter. Humans made use of this discovery by building, what else, an almighty bomb. Those of us born in the 1950s, '60s and '70s walk around with a legacy of nuclear weapon testing in our brain: Carbon-14.★ As the level of radiocarbon in the atmosphere doubled, it was taken up by plants and passed on to living animal tissue in twice the normal dose. Because cells in the neocortex are as old as we are, there the trace remains. What wasn't understood was what happened to the radioactive fallout, the tiny particles of dust and ash sent into the strato-

---

★ Radiocarbon dating uses the rate of Carbon-14 decay to nitrogen.

sphere. Guess where they ended up. In sea cucumbers. We discovered this after the Chernobyl disaster, when radioactive particles landing on the surface of the ocean immediately disappeared. It was thought, being so small, they would take a long time to sink. But no. Algae at the surface absorbed the radionuclides, which was then eaten by krill, who returned to the depths where their radioactive poo* was gobbled up by sea cucumbers.[180] Sea cucumbers – who look very much like fat cucumbers and congregate in herds – provide the ocean's waste disposal services. Once upon a time they rolled around unmolested on the ocean floor. Nowadays they are a rare delicacy of Chinese cuisine or traded on the black market for their hokum medicinal powers. Sea cucumbers, shark's fin, abalone and the swim bladders of the totoaba fish are the four 'marine treasures' in a soup called 'Buddha Jumps Over the Wall', which can cost $500 a bowl. The fisheries in Mexico who provide the ingredients are responsible for scooping up as by-catch the most endangered cetacean in the world, the little vaquita dolphin, about 5 feet long with large dark rings around the eyes (as I write, between six and ten individuals left). The growth of China's middle class drove the import of thousands of tons of sea cucumbers from the East African coast, and when that was fished out, from the West African coast. Quotas and export bans gave rise to the 'sea cucumber mafia'. Along came the ivory traffickers, then the South African abalone poachers,** and then Japan's heavy yakuza gangs wanted in on the sea cucumber action. Rival poachers took potshots at each other from high-powered speedboats out at sea.

Aside from their vital sweeping, mopping and cleaning services (a sea cucumber can process up to 37 pounds of sediment a year), they recycle nutrients, and their faeces help increase seawater alkalinity, which can protect coral reef structure by buffering acidification – which is increasing with climate change. They are host, inside and out, to other tiny organisms, *and* lunch for someone else (crabs, fish, turtles, starfish) in the trophic web. The decimation of sea cucumbers ups pollution levels and leaves a gap so much bigger than they are. This is what a bowl of snake-oil sea cucumber soup does. A short-term economic boom for a long-term bust. Sea cucumbers demonstrate the interconnectedness of the world from

---

*   The highest level of radioactivity recorded in a living organism was in a krill. And we know who eats a lot of them . . .
** By early 2000, 55,000 tons of abalone had been smuggled out of South Africa, according to Traffic, the organisation that monitors wildlife smuggling.

the stratosphere to the bottom of the sea. So, we might as well mention what Russian dissident Alexander Litvinenko had in common with a blunt-tail prawn (*Gennadas valens*). Too much polonium-210. The prawn's digestive gland registered a concentration of polonium a million times higher than the surrounding seawater in the Mediterranean after the Chernobyl meltdown; and Litvinenko's lead-lined coffin, after he died from polonium poisoning, was advised not to be opened for 22 years.

For crimes against Nature, this is perhaps small fry compared to the brainwave of dropping a few nuclear bombs on the Arctic icecap to create a warmer, nicer place. In 1946, Julian Huxley (whom we met when offering a home at London Zoo to Denis Harvey's otter), then the first director-general of UNESCO, was all for the idea. As the world slumbered, the US weather bureau developed the proposition, concluding 10 'clean' hydrogen bombs should do the trick. The Russians were *very* keen. Improving the weather would cheer up Siberia; shipping lanes would open; the tundra could be cultivated; and rain would fall on the Sahara. Gas engineer Petr Borisov published his proposal to create a Polar Gulf Stream: build a 55-mile dam across the Bering Strait. Another scheme to melt the ice was to cover it with coal dust to darken it and trap the heat. Less costly, atom bombs were seen as a kind of two for the price of one, with mining services. Like, blow up Alberta's tar sands to release 300 billion barrels of crude oil.[181] I am a child of the atom bomb deep anxiety era. It wasn't *if* a despot's finger landed on the button, it was when. Living in the 1960s and '70s was a close shave. And with the 2022 war in Ukraine, we are back there again.

GOOD GRIEF

Heaven knows we need never be ashamed of our tears, for they are rain upon the blinding dust of earth, overlying our hard hearts. I was better after I had cried, than before – more sorry, more aware of my own ingratitude, more gentle.

Charles Dickens, *Great Expectations*

Tears are a physical manifestation of compressed time, when the past is squeezed up with its final consequence in a storyline we are unable to change.

There is a term, solastalgia, for the type of homesickness that you can feel when you are still at home. It is a kind of existential melancholy for something lost, a forest, or a favourite path, a nightingale thicket or a badger sett. The Australian philosopher Glenn Albrecht coined it in 2003 to describe the effect of mining activity on communities in New South Wales. Home can be your street and your planet. Many of us, seeing the devastations of our doings, dispirited from writing letters and signing petitions, feel eco-anxiety more and more. It is recognised as a specific condition, a helplessness at the relentless onslaught against the environment. A bereavement loop. A beautiful meadow outside London has lost its fight not to become another shopping centre for more plastic nick-nacks we don't need and cannot afford; the beloved 119-year-old allotments in Watford, Hertfordshire, levelled for another car park; a huge 'ice centre' gets planning permission to be built on precious Leyton Marsh on the outskirts of London; the fabulous jumping spider on Swanscombe Marsh in Kent – home to 1,992 species of invertebrates – must leap aside for a £3.5 billion theme park;★ the precious 250-year-old Cubbington pear tree, an eye-full of loveliness and fruitfulness, cut down for the HS2 high-speed rail line which was good enough to bow around a golf course; it

---

★ With a water park and rides, six islands, 13,500 hotel rooms, an 'enchanted fantasy realm, a long-lost civilization', and 'home to mythical creatures' (oh, the irony). Swanscombe Marsh was given Site of Special Scientific Interest status (SSSI) in March 2021.

goes on and on. The marsh is *already* a theme park. How about jobs for naturalists to excite children with the wonders of the jumping spider? Every morsel of land has a greedy human eye on it. It is as if *no one* in a position of responsibility has read a single word of the UN's IPCC (Intergovernmental Panel on Climate Change) reports; or has the faintest clue about ecology or what an ecosystem is, let alone how it functions. Tim Gordon, a marine biologist at the University of Exeter, describes the sadness of documenting declines:

> A lot of the time, you're kind of numb to it . . . you know, you've got a job to do. But then occasionally, for no particularly good reason, it'll strike – you just float into the middle of the water, look around you and think: Wow, it's all dying. There's been times that you cry into your mask because you look around and realise how tragic it is.[182]

Last night we watched David Attenborough's 2020 film, *A Life on Our Planet,* which he calls his 'witness statement'. There is a moment when this vitally young 93-year-old man grows suddenly old, unable to speak; it is as if every astonishing creature in every wonderful place he ever saw comes rushing back to him. He casts his eyes down as the pain of their loss overwhelms him. Then anger spills out.

'So, the world is not as wild as it was. Well, we've destroyed it, not just ruined it. We have completely destroyed *that* world. The non-human world has gone.'

Words scramble in his mouth. It's not easy to watch.

'Human beings have overrun the world.'

He shuts his eyes. His face is tormented. It's not a gimmick. He is doing everything he can with every last breath to warn us of the peril of the extinction of the natural world. He says he wishes he wasn't involved in this struggle, because he wishes the struggle wasn't there. In 2019 he showed the IMF World Economic Forum in Davos the awful sequence in *Blue Planet II* when hundreds of 2-ton walruses, who would normally haul out onto ice, were shown falling from rocky cliffs in northern Russia to their deaths; I mean, this is hell on earth.* The dark

---

* The scene was contentious. It's not understood why this happens; the loss of ice forces herds to haul out onto overcrowded beaches where they are less adapted to move, and head off in this dangerous direction. They are easily spooked, and with a herd mentality can suddenly stampede.

cinema scans the faces and they are, to a man and woman, in shock, hands to mouths (I cry out even though I have seen it before). The helpless pain of seeing can be unbearable. The guilt that it is wholly human fault; the bafflement that we have allowed the human project to go so horribly wrong; the dismay that humans have been so wilfully dumb; the exhaustion at the same-old same-old platitudes from the mealy mouthed politicians; the fury at the hoodwinking from corporate profiteers; the frustration with the vested interests of the meat barons and plastic tycoons; the anger that we do not all mourn together; the hatred for the despoilers of the world; the disenchantment as our leaders fail to respond. It is not mentally healthy for us. Who wants to join *that* club? Yet how can we not? The fate of our fellow animals is not an *opinion* any more. We have killed them, taken their homes, their food, the conditions that made it possible for them to survive. The bloodied walruses cartwheel over our sharp shoulders. The whites of the world's most powerful eyes glint tearfully in the darkness, but let me tell you, they will do nothing to turn the tide of pressure on the planet's ecosystems.

★

Solastalgia is the disease of the Anthropocene, the era of human impact on geological time, climate and ecology, which has also been called the Age of Loneliness. I want to go home, I can't go home. The daily bombardment, from the burning rainforests and pink lagoons of toxic slurry, to the flayed hedge, the pitchforked fox, the strimmed bank of cow parsley. A tinnitus drone of the ubiquitous anthropogenic soundtrack from leaf blowers to bulldozers, hack, saw, chop, strim, flail, mow. We have gone so far that our very presence in animals' lives provokes a skewed genetic response that could undermine their long-term survival.★ We cannot live with them, and we cannot live without them. Sometimes, it feels like I am in the middle of an enormous mistake. Sometimes I am overwhelmed. So I decide to pay attention to something small. Moss, for instance. Today I am only going to care about moss. What a nice day I will have. No fretting about moss around here.

Moss, I discover, is home to a fabulous creature called a moss piglet,

---

★ For example, the shy seabirds who survive to pass on their genes because the bolder birds have been caught by our longlines.

or water bear, or tardigrade. Tardigrades ('slow tiptoers') clamber through moss and all sorts of other places. They have been found in glaciers, on garage roofs, in tree canopies, in the mid-Atlantic trench, at the top of giant redwoods, in mud volcanoes, in hot springs, on top of the Himalayas, wherever moisture collects. They are less than half a millimetre long and lively little fellows under the microscope. They look like a stack of inside-out puffer jackets with eight squat legs and tiny claws to hold on with, two eyes and a piglet-funnel mouth; they have a digestive system, a ganglion and a ventral nerve cord. There are 1,300 known species. But their life-defying trick is to stop living, if they need to, with no detectable metabolism whatsoever, for up to 30 years, and then come back to life with a splosh of water. They can dry out to a speck of dust, which is the tardigrade rolled into a ball called a tongue, in a process called cryptobiosis, which replaces water in the cells with 'disordered' proteins that produces a vitrified state. Tardigrade dust can blow in the wind to land anywhere in the world. They can withstand pressures of 6,000 atmospheres, temperatures as low as −272°C (almost absolute zero, the lowest limit of the thermodynamic temperature scale) and as high as 150°C; they can be stored at −20°C for decades, survive radiation, starvation, gases and the vacuum of outer space. Which is why they are called a pioneer species. They are the most resilient creature alive (or not alive). They survived all five mass extinctions.

The capacity to exist in a dehydrated state and be reconstituted back to life (like 1970s instant mashed potato) would be incredibly useful to humans, particularly for pharmaceuticals and vaccines that need to be frozen. Instead of complicated delivery to remote places requiring refrigeration along the route, they could be posted in an envelope. Just add water. Now we know the moon has water, maybe any tardigrades who stowed away on the 1969 Apollo 11 mission's landing craft, Lunar Module Eagle, have begun a little colony. Indeed, tardigrades officially went on the final flight of Space Shuttle *Endeavour* in 2011. In 2019 a capsule of tardigrades were in the Israeli lunar lander, *Beresheet*, which crash-landed on the moon in April 2019. We have sent all sorts of creatures into space: fruit flies, snails, carp, sea urchins, swordtail fish, silkworms, bees, ants, geckos, cockroaches, moth eggs, stick insect eggs, quail eggs, shrimp and scorpions. Fancy sending scorpions. Nematodes from Space Shuttle *Columbia* were found alive in the debris after the spacecraft disintegrated when it re-entered the atmosphere on 1 February 2003, killing all seven crew. A female cockroach on a 2007 Russian space mission became the first Earthling to conceive in space, producing 33 offspring; they called her Nadezhda – which means hope.

Love and madness

The English poet John Clare, witnessing the desecration of his beloved countryside in the land enclosures of the 1800s, suffered so much he went mad. Small fields were swallowed into the black hole of England's land-owning pockets, and swallowed too were ideas of common use and access. Where *silver springs* become *naked dykes*, and *tasteless tykes* grub up the trees and banks and bushes, where nothing living can thrive, and *bees flye round in feeble rings / And find no blossom bye, / Then thrum their almost weary wings / Upon the moss and die.*[183] Clare is turned inside-out, picked to bones, and like the beetle finds scarce a nook to hide beneath.

The American poet Robinson Jeffers found the great outdoor America too busy for him, so he looked away from the land out to the Pacific Ocean for the wildness he could not find anywhere else. In 1914, he built a stone house on a granite outcrop of land south of San Francisco that reminded him of a prow of a ship. Beside the house he built a tower on which a hawk landed every day. It was here, at Hawk Tower, that he became one of America's most successful poets. In response to a world disappearing, Jeffers developed his philosophy of 'inhumanism' and turned to wild animals for their ferocity and strength to express his rage at the reckless destruction of an irretrievable natural beauty. In his 1954 poem 'Vulture', he is lying on a hill and sees a vulture wheeling above him. Lower and nearer the bird flies, until Jeffers realises he is 'under inspection'. He lies death-still, hears the flight feathers rip through the wind, sees the naked head between the creature's great wings. So beautiful is the bird, gliding down, that Jeffers is sorry to disappoint him.

> . . . To be eaten by that beak and become part of him,
>   to share those wings and those eyes –
> What a sublime end of one's body,
>   what an enskyment; what a life after death.

By the start of the twenty-first century, we had dispatched most of the wild animals and domesticated the rest; our weight accounts for 36% of the biomass of all mammals on this planet, the animals we eat take up 60%, which leaves the mammals of the wild world just 4%. Plato lamented the destruction of soils and forests in ancient Greece 2,500 years ago; pollen analysis reveals a fertile land of rich soil long since washed away after trees were felled – many turned into ships for war and colonisation. War and power have always taken precedence over nature. Did Plato suffer from solastalgia? Elon Musk's ambition is to colonise Mars. A backup plan for when this place is dusted and done. He plans to get a million humans there by 2050. Good. We have all *seen* Mars. It's not very nice, Elon. In a run around and frolic sense I mean. No birdsong on Mars. No sea to swim in. No life because there is *no breathable oxygen*. Water supply, not good. The temperature is bloody freezing with no UV protection, and if that doesn't get you, a Martian dust storm will. But, as clever NASA explains, 'the key to survival will be technology, research and testing'. The very tools which were going to fix the fuck-ups on *this* astonishing planet. 'We can duplicate the soil on the Earth,' says futurist Michio Kaku. If you can duplicate the soil on Earth, fantasist, *duplicate* it for depleted Earth.★ Musk's goal is a multi-planetary human civilisation and he's made a multibillion-dollar snake-oil company out of it called SpaceX. Why is NASA going along for the ride? Money. There are loads of nutcase investors out there.★★ And there's the rub. Why can't they fix *this* place? Why can't hedge-funders fund the blooming hedges? Musk's Interplanetary Transport System will (he hopes) take 80 days to get his million people to Mars in three (reusable/tick) spaceship rockets every day, creating 'a lot of jobs' (tick) in the city he's gonna build on the red planet, and he won't have to cut down any ancient forests in the way (tick). The moon will serve like a motorway station: Deep Space Gateway; for Musk, Mars is just so much more *out there*. An 80-day flight to a freezing dusty rock where you can't breathe or get a drink of water. Without life, without creatures. How do you sell that? Well, he does. That's what makes a giga-billionaire. Oh, and thanks, Elon, for the SpaceX Starlink project which plans to deploy 30,000 internet-serving satellites into orbit. A 'mega-constellation' that will not only give astronomers nightmares but will interfere with billions of migrating birds' ability to read the sky.

★ Worms don't do a bad job if we let them.
★★ Although I should mention the prospects of mining . . .

As I've said, four hours in a plane does me in. But, oh, how could we leave a home so unutterably beautiful and complex and strange, with life as wondrous as any alien could possibly be? What alien could change shape and colour, regenerate legs that can think for themselves, mimic the surroundings, fake seaweed, grow horns, pop up warts, glow in the dark, pour themselves down a plug, play, dance, squirt you with ink, take off by jet propulsion and taste you all over with their embrace? What aliens can transform into little flying hankies with velvet kaleidoscopic wings and a long lasso tongue to help the flowers bloom? What more wonder did we want? I mean, what I am trying to say is we have a lot of work to do *here* SO STOP PISSING ABOUT WITH SPACESHIPS TO MARS.

Here on Earth the human brain has shrunk. We've lost around 150 cubic centimetres in the last 20,000 years, about the size of a tennis ball.[184] The ratio of brain to body mass is called the encephalisation quotient, or EQ, but the human brain has shrunk at a faster rate than our once bigger, beefier Cro-Magnon body. It could be just a more efficient brain, or maybe, like our domesticated beasts, we have lost the need for the fast-thinking chutzpah of our wilder ancestors. Cognitive scientist David Geary suspects that Mike Judge's 2006 cult film, *Idiocracy*, hits the nail on the proverbial head. Man in a hibernation experiment goes to sleep for 500 years and wakes up to discover he is the smartest guy on the planet. We have dumbed down. We don't need to outwit wild boar or aurochs any more. We go down the shop. Water comes from a tap, light from a switch and cooked dinner from an oven. We don't need to make a shelter, or see in the dark, or even to jumpstart the car. Since the Bronze Age the shrinkage in European male human craniums is the most pronounced. Heh, heh. For an idea of proportional scaling, anthropologist John Hawks suggests, 'For a brain as small as that found in the average European male today, the body would have to shrink to the size of a pygmy.' An upbeat interpretation is that a smaller brain is the result of selecting traits against aggression, which is actually back to the downbeat dumbing-down scenario, a dependent adult with juvenile characteristics who can't do everything for himself.

THAT'S NO WAY TO SAY GOODBYE

Dammed

The history of soil erosion in China tracks the evolution of an eye. As the great Yangtze River became muddier and murkier from millennia of cultivating terraces along its banks, its most sacred creature, a freshwater dolphin called the baiji, morphed with the circumstances. The baiji's eyes responded to what minuscule light there was and moved upwards towards the top of the head.* As visibility decreased, the baiji's eyesight atrophied with disuse and was taken over by acutely sensitive hearing. The baiji's natural sonar and fast clicking developed to echolocate the size, shape and speed of objects around them, and to find and communicate with each other. The baiji successfully evolved from a freshwater dolphin to a dirtywater dolphin. In time they found themselves living among sailing craft and fishing boats, cruising in pods of a dozen or in mother-and-calf pairs, pale blue and cream, about 7 feet long, known to the fishermen as the Goddess of the River. Then came sparrow-smashing Chairman Mao with the founding of the People's Republic of China in 1949. Out goes eco-ethical Confucian teaching of harmony with nature, *tianren heyi*; in comes Man Must Conquer Nature, with Mao's 'Great Leap Forward'. After the extermination of sparrows in the Four Pests Campaign (alongside rats, flies and mosquitoes) the resulting famine understandably put baiji on the menu.

Then came noise. The Yangtze was China's main highway. Sail gave way to diesel. Steamers, container ships, ferries, junks, barges, oil tankers, freighters chugging tons of cargo up and down. In the 1980s came Deng Xiaoping's reforms, a rising population and consumer expectations. He didn't actually say, 'To get rich is glorious', but many thought he did. A thousand propellors grinding through turbid waters . . . what chance did

---

* The movement of the baiji's eye was mirrored in their foetal development, beginning in normal dolphin eye position – for wide horizontal vision – to graduate upwards as the foetus grew.

baijis have? Sound waves travel further, louder and five times faster in water than air. In a few short decades the Yangtze became a terrifying world of noise far louder than the acutely noise-sensitive baijis were prepared for, or had ever known. Like living in the median strip of a 10-lane highway, alongside an airport runway, by a sewage overflow pipe after a toxic spill (400,000 chemical factories operate along the Yangtze), in an electrical storm (and don't forget you're almost blind). The bombardment confused their sensitive radar and so disorientated them that when they made a long dive to avoid a boat, they often surfaced too soon. There was so much noise they were unable to find mates, or food. The thing about being designed to live in a river is you can't pack your bags and go somewhere else. Dolphins began to be washed up along the shore. Then fishermen stopped seeing them. The Cultural Revolution had purged any study of the species so by the time China began to worry about its iconic Yangtze goddess, little of her was understood. In 1988, galactic hitchhiker Douglas Adams and zoologist Mark Carwardine went to search for them.[185] From a 1950s estimated population of 6,000 living along the river's 6,300-kilometre reach, there were barely 200 left. Now the baiji was so rare she became precious. And a good marketing device. There was Baiji beer, Baiji cola, Baiji fertiliser, Baiji loo paper. A baiji reserve on an elbow of the river in a channel between two islands was staked out, 1.5 kilometres long by 5 metres deep, with a baiji hospital, and a fish farm to feed them. And no doubt plans for a tourist attraction. Douglas and Mark stayed at the Baiji Hotel. But they didn't find them. The only one they saw was in the Institute of Hydrobiology in Wuhan. The baiji took more than 20 million years of evolution to refine, and 50 years of grand communist-capitalist ideology to rub out. First we took their sight. Then we took their hearing. Then we took their river.

In 1956, the poet Chairman Mao wrote a poem called 'Swimming' that was thunderous in tone and characteristic of his firebrand assault against nature. A great dam, he proclaimed, would stride 'athwart' the impassable gorges of Wushan to hold back the waters and transform the sheer ravines into a serene lake. And the goddess, if still alive, will be awestruck at the changed world.[186]

Athwart, it did. Alive, she ain't. In 1994, 80,000 sticks of dynamite were detonated along the Yangtze River in central China, at Qutang, Wu and Xiling gorges, where the biggest dam in the world would be constructed. Situated, weirdly, on a seismic fault, Three Gorges Dam increased landslides and displaced 1.3 million people. In company with the baiji was another Yangtze native, the fabulous 200-million-year-old Chinese paddlefish, up to 23 feet in length, officially extinct after the dam blocked their migration. Big dams don't tend to have a great track record. China is the most dammed country in the world, with 85,000. The dangers, ecological damage and costs of mitigation in the long run tend to outweigh their benefits. In January 2020, China announced a 10-year commercial fishing ban in 300 zones along the Yangtze to help the river's biodiversity bounce back. But where from?

Who-are-you, who-are-you, who cannot fly who comes to my home with my feathers in your hair?

There exists a sound recording of a Māori bird tracker imitating the call of the sacred huia bird of New Zealand.[187] Henare Hámana, a young man on the 1909 expedition to search for the lost birds in the Ruahine Ranges of the North Island, sang the huia song to lure it out of the forest into a clearing. The bird was named after the sound of his call, a clear-noted undulating whistle, *Whooouuuu-eea Whoooouuu-whu-wu-wuu-eea, Whoooouu-uee-eea*. There was a sense of Shakespearean tragedy in their departure. Theirs was a fatal song. The female bird attracted by the notes, looking for her mate, came out into the clearing. And into the trap, where a noose on the end of a carved pole was easily slipped over her head. The male bird, attracted by the distress calls of his mate was caught in the same way.

Without mammalian predators, the huia was not a fabulous flyer, but gracefully hopped and bounded from branch to branch, rarely flitting above the forest canopy. A bird about 18 inches long, of black-blue-purple lustre with glowing orange wattles, and a fan of white-dipped tail feathers. Because of their close pairing, if there was a female, the male would not be far away. They were observed touching their heads and bills, in apparent caress, while making a low cooing twitter. The remarkable thing about them, which put nineteenth-century biologists into a spin, was the radically different bills of male and female. The hen had a slender downward-curving bill about 10 centimetres long; the cock, a straighter, stouter 6-centimetre pick-axe, like a woodpecker. Their sexual dimorphism was so extreme that taxonomists had initially identified them as different species. To get at the huhu beetle larvae on which they fed, the male beak had evolved to lever open fissures in decaying wood with force, in a rare technique known as 'gaping', whereas the female could reach more deeply embedded insects. It was thought this enabled a kind of collaborative feeding because partner birds foraged together. More likely it allowed them to exploit different niches in the same territory.

Declare something sacred and you can seal its fate. Huia tail feathers were highly prized by the Māori; collected from May to July when the plumage was at its most ravishing, and so precious they were kept in a carved canoe-shaped box called a *waka huia*. Most portraits of high-ranking Māori will show a huia feather tucked into their hair. The skins, with head and beak intact, were worn as ear ornaments. A *pōtae huia* was a special headdress of huia heads, with the curved beaks hanging down, bills rattling against each other. And so it was, on 14 June 1901, when the Duke of York (later King George V) arrived at the sacred Māori meeting place, the Te Papaioururu marae, at the start of a royal visit to New Zealand, that a high-born Māori woman took a black feather with a striking white tip from her hair and slipped it into the duke's hatband as a token of friendship – which he honoured by wearing it throughout the whole visit.

The burning of forests, the arrival of cats and other agile European predators pushed the huia to the edges of their territory, but it was the desire for specimens and feathers that finished them off. The rarer they became the faster they were collected. In 1892 the ornithologist Sir Walter Buller reported he had seen only one (which he shot) whereas five years before they'd been plentiful. Once the duke sported his huia feather, the whole world wanted one. In 1883, the year my Irish grandmother was born, a Māori hunting party took 646 skins in a single month.[188] The Natural History Museum in Vienna took 212 pairs. The Ngati Huia tribe tried to obtain protection for their sacred bird, forbade hunting and entreated Governor Onslow – whose baby son, photographed wearing a pair of huia feathers, was baptised Victor Alexander Herbert *Huia* Onslow in 1891 – to give the bird legal protection. Onslow added the huia to the protected species list but it was too little, too late. Paltry fines – less than the birds fetched on the market – were rarely dished out, and in 1903 the Solicitor General ruled that while the birds were legally protected, the feathers were not. A pair were caught to send to a sanctuary in New Zealand, but Buller interceded to send them to Lord Rothschild in England instead. Buller promulgated an attitude of skewed Darwinian logic that believed New Zealand's indigenous fauna and flora (including the Māori) were destined to be overtaken and replaced by 'fitter' European species. Here he is describing his Māori guide whistling to attract a huia:

> . . . a huia came bounding along, almost tumbling, through the close foliage of the pukapuka, and presented himself to view at such close range that it was impossible to fire. This gave me an opportunity of watching this beautiful bird and marking his noble bearing, if I may so express it, before I shot him.[189]

Had the duke been given a different gift, or had he passed it to his minder, the huia's fate might have been different. Yet just six years after the future British king wore his huia feather, the last bird recorded was shot on 28 December 1907. With the huia gone, so too vanished their specialised parasites, an unmourned louse, *Rallicola extinctus*, and a species of feather mite, *Coraciacarus muellermotzfeldi*, discovered on a museum corpse in 2008. In 2010, a single huia feather made the record books by fetching NZ$8,000 at auction.

I gaze at the stuffed huia perched on a dead branch in the National Museum of Scotland in Edinburgh. The male leans in front of the female. And that's it. Two husks of silent bird, 11,000 miles from home.

That his fateful song exists solely in the mouth of his tracker is a fitting record of human and bird and memory.

In the book *Last Chance To See*, in which Douglas Adams and Mark Carwardine describe their quest for the baiji, there is a tragic sentence: 'The point is that we are not too late to save the northern white rhino from extinction.' That was 1989. In 2021 the last two northern white rhinos on the planet are both female, mother and daughter, Najin and Fatu. In 2018 China legalised the sale of rhino horn and tiger bone for medical purposes. Hocus pocus Chinese medicines drive other nation's animals down the same road as their sacred baiji, when they could just as well chew their own keratin fingernails. The great herds of millions of elephants on the great African plains are down to just 400,000, many under armed guard; what an indictment: 20,000 lions, 4,000 tigers, 2,000 giant pandas, 500 Ethiopian wolves, not a dozen vaquitas . . . The term 'endangered species' has lost its potency from overuse. We sink our shoulders, hang our heads. As we begin to understand their richness we lose species faster than ever before. What a thing, to wipe out a species. And what an indictment for it not to be a crime.

A close cousin of solastalgia is eco-furiosity, an eco-tear-your-hair-out solastalgia on steroids. It is the long loud desperate cry of the human heart.

THERE WAS AN OLD LADY

There was an old lady who swallowed a cow,
I don't know how she swallowed a cow;
She swallowed the cow to catch the dog,
She swallowed the dog to catch the cat,
She swallowed the cat to catch the bird,
She swallowed the bird to catch the spider,
She swallowed the spider to catch the fly;
I don't know why she swallowed a fly – Perhaps she'll die!

Isolation has been New Zealand's unique path in the evolution of animals not found anywhere else on the globe. Without mammals (aside from two species of bat) the bird life went its own charming way, with no defence against human invasion. The Māori finished off the moas, but when the Europeans arrived, rats jumped ship; possums were brought in from Australia to start a fur industry; stoats and ferrets were released to get rid of the rats and the introduced rabbits.

Modern conservation remedies can seem like hell itself. The measure used to protect paradise in New Zealand is a biodegradable poison called 1080, sodium fluoroacetate. In the Land of the Flightless Bird, the many vulnerable species like the kiwi, and the world's heaviest parrot – the endearing, owl-like kakapo – have been pushed to the outer margins by bedazzled incomers – possums, rodents, cats, mustelids and hedgehogs. They now only survive by being translocated to islands offshore, or remote protected pockets and fenced-off peninsulas. To give them half a fighting chance these places are cleared of all invasive species. New Zealand chews through about 80% of the world's supply of 1080 and swears by it. It is dropped by plane across more than a million acres of land every year, and there are armies of native nature lovers who say conservation for them 'is all about killing things'.[190] Predator Free 2050 is New Zealand's ambitious official plan – for what some call the Battle for Our Birds: to rid the country of all invasive mammalian predators by 2050. The pest control costs the government $70 million a year, which is nothing compared to the estimated $3 billion of lost crop productivity from rat damage alone.* Whole communities are behind the mission with a passion that taps into deep national pride. It's a crisis they see as now or never and forever lost. Killing for conservation has become a national pastime. Bait stations, trapping, poison drops. The thing about New Zealand is the no-messing. Pet dogs go on kiwi-aversion training courses. In 2018 the residents of Omaui in the south of the South Island were told they would not be allowed to replace their cats. The council eventually buckled, but the point pushed home a general awareness of pet owners' responsibility to mitigate the hunting sprees of their beloved moggies on the bird life.

On a clear day from Jonathan's family home we can see Little Barrier Island, the protected redoubt of the kakapo and kiwi. Te Hauturu-o-Toi, its Māori name, means resting place of the wind. In 1895 this small volcanic island, 28 kilometres square, became New Zealand's first nature reserve. Cats were eradicated in 1980, followed by rats in 2004, and now it is the closest it's been to its primeval state in hundreds of years, with two-thirds of its virgin forest intact, including some kauri trees that were alive when the Māori ancestors first disembarked their canoes 1,000 years ago. From the driftwood-strewn shore to its 700-metre pinnacles of cloud forest veiled in mist, Little Barrier is home to 117 species. Kokakos, kakas, kakapos, kiwis, keas, kakarikis, kererus, kororas (little blue penguins), rurus, hihis, tuis, bellbirds, fantails, saddlebacks, silvereyes, tomtits, warblers, North Island robins (once down to just five individuals), shearwaters, petrels, gannets, skuas, albatrosses, spoonbills, gulls, terns, shearwaters, frigate birds, ducks, teals, rails, stilts, crakes, oystercatchers, dotterels, lapwings, plovers, curlews, godwits, sandpipers, snipe, knots, shags, cormorants, herons, kingfishers, starlings, thrushes, myna birds, blackbirds . . . and skylarks.[191] The New Zealand storm-petrel was thought extinct for 100 years, until it turned up in 2003.

These days New Zealand exports its poison know-how. Sixty years of tests, trials and research give their conservationists the tools, which most New Zealanders support, where the recovery of native species outweighs the risk of by-kill. Companies, backed financially by the government,

---

* It's a mighty task. New Zealand's total area is 268,000 square kilometres, with cities and sewer systems, and plenty of places for rats to hide.

promote the whole eradication package: consultation, planning, logistics, poison, grain base, bait stations, monitoring and helicopter hire. *Rattus rattus*, black rat, ship rat, roof rat, house rat; *Rattus norvegicus*, brown rat, common rat, street rat, sewer rat, wharf rat; our ubiquitous companions from the dawn of human civilisation, wherever we are – in our homes, our crops, our stores. Colonisers, stowaways and resilient reservoirs of 60 different zoonotic diseases, these intelligent, omnivorous mammals have acute hearing and a highly developed olfactory sense. They are good swimmers, good climbers, excavators, exploiters, adaptors. They can survive a 50-foot fall and chew through metal. Colonies on the banks of the Po River dive for molluscs. Gestation period: 21 days; litters: up to 14. Extermination in a population increases reproductive rates. Rats are a formidable 'enemy'. Uninhabited islands are fairly straightforward; the problems come with people.

Lord Howe Island, some 600 kilometres east of the New South Wales Australian coast, is a World Heritage Area encompassing a coral reef lagoon with sandy bays, a turquoise sea, cloud forested hills and a unique biota. And 150,000 rats. It is jaw-droppingly lovely. But there has been trouble in paradise. The 350 resident humans were at war over the rat-eradication programme. Fifty tons of the New Zealand-manufactured anticoagulant brodifacoum, distributed by helicopter and hand by 60 field staff employed for over five months; 19,000 external bait stations, 3,500 internal bait stations, 9,500 hand-broadcast points and 2 tons of slug pellets to limit bait damage. All milk from cattle on the island was destroyed during the baiting period, no fish could be eaten from the sea, pets and chickens were sent to the mainland. The two species protected from the bait, the endangered Lord Howe woodhen and the Lord Howe pied currawong,★ birds native to the island and whom this was all about, were captured and kept in a zoo. Anything else was collateral damage, including the popular owls introduced to prey on the rats, for without rats the owls would prey on chicks, so they had to go anyway. While residents were happy to make the island rat-free, it proved hard to bring everyone round to the idea of helicopters spraying poison everywhere. What about ground water contamination, residual poison, the seabird population and coral reef community? And what about the prolonged suffering, for anticoagulant poison is a slow and nasty way to die. The anger, distrust and

---

★ A black magpie-sized passerine bird with yellow eyes and white flecks on the wing.

arguments went on for 20 years. Then the money came in.★ Any failure
to allow entry to property was overcome with the control orders of the
Biosecurity Act and threatened fines. In 2020 the currawong and woodhen
are back home and breeding. As one invasive species is assumed gone,★★
another has plans to lengthen the airport runway.

On the other side of the world, as far as you can go before you start
coming back again, on the filigree of Scottish islands in the Outer Hebrides
called the Uists, another almighty row about pest eradication was at full
throttle. This time it was hedgehogs. But hedgehogs are Britishers' beloved
gardeners' friends. Which is why, in 1970, seven were introduced in South
Uist – where there were no hedgehogs – to eat slugs. Soon there were
5,000 (few roads, no pesticide spray) enjoying the smorgasbord of ground-
nesting seabird eggs and chicks. Scottish Natural Heritage began a cull
by live trapping and lethal injection, rejecting requests from animal welfare
groups begging to translocate them to England where the native hedgehog
was in massive decline. Which *was* perverse, for if you can get close
enough to inject them, you can pick them up and put them in a box.
Their reason: it was more humane to kill them. A Uist Hedgehog Rescue
group formed and sprang into action; reporters filed stories which ran
on BBC *News at Ten*; a Somerset councillor raised the issue in the House
of Lords. It was then that the soft-spoken Andy Christie and his wife,
Gay, stepped in. Andy and Gay have devoted their lives, and their home,
to wild animals at Hessilhead Wildlife Rescue Centre in Ayrshire. Andy
tells me they committed to take all rescued hedgehogs to acclimatise
before being released and monitored in carefully chosen sites. And so, on
the 690th hedgehog lethal injection, SNH did a U-turn; the islanders
mobilised and the translocation began. Two thousand Uist hedgehogs
passed through Hessilhead, with no shortage in offers of homes, and an
80% success rate after eight weeks. Then everything ground to a halt
when SNH inexplicably pulled out of an EU grant application necessary
to finish the job. So there are still hedgehogs on the Uists, and still the
dunlins decline. And inexorably the hedgehogs' home in England is flailed
more mercilessly by bigger and more savage machines, ripped and shaved

★ $16 million from the NSW Environmental Trust, the Lord Howe Island Board and
   the Commonwealth Government.
★★ Two rats were trapped on Lord Howe Island in 2021. An autopsy revealed they were
   a male and a pregnant female . . .

of all food and shelter into wind-whistling silhouettes of Tidy. We all lament our disappearing friend, but do not notice the clue is in the name.

The killing industry marches on. Dingoes, cats, foxes, horses, camels, donkeys. On the Galapagos Islands, some 140,000 goats were shot from a helicopter to restore the vegetation for the beleaguered giant tortoises, but this did not deal with the remaining elusive 10%. The ingenious solution was to release sterilised male 'Judas' goats with radio collars to lead hunters to the wary females, along with radio-collared female 'Mata Hari' goats, who were chemically induced into oestrous to lure out the wary males. The sterilised Judas goats were reprieved to live out their now unreproducible natural lives. The vegetation responded, plants and bird life bounced back. The Española tortoise, once in their thousands, and down to just 15 individuals in 1960, have bounced back too. After a successful 55-year breeding programme with the last 15 tortoises brought into captivity, there is now a wild population of more than 2,300. And how about this: on 20 June 2020 the original 15 adults who saved their species from extinction were all returned to their home island, by tractor-trailer, then boat, then backpacked like colossal Stone Age coracles by 15 humans to the interior, to join their large family and live out their long lives.[192] We are the gods. Case by case, we evaluate the disarrangement we've caused. Short-term industrial cruelty for long-term good . . . or not. Nature in aspic, or nature reborn? This is the paradox in the conflict of different camps who want the same thing: a world where native wild-life can flourish. Alongside us.

At Little Barrier Island, the avian orchestra grows ever louder, a crescendo of booms, whoops, whistles, piping and bells, the great singing of the forest, every wavelength taken, each niche loaded with rich, bubbling bursts of calls and replies in an acoustic tangle of otherworldly sounds that are not otherworldly after all. I have stood many times at the tide's edge looking out to sea, watching Little Barrier appear and disappear on the horizon, imagining. The daily business of its creatures in heavenly oblivion of all the effort that was expended to take our human footprint away.

DANCING WITH WOLVES

Sometimes the conflict between two sides with the same interests can get nasty. And political. And weird . . .

When Joanna Kossak left home to live with her aunt, Simona, in Białowieża, in the 'snowy fairytale winter' of 1993, there was a knock at the door. A park ranger needed help to release his colleague, who had been caught in an iron-jawed leg trap in the protected part of the forest. After they released him, Simona took the trap and reported it to the police. A young scientific researcher, monitoring the wolves, arrived and claimed it as a scientific tool stolen from his research area. A huge row ensued about the safety of the trap, which culminated in Simona being taken to court. The researcher submitted the trap was safe. 'You put *your* hand in it,' responded Simona. Joanna showed us a photograph of the metal jaws and foot-plate stamped with a picture of a wolf's head above the words 'Alpine Texas Livestock Protection Snares'. Simona lost the case but did not back down in her fight against the researchers' method of trapping the wolves and the heavy battery collars they used to track them. Seven collared wolves and three lynx had died during this monitoring study, largely, Simona was convinced, because the 'research apparatus' and brutal procedures caused risk to the animals. It was painful for her to see the protected reserve turned into an experimental laboratory by funded graduates who scared off the wildlife and damaged the undergrowth and trees with paint and plastic flags. But here she was, a scientist herself, regularly in conflict with the establishment, living with animals and a man to whom she was not married, seen walking with a boar and a raven, a witch in the woods, all things deeply scandalous in such a remote Polish village at the time.

'Then a film crew arrived,' Joanna said, 'Hugo Smith from Yorkshire Television to make a documentary about the wolf-monitoring project using British volunteers as a tourist idea where they pay to come and help.' She laughed ironically. 'I don't think the film turned out the way they intended when they saw what was going on!'

When we got home I tapped Hugo Smith's name into Google. Up he came, now a Senior Lecturer in Broadcast Media. I sent an email and within two hours I received an intriguing reply: '. . . the film was both significant for me and a really weird trip for everyone involved.'

The camera tracks through a misty midnight-blue forest. Tree trunks in close-up flash by. The music part human, part animal, part mineral, cave-deep, echoing. Dark shadows, flashing lights. A banging, beating, clashing. A sound-stream confusing the cochlea, the tolling of church bells, ominous yet alluring, distant cries, primal, otherworldly. Men running through the forest with sticks.

Voice-over: 'Europe's last primeval forest has become a hunting ground for scientists.'

Men in a tight circle. They are carrying something by the hind legs. Wolf.

'The supreme predator is now the quarry.'

Trees. Men. Echoing bells, symbols, howls, rattling.

'The wolf, once the most widespread mammal in the northern hemi-sphere now lives on in just a few areas of wilderness. The Białowieża Forest in Poland is the last great habitat of the wolf in Europe.'

A superimposed photograph clicks into frame, we zoom in. A wolf in a net. Her mouth forced open by the tangle of knots tightening around her nose, a sabre canine, bloody gums. Back to the action; the men hold the wolf down with pitchforks. A single wild wolf-eye shines laser-white into the lens. Up comes the title: *DANCING WITH WOLVES*.

It is 1998. Hugo Smith and his film crew are following a group of British volunteers who had each paid over £700 to help with the wolf-monitoring project run by the tall young leader of the Predator Ecology Group at the Mammal Research Institute in Białowieża; let's call him Henryk. In the film, Henryk is explaining to the group that they will be tracking a wolf called Bora, the dominant female of a pack, because her tracking collar needs to be replaced before the batteries run out. If it fails before she is caught, 'her pack will be lost to science'. This was going to be a feel-good film about eco-tourism, conservation and helping wolves. But there is conflict. The problem is the well-being of the wolves. No longer using the jaw traps discovered by Simona, they are now using the 500-year-old practice of driving wolves into nets. Once a VHF signal from a collar is picked up with a portable antenna, the area is cordoned off with fladry lines (long ropes strung with red flags, like bunting) which the wolves won't cross, to help funnel them into the catching nets. But now the British volunteers are here (and have paid their money), Henryk doesn't want to let them or the film crew near the capture.

The voice-over tells us wolves have twice the bite pressure of domestic dogs, a 20 times more powerful sense of smell, and can run 70 kilometres an hour.

Wolves were given protection in Poland in 1998, the year this film was made, but continued to be poached; the research team had lost four wolves that winter. The camera goes into a police storeroom full of wire snares, gin traps and iron leg jaws. We see a photograph of a wolf strangled by a wire snare, head lolling. Then a close-up of a heavy leather tracking collar with a steel core, transmitter case and battery pack. A dead lynx is carried into frame, her emaciated body and scraggy neck dwarfed by the giant collar, the large cylindrical cell batteries exposed, rusty and bubbling, in the deteriorated case. Professor Simona Kossak bends over the dead lynx, photographing it with her camera. Her suspicion that the lynx starved to death is confirmed in the post-mortem. Kossak opposes the project's methods; she suspects the heavy collars impinge on the animals' natural behaviour and ability to hunt effectively. Expert witness Dr Claudio Sillero, Oxford professor of Conservation Biology, rules out a negative effect on the animal, but says the collars should not stay on after the study. He says they can be simply adapted to drop off after a period of time. However, hunters *have* noticed changed animal behaviour in collared wolves. They say the wolves appear more nervous, as if someone is always following them. A faint radio blip from Bora's collar is picked up by the British volunteers combing the forest with the antenna. They are closing in on the pack. The tension between the film crew and the lead scientist increases. Henryk refuses all access.

Henryk challenges Smith. 'I don't trust you; you promised many things and you didn't observe.'

'Please give me an example of one thing I promised and I haven't observed.'

'I will tell you later.'

'You'll tell me later? If you want me to move you should tell me now.'

'Yes, please move now.'

'But you have no reason for me to move now . . .'

Henryk is tall and thin, his pale face impassive. 'The problem is that you disturbed me. Thank you.' He walks off, turns to look over his shoulder. 'I will not start until you go away.'

Smith is tipped off to go and speak to 'Mrs' Kossak. So he does.

Simona has possession of 'unauthorised' photographs taken during a previous wolf capture. She deals them out like a hand of cards. She points to the bloody gums of the wolf trussed in the cat's cradle; to the five men with pitchforks pinning her to the ground.

'It's hard to conceive how much physical strength a wild animal has when trying to save itself from death . . .' she says.

We zoom in again, the twisted neck, the mouth wedged open, the bright light of a terrified eye. Kossak says the anaesthetic is injected by hand. The amount of strength needed to subdue a terrified wolf in order to give it an injection is immense. 'How this is considered to be good for its health is hard to imagine.' We are left to wonder.

Dr Sillero confirms that the photographs show the wolf under a great deal of stress. He suggests a blowpipe would be better to inject the animal without the need of pronged forks. Smith asks if that's affordable for the poorly resourced Polish scientists.

'We can make one for a few quid,' Sillero says.

Next up, an ambulance has rushed the film crew and British volunteers to hospital. Everyone is violently sick; one person is on a drip. The sound recordist has to be flown home. A mysterious case of food poisoning. Smith nearly aborts the film but recovers enough to get a camera back into the forest. We see the fladry lines swagged between trees, we hear the loud beating of metal pans, a glimpse of wolf flashing past, men shouting, sticks waving. We don't see the moment of capture. But now we zero in.

A hand looms at the camera. 'Impossible to take pictures. Stop! You make a problem for them.'

The wolf, Bora, is upside down in the net surrounded by men.

'Do you know that you are not allowed?'

We see Bora being carried into the wood. She is laid down asleep, her new collar bigger than a man's fist.

Twenty years on, Smith remains *very* suspicious. After the poisoning one of his crew was ill for months. Nothing looked right to Smith and his team: the use of volunteers to fund the project as eco-tourism was a scam, and the techniques were so brutal and traumatising on the wolves he felt it must impact their behaviour. He was a young filmmaker then; he says he wouldn't be so easy on the project leader now. He thinks what he witnessed was bad science. He'd come across Simona by asking around if there were any challengers to the project. Henryk told him not to take the crazy witch seriously, but Kossak's position seemed grounded in better science, common sense and rationality. Smith concludes that he was dealing with a kind of post-Communist patriarchal mentality, in the form of a scientist who did not want any outsider eyes on him. The experience remains vivid and ominous in his memory. He still sounds rattled by it. The deep dark brooding forest as a backdrop, the banging saucepans, the white of a wolf's terrified eye.

These days, in Yellowstone, they use GPS satellite collars. If the scientists want to, they can send a signal and the collar drops off. With many years of research we know a lot about wolf behaviour and pack dynamics. Radio collars teach us about wolves by telling us where they are, but they can tell other people where they are too. It didn't take long for local hunters to track signals from radio collars, waiting for a wolf to leave the protection of the national park. In 2018, 'Spitfire', a legendary female and leader of the Lamar Valley pack, was killed a few miles outside the park entrance. Beloved 'O–Six', Yellowstone's most watched wolf, a once-in-a-generation hunter, granddaughter of legendary 'Twenty-one' from the first Yellowstone litter, was also killed by a trophy hunter. Her collar showed she spent 95% of her time inside Yellowstone park boundaries. The destabilising effect of her death as a high-status adult with experience rippled through the pack for a long time. The roll call is long and heartbreaking for the wolf watchers and conservationists who know them individually. Yellowstone wolves, accustomed to people, are less cautious and easy targets. Hunters take the body for the pelt; when the pack return in search of their family member they are likely to lose another. Good American taxpayers' money hanging on Billy Bronco's wall.

The paradox is that to track an animal's natural behaviour we fingerprint their wildness. If you see a wild wolf with a collar on, it isn't the same, it is against the nature of the wolf, his stealth-secrecy transcribed to a *bleep-bleep*. Necklaced, tampered with, he and we must give up the old-wild for a world of surveillance with the reminder of why – good reasons and less good reasons. In Białowieża, just seeing wolf tracks and scat was enough. To feel them, wild, keeping away from us. Fresh bear dung certainly enlivens the forest air. The Scottish naturalist Jim Crumley despairs of the many gizmos and aerials attached to our wild creatures: collars, bracelets, anklets, implants, glue-on tags, harness-mounted units. The intention is to study with minimum disturbance, but capturing and tagging causes stress, and can trigger psychological effects, changes in cortisol levels, altered behaviour and sometimes death. 'Capture myopathy' is just seizing up and dying from being caught. How much data is enough data? To know what we need to be doing. How much of this data affects outcomes? And how much is just tracking decline? How much is locked in a treadmill of funding research to tell us each year that the numbers are going down, and down – and that we need *more* data? As Crumley points out, 'We can take a string of far-reaching initiatives to the landscape of Scotland, and almost all of them involve nothing more difficult than that we should stop doing certain things, give nature its head and let wildlife manage wildlife.'[193] Help can become hindrance. Human presence disturbs wild animals. Animals avoid paths of recreational use. Human-dominated landscapes affect foraging and reduce reproductive success *more* than the natural predators do. Roads dissect habitats into little pieces – the chopped-up carpet – that separate wild animal populations and inhibit gene flow. In England, national parks are leisure areas for people, and for that they are creature-poor. We want to be in nature, which is a wonderful thing until the golden goose is gone.

Bringing back the wolf to Yellowstone was good business. In normal times, the local economy benefits by $35 million a year. At Creek Cut Wildlife Trail you can pay to hear them howl, or go howling yourself on the sunset howling safari: 'Dig deep and howl!' the ranger shouts. The ranger howls, the wolves howl, everyone howls. Tempting. At the International Wolf Center there are field trips, seminars, group visits: Wolves After Dark, $70; Wolves at Wolf Family Rendezvous, $75 adult, $50 child; Tracking the Pack, $135.* After centuries of bad press, Americans have begun to love wolves. For $75 a teacher can rent a Wolf Discovery Kit. (Oh, to have such a teacher.) You get: *Wolf Pelt, Wolf Skull, Wolf Scat, Herbivore Scat, Wolf Claw, Beaver Skull, Deer Leg, Deer Jaw Bone, Moose Jaw Bone, Deer Antler, Red and Grey Wolf Tracks, Radio Collar, Books, Coyote Skull, Field Guides, Magnifying Glasses, Wolf Paw Rubber Stamp, Mapping Activity, Wolf Howl CD, Video, DVD and the Teacher's Guide.* I love the kits; I don't like the live wolf exhibit. The 'ambassador wolves' who 'represent their wild counterparts'. I watch this week's video: 'Aidan, a pack member in retirement . . .' Forced retirement. He licks the camera, he walks up and down, and down and up the chainmail fence of his

---

* The 'Track Wild Wolves' webpage, where you could track wolves online, has been removed 'to minimise chances of their radio-collared wolves being killed'.

enclosure. It doesn't take a wolf scientist to see he is bored (and too fat). Aidan has a tumour; Luna has a weeping spot on her neck. The site tells you wolves can travel up to 500 miles. They *know* that. So why are we watching them like this?

In October 2020, the wolf lost all protected status in the 48 contiguous states. A fortnight later Colorado voted to reintroduce wolves into the Southern Rocky Mountains. The fate of the wolf plays like a yoyo, in and out of court, de-listings, re-listings, objections, hearings, lawsuits, court orders, injunctions. In Wyoming a protected animal becomes shoot-on-sight vermin overnight. The territory of the wolf is the territory of the cowboy – Montana, Idaho, Wyoming, Oregon, Utah – where hatred is hardwired. Wolf season in Wyoming is all year round. Guns, snares, poison, dynamite, yahooing shooters from snowmobiles, quad bikes, hunted by helicopter, plane . . . pregnant females, pups in their den. In 2016 Wyoming sheep casualties from weather and disease accounted for 37,550 animals; wolf kills numbered just over 200; domestic dogs kill twice as many sheep as wolves. As far as taking game, a wolf will always take the sick and weak first. But more than that. By keeping the deer and elk on the move, the wolf plants the aspen and willow that supports *more* deer. And a host of other animals while mid-level predators are kept in check. And so it goes. Fear is the gift of the wolf that the hunter cannot replicate. That's bias with science.

Our last night in Białowieża, we are sitting around a table in an old village schoolhouse. The building belongs to the writer, naturalist and Polish broadcaster Adam Wajrak and biologist Nuria Selva Fernandez (an expert interviewed in *Dancing with Wolves*), and has been lent to the protest group 'Camp for the Forest', campaigning against the increased logging in Białowieża. Adam has been described as Poland's David Attenborough. On the wall there is a Banksy poster of a girl looking at a chaffinch, captioned: If you get tired, learn to rest, not to quit. Although the Court of Justice of the European Union found in the protestors' favour, they remain to guard the forest, to educate and inform. They want all the Polish forest to be given national park status, to protect the mammals, trees, fungi, pygmy owl and three-toed woodpecker. Jonathan asks Adam about the people living with wolves; he says how relaxed everyone seems in comparison to the uproar at any suggestion of introducing wolves in Scotland. Adam is astonished there are no wolves in Scotland. I can feel him disbelieving us, fingers itching to check it out on Google.

'Our president was a hunter,' Adam says, 'but to be elected he had to stop hunting. In Poland, in public, it is a shame to be a hunter.'

Adam believes this is because, historically, it was the noble elite who went hunting, followed by the communist elite. Both disliked by the people. He finds it strange that the British royals hunt and are not ashamed. He thinks Poland was influenced by Jewish culture – you cannot bleed the animal properly if you hunt.

'Wolves have been protected since 1996. To kill wolves one must feed the story they are dangerous, learn to hate them, then society will accept. Killing the wolf feeds the fear.'

I ask him how he sees the future after all his years broadcasting and campaigning to protect the natural world.

'I think we have a chance to stop a lot of bad things. The catastrophic logging in Białowieża Forest in 2017 brought a lot of attention. Suddenly people coming from everywhere . . . I was sceptical, but in eight months they stopped. Why not be optimistic? We Polish people have good and bad sides, we are ready to fight for bear, wolf, Białowieża Forest, and the Biebrza Marshes, but we don't understand climate change. We are very bad on this. These are our challenges. Educate people better.' Adam looks down at his hands for a moment, 'I am more and more certain what we can do for animals is to give them space, not to kill them, this is the best we can do.'

ATISHOO, ATISHOO

Then suddenly, there we were, 2020 lockdown. The whole world in the grip of the pandemic. Another disease that leapt the species barrier, and whoosh, like dye in water, it flashed around the globe. A disease deciphered into mathematical models and a new lexicon: R values, Track & Trace, viral load, herd immunity, social distancing, PPE, circuit breaker . . . lockdown! Political urgency moves mountains, not scientific pressure. The whole place screeches to a halt. The roads are quiet. The skies are quiet. We buy vegetable seeds online; chicken coops sell out; loo rolls are fought over. Tellies flicker streaming *Normal People* and a resurrection of the 2011 film *Contagion*. Apps, face masks, lateral flow. Number crunching. Deaths by the day, by the week. Epidemiologists are our new celebratory chefs. In Britain the Tory dream of munching down the State evaporates in a virus that feels like your throat is made of crushed glass.

All because somewhere in Yunnan Province a hunter had likely walked home with a sack of horseshoe bats he'd netted in a cave in the mountains. Maybe he sold them at a wet market (what a name) to a stall selling racoons, crocodiles, frogs, monkeys, barn owls, turtles, snakes, mouse-deer, foxes, ferrets, piglets, badgers, peacocks, wolf pups, bamboo rats, porcupines, pangolins, ducks, dogs, cats or more bats. Each creature with their own ecosystems of bacteria and fungi and viruses, coexisting for centuries without causing much ado, until a chance mutation gave a particular virus the ability – when the opportunity arose – to jump ship to another host. As the opportunists viruses are, it did. Off to Wuhan to run amok. From lonely cave to high-speed train to a passenger on a 747 plane. South Korea, Iran, Italy. What a whirlwind tour this virus had tapped into.

As mammals with an ancient lineage co-evolving with an array of viruses for 50 million years, bats are fascinating vectors. They roost in large numbers, can fly long distances and have a unique immune system. The bat's metabolic rate increases dramatically during flight, which creates ions that could cause cell damage, so to counter this they have a diverse antibody repertoire that makes them resilient to viruses. Humans shouldn't mix with bats. It is not the horseshoe bats, or chickens, or swine who need to change their behaviour. It's us. All we have to do is to *stop* doing it. Leave the bats and monkeys and pangolins alone. Domesticated animals are also natural reservoirs of pathogens, and overcrowding provides the perfect conditions for viruses to evolve, mutate and transfer to us. Disease provides fast-track masterclasses in evolution. What happens in one person's backyard these days very quickly happens in everyone's backyard. We can't say we weren't warned. In 1997 a three-year-old boy died in Hong Kong of viral pneumonia. Except it wasn't. It was bird flu. H5N1. After six people died, 300 million farmed geese and 1.5 million chickens were slaughtered. In 1999, when the Malaysian villagers of Sungai Nipah real-ised they were catching a fatal encephalitis from their pigs they fled, leaving the animals for the army to cull. The Nipah virus had hopped from bats to pigs to humans. SARS virus also originated in bats but jumped to the Himalayan palm civet sold in markets for their musk and meat. Detected in 2002, SARS virus gave humans a false sense of security because it fizzled out by July 2003. Then came H7N7, then H1N1, then swine flu, then MERS (Middle Eastern respiratory syndrome), blamed on the rapid increase in the camel population in Saudi Arabia.

Covid-19 caught us unaware. It shouldn't have. We know new diseases arise from how we live with animals and are quickly transmitted by our huge mobile population. How we pack domesticated ones together in overcrowded confinements; how we push wild ones beyond the brink of their habitats; how we haul them into our world. Industrial animal farms are a jungle of petri-dishes waiting to bloom, which is why we cram them with antimicrobials. Thus factory farms drive drug resistance and erode medicine's efficacy for humans. And whose idea was it to feed vegetarian ruminants with meat and bone meal? Bingo for bovine spon-giform encephalopathy (BSE), or mad cow disease, transmitted by consuming infected tissue with the ability to manifest *very* nastily many years later in a variant of Creutzfeldt-Jakob disease (vCJD). Airborne, fluid borne or flesh borne, factory farms and wet markets are the places where pathogens like to socialise.

AIDS, Zika, Q fever, Ebola, dengue, are all zoonotic diseases which hopped the species barrier. The clever rabies virus makes dogs want to bite you. What's new is that the spillover of pathogens from animals to humans has tripled in the last decade. Nonetheless, trawl back through history. From domestic animals: diphtheria, influenza A, measles, mumps, whooping cough, rotavirus, smallpox, tuberculosis. From apes; hepatitis B and HIV/AIDS. From flea-bearing rodents: plague, typhus. Influenza took 20–50 million human lives at the end of the First World War; the Black Death saw off a quarter of the European population. Human pathogens as invisible agents from the Old World abetted the European conquest of the New World. Native Americans, Aztecs, Mayans, Pacific Islanders, up to 90% of indigenous populations were cut down without a fight. The smallpox virus had a 3,000-year run.★You can see signs of the pustules on the mummified head of Pharaoh Ramses V; original host – probably an African rodent. The last case occurred in the UK in 1978 after a medical photographer died from a sample grown for research. The World Health Organization (WHO) called for all smallpox virus stocks to be destroyed, but Russia and the US have retained their supplies – for research purposes. Smallpox scabs tip up from time to time in vials in cold storage rooms. In 2003 an envelope of the little flakes was found in an 1888 book on Civil War medicine in a Santa Fe library. In 1985, the *Lancet* published a paper proposing that smallpox could remain viable for a century in the graves of those interred in cool, dry climates. Malaria, delivered by the female *Anopheles* mosquito, the most prodigious killer of all time, originated from a bird parasite either a few thousand years ago or a few million years ago, before we split from our common ancestor with chimpanzees; scientists can't decide. The bubonic plague has been with us for millennia, identified in human DNA from 5,000 years ago. People could not imagine the swift invisible enemy in their multitudes: Rothschild's fleas carrying the bacterium *Yersinia pestis*, riding *Rattus rattus* on the overland route from China or stowing away in the bellies of ships. In the seventeenth century, plague doctors wore cloaks and occultish masks with long bird-beaks stuffed with herbs, cinnamon, myrrh and honey, to purify the sick air. In July 2020 a teenager in Mongolia died from bubonic plague after eating infected marmot meat. In the States there are around seven cases of bubonic plague each year. Lyme disease, caused by a bacterium carried by ticks, was spread by deer as American forests were cleared, exacerbated by the removal of host and tick predators. So no more lying in long grass chewing a stalk of Timothy grass without your socks tucked into your trousers.

★ Eradicated in 1980 by the Global Smallpox Eradication Programme.

'Human health and animal health are interdependent and bound to the health of the ecosystems in which they exist.'[194] *Homo sapiens'* sharp elbows have pushed other species into closer proximity by taking their homes and food sources. Optimal rates for microbe spillover occur once 40% of the forest cover disappears.[195] Less diversity means less resilience and more opportunity for disease. Greater biodiversity hinders transmission by a dilution effect, making it harder for a single pathogen to spread or dominate. Everything comes back to our pressure-cooker relationship with animals. All our woes can be traced back to this one thing. Astonished radio presenter Sarah Montague, working from home during the first lockdown, tells us a deer is watching her through the window. Reports of nature's comeback flooded our computer airwaves from around the world. Mountain goats paraded the streets of Llandudno in Wales; jackals in Tel Aviv; racoons in Central Park; buffalo on an empty highway of New Delhi; wild boar in Haifa; a cougar in Santiago. Animals came back out of the woods. It must have seemed curious for them, those weeks of no cars and silence. The quiet aided their nesting period. Pink flamingo numbers were up by a third in Albanian lagoons. How long will we remember those long months of birdsong, the peaceful skies, the kindnesses of strangers?

ATTACK DOG

The most dangerous worldview is the worldview of those who
have not viewed the world.

Alexander von Humboldt

environmentalist, *synonyms*, CONSERVATIONIST,
preservationist, ecologist, green, nature-lover, eco-activist;
informal, derogatory eco-nut, ecofreak, tree hugger.

In the olden days you could smoke on the top deck of a London double-
decker. I loved leaping on the open platform at the back as it whizzed
by; you ran like stink, grabbed the pole and hauled yourself on board as
the bus swung round a corner. Then you climbed upstairs for a cigarette.
In my twenties I loved smoking, but you could hardly breathe on the top
deck so I would open a window (you were allowed to do that in those
halcyon days of greater ventilation), and invariably someone would ask
me to shut it. I could never work out the logic of the unspoken rule:
the window-shutters always won. You could die of smoke inhalation but
not a draught. My point is, why do some things in our culture always
override others? Like tidiness. Or the economy. Or jobs. Or shutting the
window. Anything short-term overrides anything long-term. It only takes
one strimmer, one chainsaw, one bulldozer, one vested interest, one president, to destroy an ecosystem, yet truckloads of science, whole movements
of protest and more than 100,000 signatures to save one. If you're lucky.
Sheffield Council was *allowed* to cut down 5,500 healthy 100-year-old trees.
Wirral Council was *allowed* to spray the beach with herbicide. It was as if
they had never heard of climate change or biodiversity loss or pollution.
Hedges are flailed to diseased stumps beneath flocks of fieldfares flying
overhead looking for berries. As if we *hate* nature? It hurts many of us —
it really hurts, mentally, physically, in the guts, in the brain, in our hearts.
Why do we lose to the spray-strimmer-tidy crew? The English cottage
garden, once a place for bees, flowers and birdsong, is now a place for

decks, Astroturf and BBQs. Outside space becomes an extension of inside space where little can live except people. But that's allowed.

Jerks

Ecological sense runs up against a profit-junkie world and loses every time. Save the planet or the economy, it'll be the economy. *Homo economicus consumicus insatiaticus.* Our world is built on short-term interests. Sufficient is a dirty word. The ecological downside of profit (externalities★) is *inconvenient* to think about, so it is easier (and effective) to ridicule those who try to make you. Apocalypse-mongers, eco-miserablists, doomsters, disingenuous mopers, 'the environmental Taliban',[196] dippy, hippy, eco-la-las, grow-your-own-socks Utopian sentimentalists, eco-fascists, eco-Nazis, eco-terrorists, *criminal* terrorists! There are bunny huggers and predator huggers.★★ 'Watermelons' are green on the outside and red (communists) on the inside. Slap a hate label on a group, then charge your adversary with being privileged, naive, urban, elite or misanthropic. Buzzwords make it alarmingly easy. Just accuse someone of 'virtue signalling' or 'rewilding' to get them on the back foot or silence them. Somehow the term 'rewilding' has become toxic in the very places where revived ecosystems, restored nature and returned wildlife might have brought uplifts in jobs and local revenue. *Who are these people coming in and telling us what to do? They want to turn the place into a wilderness and everyone will starve.* The accusation against rewilding for taking land out of food production is something never levelled at golf courses, racecourses, pheasant shoots, grouse moors or motor circuits, and ignores the point that rewilding ventures *don't* take productive agricultural land.★★★ Of course we must eat. But why pursue destructive, toxic, cruel, dangerous,

---

★ Externalities meaning the cost of the consequences of the commercial activity, such as pollution, which the market price neither reflects nor pays for.

★★ Will Travers remembers the late palaeontologist and conservationist Dr Richard Leakey in 1997 addressing a conference in Harare, Zimbabwe, of delegates of CITES (Convention on International Trade in Endangered Species of Wild Fauna and Flora) with ranging views on the ivory trade. Leakey said, "'There will be some here tonight who think that because of the importance I place in the individual animal, the respect and indeed affection I have for an individual elephant who can guide her family to safety through thick and thin . . . they will think that I am some sort of bunny hugger . . .' He paused, reached inside his jacket, pulled out a toy rabbit and declared, "Well I am!"' (Will Travers, 'Dr Richard Leakey – A Personal Appreciation'. Born Free Foundation, 5 January 2022.)

★★★ Nature restoration schemes (rewilding) currently cover: 1% of the UK, 25,000 hectares. Pheasant and partridge shoots take 12%, 3 million hectares; grouse take 8%; golf courses 2%.

expensive, short-term, soil-exhausting, microbe-killing methods? Rewilding is not about abandoning land or creating playgrounds for the rich. It is an ecological reboot that repairs our life support systems, for free. With minimal intervention animals do it for us. Browsing, grazing, defecating, disturbing, opening glades, transporting seeds, left to their own devices and natural processes animals create dynamic habitats that teem with life. This can solve some of our biggest challenges: insect collapse, soil health, carbon seques-tration, flood mitigation. It's the most exciting environmental fix out there, with legs on! The signalling is not the critics' problem, it is the virtue. So little of nature is protected that ordinary people feel compelled to protect it themselves. The pity is the need to justify conservation on human-centred utilitarian grounds, not for the intrinsic value of the creatures with whom we share, or should share, the planet. Are we stark staring mad? Someone somewhere is missing the point. Maybe it's me.

Most of my best friends are humans

Misanthrope is a powerfully weaponised word. The *Chambers Dictionary* definition of misanthrope is: a hater of mankind. In that case, surely, few people would qualify. Misanthrope is a word bandied about too loosely. I would counter that for most so-called 'misanthropes' it is not a hatred of humans, so much as an antipathy towards the side of humans that tends towards senseless cruelty, destruction of habitats and waste of life. The charge of misanthrope changes the target. You hate humans therefore you are anti-human and care more about animals than starving babies. That is where that conversation goes. 'Misanthrope' is a gagging device. The question of whether we are compassionate and cooperative or fundamentally violent and selfish has only one answer: we are both. This spanner in the works is what the late, great E.O. Wilson called the 'Paleolithic Curse'. The dysfunction, as Wilson saw it, is innate. Competitive behaviour that advantaged us in hunter-gatherer days does not favour a global society where cooperation is more important for survival and our ability to thrive.[197] We cooperate and we compete. Because we are naturally village people we find it hard to care too much about another village, or another tribe. Harder still to care about another species. Unless they are dogs, cats or honorary tribe members. We are like any animal, stuck in our ways. Tribalism makes us easy prey to snake-oil salesmen and those religious and political leaders who even now succeed in ignoring the hard science of our potential demise. Greta

Thunberg, the young Swedish environmental activist who has access to many world leaders, speaks truth to power about their 'very, very low, much lower than you would think'[198] level of understanding and knowledge about climate change. When critics ask what her school strikes achieved, Greta is nonplussed, as if *they* expect *school children* to fix it. Human nature is just that, the genetic inheritance of our past. Its expression can be detrimental to the survival of the group and our home. The result is the internal conflict of, may we say, the human soul. We are the Janus creature, both base and principled, as aggressive as we are loving, and that tension is the wellspring of our poetic impulse, our comedy and tragedy, our Shakespeares, Tagores, Angelous, O'Keeffes, Sapphos, Picassos and more. But somehow, we must safeguard a world that our poets can sing about.

In 1968, the cover of Stewart Brand's counterculture classic, the *Whole Earth Catalog*, showed our blue planet from space. Brand understood the power of this image to stir the human heart: our home floating in the deep dark universe. The *Catalog*, a kind of prehistoric Google, appearing long before the World Wide Web, spawned an information revolution. The idea was to reconcile a vernacular technology with environmentalism, connect culture to nature with green tools. And make it cool. A compendium-cum-manual of self-sufficiency, ecology and DIY, from electronics to beekeeping with a few mind-bending drugs in between. The first sentence famously read: 'We are as gods and might as well get good at it.' Forty-two years later, in Brand's ecopragmatist manifesto of 2010, *Whole Earth Discipline*, he changed the line to: 'We are as gods and have to get good at it.' One can hardly argue. Except when it comes to extinctions, for which he espouses a depressing remedy. De-extinction is the cloning and bioengineering (at huge expense) of existing DNA to recreate animals of the past. From the passenger pigeon to the woolly mammoth. One has to ask: is this for us, or for them? Aside from the inherent health problems of being a clone, the same pressures exist, such as habitat loss, that caused most of these extinctions in the first place.★

★ By contrast, back-breeding selects genetic characteristics of a modern animal's distant ancestors to create a wilder, more robust animal for whom there are habitats available to exploit. Like the hardy Heck cattle back-bred in the 1930s, the Tauros Programme aims to bring back the mighty aurochs (6 feet at the shoulder), hunted to extinction in 1627, by back-breeding from their closest cattle relations to produce a hardy, self-sufficient wild bovine grazer as proxy to replace the ecological role of the missing mega-herbivores, as close to an aurochs as technically possible.

For E.O. Wilson, writing in the 1980s, the worst thing that could happen was not economic collapse, limited nuclear war or conquest by a totalitarian government:

> As terrible as these catastrophes would be for us, they can be repaired within a few generations. The one process ongoing in the 1980s that will take millions of years to correct is the loss of genetic and species diversity by the destruction of natural habitats. This is the folly that our descendants are least likely to forgive.[199]

Everything funnels into that single point. This book. Our skewed relationship with animals. What we know contradicts what we do. Look, if a tapeworm comes into my life, however interesting, sentient or rare, the tapeworm is gonna get it. There's a hierarchy, but it's what, why, how, and *all* the consequences.

The hopeful *GAIA Atlas of Planet Management* of 1985 shows a world heading for trouble, but one we would happily swap for today. In the global forest picture map, Borneo is coloured all-over dark rainforest-green. There was no shortage of plans and *voluntary* targets, and no force of law. Ever heard of the IPBES? It's the equivalent of the IPCC, but for biodiversity: the Intergovernmental Science-Policy Platform on Biodiversity and Ecosystem Services.[200] (The name alone will kill it off.) We are becoming immune to the word 'extinction'. We can't even protect an enormous rhino. Shark fin soup has become more valuable than sharks. Face cream more important than orangutans. Tipping points tip. Parks are magnets for poachers. Tracking collars are hacked by hunters. Super trawlers scoop out the oceans and put nothing back in. Am I being emotional? Yes. It's a *rational* response. Reason and science have not convinced the world (whoever that is) to stop damaging the biosphere.

★

White-out. A harsh blast of winter we've called the Beast from the East has covered our valley with 3 inches of snow. The ground beneath is frozen. The miracle of wild lives never ceases to amaze me. The bird table is a frenzy and I can hardly keep it stocked up. The fringes of the winter-bourne stream that threads its way through the meadow is thronged with rooks, fieldfares and – rare visitors these days – a congregation of lapwings.

The only foraging ground for miles around is where water percolates to thaw the frozen grass. At dusk, the lapwings hunker down under a slate-heavy sky, hunched in frost around the dull mirrors of uncracking ice. They cry all night like babies or hungry ghosts.

In the morning I rush around like a mad person with trays of meal-worms and seed by the sack. In the afternoon I walk to the top of the hill for the snowy view. I pass a house and notice the once laden bird table is empty. It perplexes me. I bump into the new owner, a military man, on the summit. We exchange a few words.

'I worry for the birds with nothing to eat,' I hear myself say.

'Oh, they are very robust,' the army man mansplains.

'Are they?' I ask, but not really asking. It's touch and go if I can behave myself.

'Oh, yes, they can survive at least ten days without feeding,' he assures me.

'I don't think that's correct,' I say.

'Oh, yes, believe me,' he says.

There is something so perfunctory about the conversation. I can feel my hackles rise. It seems to me the prerequisites required of his cadets are required of wild birds. They must tough it out, man-up or they have failed. I stomp back home. It depresses me that a man can know so much about ordnance and ammo and war tactics, and so little about the world he is commissioned to defend. But why *should* he know, or anyone know anything if it is not considered worth teaching at school, worth defending in politics, worth anything . . .? I fume as I walk. At the same time I feel an ethical dilemma. What is my duty as a fellow human and fellow creature? Warm human in front of telly scoffing cake juxtaposed with cold bird, crop empty, scratching the scant frosty patches of earth. Stomp, stomp. It swirls inside me. In 20 years our village boasts less garden, more parking, fewer hedges, more tidy, fewer bird feeders, more cats, and now me, the mad woman at the end of the lane. Small birds, such as tits, need 30% of their body weight to get them through a long cold night. A healthy bird should have reserves to last 48 hours without food. I should acknowledge the poor man responded to a printout of wild bird survival information with grace, though his snow-capped bird table remained bare.

★

Emptiness is painful. Attachment to place is something we all feel. It is not about owning the land. It is about loving it, paying close attention, knowing who lives there. Local landscapes mean most to us, wildlife on our doorsteps imbued with personal affection, intertwining our own life with the natural world, which is where it belongs. A special corner in a field, the particular bend in the river, where a white heart-faced owl alighted on your head, where the spirit salves; until something happens which makes you stop going. A friend told me of walking his favourite path when a Land Rover sped by chasing a hare across the field. The vehicle stopped, a man jumped out with a gun and took a pot shot. He missed. But something cut deep: as the idea of the hare dying in the field flashed through my friend's mind, the life-giving properties of the path were poisoned for him. Time for courage to turn the boat around and speak to our better nature.

LOVE, ACTUALLY

Len (Gwendolen) Howard, a concert musician who played viola, left London at the beginning of the Second World War and moved to Sussex. She gave over both the inside and outside of her cottage to the wild birds of her garden. If they had no fear of her, she believed the birds would give her a true insight into their natural behaviour. Instead of peering into a box, or out of a box, she decided to cohabit. As they gained confidence she noticed how each individual was different in character, so she decided to study and record their life stories. Summer, winter, in and out the birds flew, free to come and go as they pleased.

Miss Howard stands at her open window, hand stretched out with a great tit perched on her fingers. She is wearing an odd cloth pixie cap tied under her chin. She looks like a Plantagenet nurse, or a clothed mouse in one of those Victorian taxidermy dioramas. From 1940 until her death in 1973, Len collected the biographies of the birds who lived at Bird Cottage with her: blue tits, coal tits, robins, blackbirds, thrushes and her favourite, the great tit, who she thought the most intelligent of all. She shared her wartime rations, chased away the neighbours' cats, made roosting boxes out of cereal packets and left the fanlight open, wind or rain. The surrounding hedges and trees

were allowed to grow thick and tall, the gateway was overgrown, with
a sign saying:

NO VISITORS

NESTING BIRDS

MUST KEEP COTTAGE

QUIET

NO CALLERS

The interior was 'arranged to suit them', with her life 'more or less
regulated by theirs'. They flew around her, perched on her shoulder as
she typed, pulled her shoelaces, a wren asleep in a coconut shell, a tit in
a roost box; and of course they wrecked the place. They pulled the stuffing
out of the upholstery, tore lampshades to ribbons, shredded book covers,
made holes in her shoes. She covered her furniture with dust sheets and
spread newspapers around only to come home to torn paper strewn all
over the floor. The curious cap, I imagine, was to protect her hair.

Her work was described (by male ornithologists) as 'amateur' bird
studies, yet every hour, every day, from 1940 to 1973, Len Howard noted
genealogical trees, paid acute attention and recorded the behaviour of her
lodgers. Their births, deaths, fortunes and misfortunes read like a long-
running soap opera. Bigamy, desertion, solo parenting; three in a marriage;
lame fledglings; trapping spiders; fights; fall-outs; territorial battles; the
bold; the fearful; the excitable; the neighbour's cat wins a long-running
battle and cleans out a whole brood (the same culprit behind most trag-
edies); the local council removes a hedge (despite appeals for delay) in
the middle of nesting season. Star, a female great tit, born 1944, dies in
1953 (average lifespan for a great tit is 3 years), has four mates and a
multitude of offspring. What a legend Len Howard must have been:

On the evening of 17th June, while walking in a wood half a mile
from my cottage I saw flying towards me a great tit with her four
fledglings behind her. I held out my hand and after a moment of
keen scrutiny she came to me fearlessly although we had not met
for eighteen months.[201]

Her musician's ear was tuned to their slightest inflections and timbre. She
saw language in their movement, their wing twitches, tail-spreading,

chin-ups, in their fast or slow gapes, their beak-opening, shutting, tapping, faint sounds and hesitations. She noticed the subtle things we humans all notice with each other, which are hard to describe, such as the expression and light in the eyes. She wrote prodigiously about memory, recognition, fear, flight, intelligence and birdsong: 'a bird's heart goes into his music, so by full and intimate appreciation of his song we get nearer understanding a bird's nature'.[202] Len recorded their recreation in the form of little games and fabulous inquisitiveness: photo of great tit opening a matchbox; photo of great tit untying a parcel; photo of great tit inspecting a toy parakeet; on her pen as she draws; inspecting her shoe. When the great tits wanted to peck at her butter dish, normally forbidden, they looked at the butter, then up at her face, then at the butter. If she said *Come on* coaxingly they helped themselves, if she said *No* firmly, they kept away. An angrier *No!* could make them fly to the window. Well, I believe that.

X: GOLDEN JOINERY

The best time to plant a tree was twenty years ago.
The second best time is now.

Chinese proverb

WHO CARES

A thump on the glass door. My retina memory reel spools back. Small
brown wings flying towards me. The thump, but light, glancing, not a
death blow, surely? On the flagstones in front of the door outside, a tiny
cartwheel of feathers, its spindle off-centre, round and round, a life-grasping
pirouette. I watch as the revolutions slow, the fluttery shudders rippling
through the chaff body. My hope begins to fade. Quieter and quieter.
The little thing. Yet still alive, still quivering. And I can't work out who
it is, beak to the ground, chestnut wings outstretched, freckledy bark
moth-wing zigzags. Not a sparrow. Is it a dunnock? I don't think so. I
am confused. I will find out when I pick him up, but right now I am
kneeling on the other side of the glass door, watching, waiting, a sad
heavy feeling. I don't want a human hand to frighten his last moments.
My eyes must have shifted away for a millisecond, because in a flash the
bird is sitting upright. A tiny, tiny bird, no wonder, he is a treecreeper.
The finest curl of his bill, his white breast. He sits, not moving, but
breathing. Minute after minute. I cannot open the door. I worry he is
about to keel over. I think of him once in his tiny egg; what his beak
was like then is hard to imagine; born in the crack of a tree, his parents
coming and going with minute insects, flies, ants, beetles, spiders. Twenty
minutes we sit together; he is blinking now. His treecreeper world uncork-
screwing in his head. Treecreeper blood pumping to eye sockets. Minute
23: a *wwhwrrhh* of tiny brown wings and he is gone!

Relieved, I hang hazel branches on string in front of the reflections
of open sky on the glass door. Educated guesses estimate there are *100
million* bird strikes on windows reflecting the sky each year in the UK,
of which a third are fatal.[203] In the US the bandied-about figure is a

billion. That is barely believable.★ The mathematical assumption is that every house accounts for one or two a year. Reflecting glass is a big problem for birds. It is an awful thing, finding a warm body beneath a window, beak up, sticky legs stretched out. Later that day I find out about a kind of glass coating inspired by spiders' webs that helps prevent bird strike. If you look across a dewy meadow in the early morning you will see the flat webs of orb spiders with their concentric circles, bicycle spokes and cross threads perfectly intact; yet this is where birds poke about looking for insects. The reason birds don't collide with spider webs is because orb weavers decorate their webs with a UV-reflective thread called stabilimenta, which is invisible to us but can be seen by birds. The web needs to be invisible to the spider's prey, but visible or deterring to the spider's potential predators.[204]

A German glass company began to design glass on this principle and came up with an ultraviolet coating applied in a crisscross pattern like pick-up-sticks across the glass, which humans cannot see. They have used it to create a room on the top of the old Lookout Tower on the Holy Island of Lindisfarne, where tens of thousands of birds arrive to winter-over, and many more fly by in the great migrations every year.

Finding beauty

Imagine the electric pulses that leapt from George Bargibant's retinas to his brain in 1969 when he was examining some gorgonian coral under the microscope and saw the first tiny, *tiny* coral-matching pygmy seahorse to grace the human eye. Pygmy seahorses are almost indistinguishable from the coral where they live, which is why we didn't know they were there. When the fry hatch from the male's brood pouch, they are 2 milli-metres long. Bargibant's pygmy seahorses are either purple with pink warty bumps called tubercles, or yellow with orange tubercles, depending on the coral host. The adults grow to about 2 centimetres. By fertilising the eggs once they are in his pouch, Dad seahorse is one of the few creatures who can be absolutely sure of the paternity of his brood.

Living creatures keep the Earth habitable. 'Life' on Earth is the piston that pumps out the atmosphere we breathe, the nutrients we need, the

---

★   Three major causes of premature death for birds are cats, windows and cars. Cats kill between 1.4 and 3.7 billion birds in the US annually.

temperatures we depend on to survive. Oceans are maintained by life itself. The abundance of species gives Nature resilience, strengthens ecosystem stability and increases productivity. From bone-crunchers to rot-munchers and everything in between. Darwin's 'endless forms most beautiful' become even more beautiful by understanding the mutual interdependency and interconnections of all life. When we understand, we begin to care. A switch flicks our way of seeing. The entangled bank blossoms and froths over and finds beholders who will strim it no more. What a relief to see nature not as adversary, but as friend. So what could be more important in our education than how these life support systems function? How can we look after something if we have no understanding of how it works? At school I learnt the names and fates of Henry VIII's six wives – divorced, beheaded, died, divorced, beheaded, survived – but not the fate of the Steller sea cow. Which history lesson would serve us better now? Human knowledge is the astonishing sum of all that has gone before us, the accumulative brainpower of a multitude of geniuses. Information is crunched and stored and assimilated and can find more eyes and ears than any time in history – hey presto, pygmy seahorses at lightning speed up on our screens. Beliefs only change if we understand why they must. There's a log jam because some things are inconvenient (to the few) for the many to know. From knowledge we reap beauty, beauty inspires caring, by caring we can do stuff, and pride feels better than shame. More and more humans are thinking about our relationship with the natural world, and many – courageous mavericks, ordinary people – are devoting their lives to helping the lot of other inhabitants. Slow burns and epiphanies. People who have changed their lives by changing their minds.

For 40 years, Ratanlal Maloo, a devout Jain, fed sacks of grain to evergrowing numbers of demoiselle cranes alighting in his Rajasthan village of Khichan in northern India, on their migration south from Eurasia and Mongolia. Ratanlal just decided. He was entranced by these small cranes in their elegant grey plumes with long black neckties that ruffed over their chests. First there were 12 who joined the sparrows and peacocks as regular dinner guests, then there were 80. They arrived in September and disappeared in February. Each year the numbers increased. When the local dogs cottoned on, Ratanlal built a Feeding Home or *Chugga Ghar* surrounded by a 6-foot fence to protect them, and then he persuaded fellow Jains to build a granary to store the grain. After 40 years 15,000 cranes were visiting. The numbers created a spectacle which brought tourists to the village. Ratanlal's kind soul migrated in 2007, but his legacy of a village that feeds migrating cranes at Khichan lives on.

Jay Wilde was born and raised on the 173-acre cattle farm in Derbyshire which he inherited from his father. 'When you're young you copy what other people do,' Jay says, 'particularly when you have a strong father figure.' Nevertheless, Jay switched from dairy to organic beef farming because he found it too hard seeing his cows and calves become so distressed when he separated them. He says he began to realise the cows had feelings and personalities, and it became increasingly difficult to disconnect the feeling of having to get the job done from the fact that they were individuals. Taking them to the abattoir felt like a betrayal. It was soul-destroying. When Jay's father died, he was free to change but that too seemed threatening. His farm became nicknamed the Funny Farm. Then came the day in 2017 when he just couldn't send the cows he'd been caring for to slaughter. So he didn't. He turned his farm over to growing vegetables. Miraculously, he found a home for his herd at Hillside Animal Sanctuary in Norfolk, but kept 12 to graze and manure his ancient pasture. At Hillside all Jay's cattle stayed together, mothers, calves and family units, to live out their lives. Their reprieve cost him his year's earnings, but the burden of his anguish was lifted. Then Alex Lockwood, a young filmmaker, showed up and made a short no-budget documentary about Jay's story: *73 Cows*, which won a BAFTA. Kate Wilde, Jay's wife, says they are working together to build something really good, and in Lockwood's gentle film, you can see that they are. Jay hopes the film will inform people about a more efficient way of feeding the world. Lockwood also does what he can – his latest documentary, *The End of Medicine*, exposes the link between pandemics and factory livestock farming.

In January 2019, a 60-year-old Devon sheep farmer, Sivalingam Vasanthakumar, was driving to the slaughterhouse with £9,000-worth of lambs, when he turned round and drove 200 miles to the Goodheart Animal Sanctuary in Worcestershire. Same thing: Sivalingam had farmed all his life but couldn't bear to watch the animals he raised lined up for slaughter. He too is now a vegetable farmer. But it was hard to go against the grain.

★

Mary Barton dresses up as a badger called Betty and goes to London every week to stand outside DEFRA to protest the government-sponsored country-wide badger cull. Her banners read 'Vaccinate Not Exterminate', 'Biosecurity Not Bullets', 'I Am Innocent'. One DEFRA employee thought she had dressed as a skunk, which shows how much people working in this department know about British wildlife. Mary cannot understand why badgers are singled out for wholesale extermination when reservoirs of bovine TB (bTB) are also found in deer, cats, ferrets, foxhounds, sheep and other mammals, and the primary transmission route is cow to cow. Mary advocates testing cattle for TB before they are moved with the more accurate interferon-gamma test, and vaccinating them. For the rest of the week Mary writes letters to which few of their recipients reply (except David Attenborough, isn't that telling?). Mary is under no delusion she is derided but continues her blog and single-pawed protest all the same. George Eustice, then the Secretary of State for Environment, normally avoided Mary but one day the rear entrance was blocked by Extinction Rebellion, so he had to walk past her:

Mary: 'Oh it's you.'
George Eustice: Sickly smile.
Mr Eustice didn't stop to talk to Mary so she followed him.
Mary: 'Why are you killing Badgers?'
Eustice: Sickly smile.
Mary, now rushing to keep up: 'Why are you not vaccinating cattle?'
Eustice: 'We are.'
Mary: 'You are not. You are doing an unnecessary vaccine trial. We have the vaccine ready to use now.'
Eustice: Sickly smile.
Mary: 'Why are you not testing cattle properly using the Actiphage DIVA test?'
Eustice doesn't reply and dashes through the door to escape a very irate Mary. Mary burst into tears in rage and frustration . . .[205]

Bovine TB is a devastating disease that can wipe out a farmer's livelihood, and 30,000 cattle a year. No one wants that. It costs a fortune. The badger cull also costs a fortune, but shows willing to act, even if the science doesn't support it. The studies show that culling is at best ineffective. At worst, it can increase the spread of TB by scattering populations of escaping badgers far and wide. The cattle skin test for TB is notoriously unreliable, with false negatives which allows infected cattle free movement to infect other cattle. Poor shed hygiene, infected slurry and intensive farming of high-density herds are factors which increase the bTB risk to herds. Indoor silage-fed cows are twice as likely to catch TB as grass-fed animals. Healthy, thick hedges where wildlife will choose to defecate decrease TB risk by 37%. That doesn't matter, because the NFU has the mindset to cull badgers as the solution, which is popular with their base, who believe the badgers are the major culprits.* Neither the best scientists nor even David Attenborough can persuade them. To accept that the main problem is in the farm reservoir and cow movement means a change of practice – a huge undertaking for which there is no appetite. Vaccination is showing more than 70% efficacy, and better testing is available, but nothing will satisfy the NFU until the badgers have been culled. In seven years an estimated 164,000 badgers have been culled, costing over £80 million. Under 1% of badger carcasses sampled in three areas had infectious bovine TB. The heavily culled area of Gloucestershire showed a drop in bTB of 60%, but then in 2018 it went up by 130%. There is no requirement to test any of the badgers killed in the programme.[206]

Mary caught Covid-19 in 2020 and thought she would die, but she survived and fights on.

LOVE, ACTUALLY

The unlikely conservationist

A log cabin sits on the rim of a silver lake sheltered by pine and birch sentinels a day's journey from civilisation. The southern end touches the water's edge, in which the cabin mirrors itself. Against this gable a pile of brash has climbed out of the water looking remarkably like a beaver lodge – because it is. There is a corresponding pile of brash on the inside of the cabin which is connected by a tunnel under the dirt floor to lead directly to the lodge in the lake, for its occupants to come and go. We are at Lake Ajawaan, deep in the northern boreal forest of Saskatchewan's Prince Albert National Park. 'Beaver Lodge' was the last home of the inscrutable Grey Owl and his two charges, Jelly Roll and Rawhide.

Grey Owl – fringed buckskin jacket and broad-brimmed hat – paddles

---

* While spikes in the badger population correlate with maize crops grown for pheasants.

a canoe towards us from around the bend of a narrow inlet in this short silent film that was the first to be made about beaver behaviour, nearly a century ago.[207] Lush vegetation spills over both banks into the jagged reflections of tall pines. He slaps the surface of the stream with his paddle, cups his hands to his mouth and noiselessly calls. A wake curves through the white water led by a black nose. Grey Owl grabs a branch to hold the canoe steady, leans the lip of the canoe towards the water and a beaver hops in. Cut to Jelly Roll eating a branch; Jelly Roll and Rawhide on the bank; plop, into the river; into the canoe; swimming; stealing an apple from a cardboard box; grooming, scratching, standing up, lodge building; little busy hands, fish leather tails.

It would transpire, years later, that this striking man with long braided hair was not of Apache and Scottish descent as professed, but Archibald Stansfeld Belaney, born in Hastings in 1888. Archie had left home at 17 years old to realise his dreams of living in the wilds of Canada. He eked out a living as a fur-trapper among the Ojibwa people and began to reinvent himself. His trail left 'wives' and children along its way until his life irrevocably changed in the summer of 1925, when he met a fiery 19-year-old Mohawk girl, nicknamed Pony, whom he would call Anahareo. He was 36 years old. Anahareo followed him on snowshoes deep into the hinterlands of northern Quebec but seeing first-hand the senseless cruelty of his profession, she was appalled. What transformed Archie Belaney, trapper, outlier, bigamist, drinker and drifter into Grey Owl, the writer, speaker and campaigner, were the two beaver kits rescued by Anahareo after their mother had been caught in his trapline. The young beavers, McGinnis and McGinty (a nod to the Irish workers on the railroads for their industrious nature), miraculously survived on tinned milk and porridge, and worked their way into Archie's heart. He was disarmed by their gentleness, naive trust and insatiable curiosity, and beguiled by their expressive chatterings and riotous delight when they romped around the camp. Their comedy and demonstrative affection drew affection back. Before Archie knew it, the four of them had become the closest family.

I found it strange and a little disquieting that these animals, that had seemed heretofore to have only one use, and that I had destroyed by hundreds, should turn out to be so likeable, should so arouse the protective instinct of a man who was their natural enemy . . . These beasts had feelings and they could express them very well; they could talk, they had affection, they knew what it was to be happy, to be lonely – why, they were little people! And they must be all like that.[208]

Archie began to see how the beaver stood for something essential in this wilderness, yet 100,000 square miles of Ontario country was dry of beaver. Save for their deserted lodges, it was as if they had never been there. Archie began to write about Canada's vanishing wilds, but believed only under his native persona of Grey Owl would ears be open to him. *Canadian Forest and Outdoors* and, in England, *Country Life* magazine began to publish articles he sent them. It transpired there was a great appetite in the late 1920s for the words of an eloquent 'Indian' living with wild beavers. Grey Owl's stories lifted spirits and the imagination.

But it was the disappearance of the two young 'Macs' in the spring of 1929 that hardened Grey Owl's resolve to devote his life to saving their brethren. They had set out for their evening swim in Lake Touladi; Grey Owl had called out to them and they'd replied with a long, clear, happy note, two Vs forging ahead sending 'rippling bands of silver' in their wake until the dusk absorbed them. Next morning they had not returned to camp. Grey Owl's account of his search for them in *Pilgrims of the Wild* describes wading through tangled, mosquito-infested cedar swamps in pouring rain, his ears ever alert for the beavers' greeting, every lap of water, every rustle in the trees . . . but nothing. Days passed, then months. In her sleep one night Anahareo said, 'They loved us.' All Grey Owl wanted was to bring McGinnis and McGinty home to Anahareo. He describes a 'hostility' aroused inside himself, knowing their trust in humans, and that humans were the ones not to trust. Each day he searched but did not find them.

As some kind of atonement to the Macs, Grey Owl's sole aim was to protect and nurture a thriving beaver colony – to repopulate their former range, and show others the value and fascination of these natural architects. 'From death springs life,' he wrote, for along came Jelly Roll, an orphaned kit given to them to foster, soon joined by Rawhide, a male beaver with a damaged foot: the two beavers who would become international celebrities in their own right. Following the fame Grey Owl's articles had brought, in 1930 National Parks of Canada offered to build him a cabin in uninterrupted wilderness to share with Jelly Roll and Rawhide and raise his beaver colony. They would pay a salary, and in return receive the invaluable publicity for their cause. 'And so I chose aright for once,' Grey Owl confessed. Beaver Lodge, built to his own specification (with its subterranean passage) would be the haven for Jelly Roll, Rawhide and their offspring. It was here that Grey Owl became Canada's most celebrated 'native' conservationist.

Despite the challenges that living with two 'ambulating sawmills' presented – table and chair legs chewed to a fine pencil-point, gnawed doors, sticks and brush stacked into corners of the cabin, staged wrestling matches, wet furry bodies bounding into the bed – Grey Owl took to the beavers' nocturnal hours to spend time with them, writing articles about their habits, characters and their lives together. As a natural property owner, Jelly Roll's proprietorial airs inspired her nickname, The Queen:

> . . . she moved and talked and did things, and gave me a response of which I had not thought an animal capable. She seemed to supply some need in my life of which I had been only dimly conscious heretofore, which had been growing with the years, and which marriage had for a time provided. And now that I was alone again it had returned, redoubled in intensity, and this sociable and home-loving beast, playful, industrious and articulate, fulfilled my yearning for companionship as no other creature save man, of my own kind especially, could ever have done.[209]

In 1932 Grey Owl and Anahareo had a daughter, Shirley Dawn, but did not weather their own stormy relationship though a complicated love for each other remained. As their lives separated, Jelly Roll became like a person to Grey Owl, and behaved like one, resolutely on her own two legs, busying herself with jobs: haul, pile, stash, modify, construct; or laying her head on his knee as he wrote, talking her 'uncanny language' in her dozing. 'I could in no way see where I was the loser from this association, and would not, if I could, have asserted my superiority, save as was sometimes necessary to avert wilful destruction.'[210]

The English fur-trapper had transformed himself heart, soul and physical body into Grey Owl, who would devote every scrap of his being to the beavers' plight; and Lake Ajawaan would become the heart of the range for new generations striking out into the park's territory. Jelly Roll had become a film star, the most famous beaver in the world, whose

antics and those of her family – Happy, Hooligan, Buckshot, Wakinoo,
Sugar Loaf and many more – changed hearts and ways of thinking.

As Grey Owl's fame spread he took on an exhausting itinerary of talks
to get his message over, packing out halls with audiences of 3,000 or more.
In white-fringed buckskins, a single feathered headband and his long dark
hair in braids, his commitment to his beaver cause was undiminished. He
travelled throughout Canada and the States. A British tour included a royal
audience with King George VI and the Princesses Elizabeth and Margaret.

Grey Owl's story is of a man who made himself up. Or put himself together.
The point is, the 'lie' doesn't matter, it is what he became and what he did.
His truth to himself. An identity inextricably entwined in the discovery of
the individual identities of the animals he was trying to protect. The last
sight of his two Macs disappearing into the horizon were the abiding spirits
who inspired his every breath to save the Beaver People. Grey Owl died of
pneumonia and exhaustion on 13 April 1938 aged 49. He is buried behind
Beaver Lodge in this wild place of beavers, beside Lake Ajawaan, the 'splash
of quicksilver', where the descendants of Jelly Roll and Rawhide live on.

*

When Grey Owl cursed the stumps and slash of graveyard forests levelled in the rapacious lumber rush, he not only pre-empted Aldo Leopold but rewilding science of the twenty-first century: 'There were not even any wolves here, and a country without them seemed lacking in some strong ingredient. I missed their wild cantatas, and the lazy inbred deer, lacking any incentive to move around and improve themselves, died like rabbits all through the woods.'[211] A decade later Leopold wrote that the penalty for an ecologist was to live 'alone in a world of wounds'.[212] Today there is a growing movement of nature restoration across the globe. Giving back, where we can, is slowly spreading across nations. Understanding how ecosystems function with the vital role of animals to charge the nutrient cycles we all rely on. Time for our leaders to get with the programme. Where politicians obfuscate, ordinary people are getting on with it. In a few places, it just does it itself . . .

NO MAN'S LAND

Few around in the 1970s will forget the medal-jangling, insignia-bristling, gold-braided, enormous figure of the tyrannical military dictator of Uganda; or, to use the title he preferred: His Excellency President for Life, Field Marshal Alhaji Dr Idi Amin Dada, VC, DSO, MC, CBE, Lord of All the Beasts of the Earth and Fishes of the Sea, and Conqueror of the British Empire in Africa in General and Uganda in Particular. This sinister syphilitic megalomaniac killer-clown, responsible for up to half a million human deaths and known for keeping his enemies' heads in the palace fridge, was very fond of killing African animals too. To which end he commandeered a grand hotel, the Pakuba Lodge, on the banks of the River Nile, downriver from the Murchison Falls, as his own personal safari State Lodge. He particularly liked hunting the crocodiles whom he had reportedly fed on disabled citizens and ministers who had fallen out of favour. After he was overthrown in 1979, Pakuba Lodge fell into disuse and was slowly reclaimed by the jungle and its wildlife. Among the crumbling stones of the terraces, lounge bar and restaurant, packs of hyenas roamed, baboons climbed through windows and a family of warthogs took over the old kitchen. Camera traps revealed visiting lions, porcupines and pangolins. The BBC heard about it and in 2018 filmed the most extraordinary scenes of peaceful coexistence between animals we think of as mortal enemies.[213] In the most astonishing sequence, a leopard wanders into the kitchen where a baby warthog lies alone asleep on the floor. The leopard sniffs the hoglet, who wakes and runs squealing to the corner of the room. The leopard watches what is a very easy meal, then wanders off. Next morning, camera still rolling, Mama Warthog comes back, all her hoglets present and correct for breakfast.

★

Places with violent histories across the globe have rewilded themselves in our absence. The strip of land dividing the Korean peninsula – 243 kilometres long by 4 kilometres wide – is pimpled with mines and has been empty of people since 6 September 1953. Between North and South Korea, the Demilitarised Zone (DMZ) passes through mountains, river valleys and overgrown rice terraces. This, once one of the most dangerous places on Earth, guarded by watch towers and hemmed in by a double fence of looping razor wire, has become a refuge for creatures pressed by the densely populated countries on each side.* South Korea's Ministry of Environment has identified 5,000 species of animals and plants in the zone, including Asiatic black bears, Eurasian lynx, Chinese water deer, yellow-throated martens, an endangered wild mountain goat known as the long-tailed goral, white-naped cranes, red-crowned cranes, cinereous vultures, sea eagles and black-faced spoonbills. The legacy of human killing fields is a sanctuary for wildlife. The biggest threat to the DMZ is peace.

Spanning tens of thousands of miles, the Pan-American Highway connects Prudhoe Bay, Alaska, to Ushuaia on the southern tip of Argentina – except that it doesn't. For about 80 kilometres on the border between Panama and Colombia, the road disappears into the most inhospitable (for us) rainforests, swamps and mountains in the world. The lack of a road means weak central authority and so the 5,750 square kilometres of Panama's Darién National Park, a UNESCO World Heritage Site, has become a refuge for guerrillas fighting the Colombian government and drug traffickers alike. The park shelters endangered species such as jaguars, harpy eagles, brown-headed spider monkeys and great green macaws. 'It helps that the road has not been completed,' says Dr Ricardo Correa, conservation programmes advisor at Panama Wildlife Conservation.[214] Crime and poverty are impediments to development and protect the biodiversity of the region.

Similarly, the Iron Curtain that divided Europe with ideology and barbed wire after the Second World War was also a wilderness corridor. The border zone cleaved the continent, splitting the West from the Eastern European communist countries, stretching 12,500 kilometres from the Barents Sea in the north to the Black and Adriatic seas in the south. In this no man's land of minefields, watchtowers, bunkers, machine-gun nests and barbed wire, insect populations thrived in the absence of farming or

* North Korea's hermit kingdom of almost 26 million; South Korea, over 51 million.

pesticides, ponds filled mine craters, swamps remained undrained and river
landscapes were left alone, enabling black storks, beavers, bears, bee-eaters,
wolves and lynx to take advantage of this heavily guarded redoubt. Imperial
eagles even nested on the Bulgarian/Greek border. When the Iron Curtain
fell in 1989 and the area was threatened by development, the 24 countries
on the borders came together in 2003 and launched the Green Belt
Initiative to protect what is the longest ecological network of its kind,
linking natural areas to parks and biospheres. Of course, some sections
are threatened by development, but this is what humans can do. It was
noticed that 18 years after the barrier between Bavaria and Bohemia was
removed, deer still refused to cross the frontier, for that hefting geography
(learned boundaries) had become part of their culture.

Kintsugi, or golden joinery, is the Japanese art of mending broken pottery.
The breakage is not disguised but repaired with lacquered gold seams,
making the scars highly visible yet beautiful and the vessel stronger. *Kintsugi*
is the acceptance of change, where restoration celebrates transience and
humility. So comes a way of looking at broken things as an opportunity
for something illuminated and different. *Kintsugi* flips our way of seeing.
Sites of fracture become sites of mending, in the same way we might
imagine our blue planet, girdled in seams of green along the fault lines:
the old Iron Curtain of Europe; the cinched waistband of Korea's DMZ,
the bruise of Chernobyl (we hope); and all the broken ecosystems waiting
in the wings to blossom. We could map the world with *kintsugi*. From
old quarries and brown-field sites to bigger dreams.

Green joinery becomes story. With only 2.9% of the world's terrestrial
ecosystems faunally intact, it's all hands on deck.* By bringing back animals
as ecosystem engineers our work is easier and cheaper, with far better
results. Beaver, boar, bison, tortoise, Exmoor pony, greylag goose or termite
trigger dynamic nutrient cycles, create habitats and increase the variety
of life. More is more. At Underhill in early summer there are so many
toadlets you cannot walk around the lake, a grass snake threads across the
surface, a dabchick dives, a hobby jets over, swooping on the dragonflies.

---

* E.O. Wilson's 'Half Earth' proposal identifies the viable natural spaces that we have left
in order to safeguard them. The International Union for the Conservation of Nature
(IUCN) reports 44,000 substantial wildlife reserves over 14 million square kilometres,
with hugely varying levels of protection. This is 15% of the land and 10% of the sea
but includes the most inhospitable places.

National Geographic launched the Last Wild Places initiative to protect 30% of the planet by 2030, including American Prairie, which is gathering land in Montana to restore the grassland habitat of America's Great Plains, bringing back buffalo and protecting prairie dogs. In northeastern Russia, Pleistocene Park, a family-run enterprise founded by Sergey Zimov and run by his son, Nikita, has introduced Yakutian horses, Kalmykian cows, moose, musk ox, yaks, reindeer and bison to graze and fertilise the Arctic tundra grassland ecosystem to replace the role of mammoths and protect the permafrost carbon reservoir. The World Land Trust partners projects across the world, so far protecting 2,222,247 acres of threatened habitat in 20 countries. In the great marshlands of northern Argentina the Iberá National Park protects 13,245 square kilometres, extended and rewilded thanks to the vision of Kris and Doug Tompkins acquiring land with philanthropic funds and investing in nature tourism. Tapirs, giant anteaters, peccaries, jaguars (absent for 70 years) and red-and-green macaws (extinct here since the nineteenth century) have been brought home. There's Peace Parks Foundation, Panthera, the Gondwana Link, all restoring ecosystems and connectivity. There's the entire country of Costa Rica, for goodness' sake, home to 5% of the world's biodiversity, with nationwide sustainability policies. The 62,000 acres of the Lewa Wildlife Conservancy used to be a cattle ranch in Kenya. Berlin is full of nightingales because they let the wild in. Even broken attempts of mending one way, can find another way. In Africa, the Great Green Wall (GGW) had the epic ambition of planting trees across the breadth of the continent as an 8,000-kilometre buffer to desertification. It was a poorly executed, impractical plan which lost most of its trees within months of planting. So the focus shifted to regenerative practices, old knowledge and a return to indigenous farming techniques. A method used by Yacouba Sawadogo, a farmer in Burkina Faso, is a practice called *zaï*, which involves digging and filling shallow pits with manure to lure back termites to break up and irrigate the soil with their tunnels, and so preserve its vitality.[215] Over 20 years, Sawadogo has created a 62-acre forested area visible on satellite images, called Bangr-Raaga, meaning Forest of Wisdom.* Slowly the GGW is evolving into a mosaic of landscapes with the empowerment of its communities. There

---

* Sawadogo, using this method to restore soil damaged by desertification and drought, was awarded the UN Environment Program's Champions of the Earth award in 2018. The use of *zaï* has led to the water table levels rising about 5 metres on average and as much as 17 metres in some areas.

are billionaires beginning to realise that restoring habitats is more fulfilling and interesting than huge yachts or fast cars. The Danish clothing magnate Anders Povlsen is now the biggest landowner in Scotland, reversing the deer-sick landscape encouraged by shooting estates, and letting Scots pines and rowans march up his side of the River Feshie. Wolves and (the no-brainer) lynx would make work easier, and the Cairngorms would be the natural place for a wolf pack to bring back dynamic life to these islands. But alas for the brave British, once so innovative and forward-thinking, medieval fear prevails.

One of the most uplifting things in my lifetime has been the return of large fauna to their European range, for wolf and bear prints to mark the forest floors again. There are around 17,000 brown bears, 12,000 wolves and 9,000 lynx across the densely human-populated continent, making their way to France, Germany and into the Netherlands. No more a lone wolf howl on the Spanish–Portugese border. It's not all happy families, but public opinion is coming round to wolves as eco-tourism brings greater prosperity and work to poor rural areas. Rewilding is taking off with the helping hands of humans who get both the advantages and imperatives; from scraps of marginal land like ours at Underhill, to unprofitable farms like the 3,500-acre Knepp estate in Sussex, to big ambitious projects like Rewilding Europe, a non-profit organisation, working with landowners and rewilding agreements on 2.3 million hectares over eight operational sites. The 580,000-hectare Danube Delta, Europe's greatest wetland, has been returned to its former function and natural processes, providing clean water, fish and flood defence by removing dykes, dams and irrigation channels that cut off lakes, the main river and tributaries from the flood plain. Here live the largest number of fish species in Europe, including four species of sturgeon. Here, to this vital migrant staging-ground for pelicans, storks, glossy ibis, spoonbill and cormorants, one of the world's largest flying birds – the endangered Dalmatian pelican – is returning. Here Konic horses, Tauros cattle and water buffalo now graze a mosaic landscape where beaver can steward and prosper, as everything begins to flourish once again, attracting tourists (and great swarms of flies!).

How thrilling the world could be for its younger inheritors. To realise a super-charged and properly green economy, where technology meets nature on her own terms, where jobs not only become meaningful and exciting but *feel* worthwhile. Where agriculture is regenerative and goes vertical instead of horizontal. Where waste becomes history. Where we

don't expect 'pent-up' savings from lockdown (lucky them) to boost
short-term economic growth in a consumer frenzy, but rather pent-up
energy from lockdown to boost ecological literacy and lesson-learning.
There are no brakes on how innovative and cooperative *Homo sapiens*
can be. Imagine if Wales went Costa Rican green. Imagine, 4 million
acres of restored ecosystems and regenerative farms; how globally celebrated would that be? What a destination; what work, what income; what
imagination; what enlivenment; what prospects. Wildlife in, noise out; a
place of silences and singing. Wreck the economy? Would it?

Common wealths

While we are on that uniquely human invention of trade, should we not
stump up collectively for the safekeeping of forests that provide the useful
oxygen we breathe? A global fund for a global commons? Instead of
expecting the poorer peoples of this world not to do what we did centuries ago, to do the right thing for nothing (or too little)? A global commons
of ocean protection zones that is bold and big and serious, that seeds the
seas around them. Because however much I'd love there to be zero industrial trawling, zero dredging and long-lining, I can't quite see it. How
about 50% no-go zones of blue joinery busting with fish, *without* anthropogenic noise? Imagine. In smug New Zealand blue joinery has long
proven benefits. Inspectors will board your boat and confiscate it if they
find a catch that breaks the rules. No messing. Although we won't let
New Zealand off the hook for renaming the slimehead – a long-living
(250 years) deep-sea fish – as the more verbally palatable orange roughy,
and then trawling these ancient creatures to the brink. It just doesn't
seem right to be frying up a ten-dollar fillet who was two centuries old.
A few years ago, on holiday in Wales, Jonathan swam in goggles up the
Pembrokeshire coast north of Fishguard, an hour a day, and he didn't see
a single fish. He emailed Pembrokeshire park services to tell them (I
know), and asked why they didn't have no-take protection zones. They
said they were in a consultation and survey phase. He responded that
there was nothing to survey. They didn't reply. Blue joinery. Shucks –
oyster fisheries, mussels, clams, shellfish! The Billion Oyster Project in
New York aims to restore the oyster reefs in the harbour: clean water,
flood defence, habitat, breakfast. Blue joinery.

Next up, a common wealth of hedges; proper hedges! Hedges dressed in their finery of frothy blossom and berries and stem tips hiding hibernating bees and the invisible eggs of insects. Not those flayed, wind-whistling, good-for-nothing tidy square nonsenses. There are still 500,000 miles of hedgerows clinging on in the UK. Imagine. A connectivity that criss-crosses the land for insects and safe highways for creatures. You might be thinking by now that I am rabid about hedges, and you'd be right. Their value is underestimated. They stand in for our lost ancient woods, but only if they are healthy. Green joinery. In our chalk valley neighbouring farms have joined in cluster schemes to plant field margins for pollinating insects and farmland birds. But only effective if we stop dragging spray booms across the land. Golden joinery. These flowery margins bring swooshing flocks of chaffinches, pipits, yellowhammers, buntings and linnets in flurries over the gifts of farmers, and in return for seed heads they not only bring health back to desertified landscapes, but pride, interest and pleasure. It's a beautiful thing.

There are a million acres of gardens in the UK. Come on! A pink-yellow-orange-red-green joinery of gardens. Messy-beautiful instead of tidy-sick (and ugly). The swish of the scythe instead of the scream of the strimmer, long swaying grasses, homes for glow-worms and places to string webs; mow a path not a lawn. Who wants a sterile green baize card table? Embarrassing. Sit back; love wild patches more and use words like 'manicured' less. Why burn everything when it can quietly decay under a hedge and make larders and homes for beetles and wrens? How can ladybirds hibernate if we burn and tidy every corner? Let us stitch our fragments of Persian carpet with golden joinery. Bring back the blowsy cottage gardens of our grandparents that we are so fond of on gift cards, swaying with flowers, beans and bees. A common wealth of gardens. Let *growth* be green and for bigger ambitions. Bigger courage. Bigger care. Bigger responsibility. Bigger love. And while we're at it, what about that Born Free elephant sanctuary in Europe, in atonement for misunderstanding these great sensitive creatures separated from their own kind? (With a bit of free eco-engineering thrown in.) Imagine! Restoring nature to small remnants of land, like our 22 acres, to those large unprofitable estates, adds up. Better surely to let unproductive land work for free, than subsidise poisoning it. We have to stop knocking 'rewilding' (and blighting the word), we really do. Green, blue, golden, multicoloured joinery.

★

Beige office, beige paintings, brown desk. Woman comes in through the door. 'Dr Switzer?' 'Come in, I'm just washing my hands.' View of Dr Switzer's back in the bathroom. The patient has a phobia about being buried alive in a box. Dr Switzer explains that he charges $5 for the first five minutes and absolutely nothing after that. 'Too good to be true,' the patient laughs. Dr Switzer guarantees the session won't last the full five minutes. He says she will have to pay in cash or by cheque and he doesn't give change. She laughs, 'All right.' He looks at his watch. 'Go!' The patient tells him she has a fear of being buried alive in a box. He asks if anyone has ever tried to bury her alive in a box. 'No,' she says. But she can't be in anything boxy. 'So what you are saying is you are claustrophobic,' he says. He tells her to listen very carefully to the two words he is going to say to her, and incorporate them into her life. She scrabbles in her bag for a notebook to write them down. Most people can remember them, he says.

'Okay.'

'You ready?' He leans forward. 'STOP IT!' He leans back.

'I'm sorry?'

'Stop it!'

'Stop it?'

'Yes. S. T. O. P. New word: I. T.'

'What are you saying? So I should just stop it?'

'There you go. You don't want to go through life being scared of being buried alive in a box, do you? It sounds frightening.'

She nods. 'Yes.'

'Then stop it!'

The patient decides to take the whole five minutes but gets the same two words for all her problems. She doesn't like it, so he tells her he will give her ten words, which she can write down.

'Stop it or I'll bury you alive in a box!'[216]

Bob Newhart's counselling sketch spells out what humans have to do. We have to stop it. *Wu wei*, in Chinese, means non-doing, a noble art, according to Taoism, of flowing with nature, letting go of our need to control, as the bamboo bends in the wind. We have to leave fellow animals and their habitats alone, for our health, for the climate, for our future. And, of course, for theirs. But like Dr Switzer's patient, humans don't stop it. So far the world has spent $12.5 trillion on Covid-19,[217] and barely a dime on pandemic prevention because you can't *show* what you have prevented if you have prevented it. We *know* factory farming and deforestation drive the risk of disease, climate change and loss of fauna; and loss of fauna drives risk of disease. We *know* if we didn't do these things climate change, pollution, antibiotic resistance, health, wildlife loss and habitat destruction would be vastly mitigated. If we think factory farming is the only way to get cheap food to the poorest people in the world, we are wrong. It is a very expensive way. The poorest people don't eat chicken nuggets, beef burgers or chocolate spread. Bushmeat is taken from the forest by those dependent on it, and we live in an unequal world. But it is corporate choices that drive the destruction of ecosystems. A doctor on the radio recently suggested that if we start to make changes as consumers, the suppliers will shift. Well, they would if we did, but we won't. Until we *have* to. And isn't it just shite for the onus to be placed on the individual consumer *again*? Not buying palm oil products would spell an end to the palm oil plantations. But you try *not* buying it. Palm oil has more than 436 other names. How about this one: CBS (cocoa butter substitute) or even vegetable oil; they are not lying. Here's a trick, if there is *stear*, *laur* or *glyc* embedded in the name, it's likely palm oil or a derivative. Aluminium **stear**ate; Ammonium **laur**yl sulphate; Ethylhexyl**glyc**erin. It's in 50% of supermarket products. Poor Jonathan, no more Bath Oliver biscuits. What did we use before palm oil? Butterfat, olive oil, various seed oils such as sunflower, safflower, cotton, grape, beef tallow, pork lard, chicken and duck fats, avocado and whale blubber. It is in consumers' hands, but no one has asked me or anyone else if I want to exterminate orangutans for face cream, chocolate spread or Bath Olivers. We don't all make the right choice, life is not like that, as Dr Switzer's patient could tell us. The Irish did not stop smoking in public places out of wise sense or the goodness of their hearts. It became illegal. They had to stop. Everyone had to stop. So we did. Ditto seat belts. Ditto ozone-damaging aerosols. We stayed at home during lockdown not because we were good citizens, we had to, it was the law. Sigh. When we *have* to stop doing something for the greater good and we understand why, it really isn't that bad. What *is* bad is the consequences of not stopping. So we have to reduce our impact on this planet or Dr Switzer will bury us alive in a box.

IT'S NOT A CRIME

I am sitting on a golden banquet chair under a huge chandelier on a warm day at a summer festival the year before the pandemic. The speaker, Jojo Mehta, is describing the most pragmatic, workable proposition to safeguard the environment that I have heard. She is explaining how it would be possible to make destroying ecosystems an international crime. A law that could change the world; a law that might *save* the world. It is neat and beautiful. This is the brainchild of Mehta's friend, the lawyer Polly Higgins, who died tragically young at 50, in 2019. Higgins's proposal was for ecocide to be recognised as a crime against humanity. Without a *criminal* law to protect something, it is not protected. It is phenomenally difficult to get a new law put onto the statute books, with countless hurdles and complications. The odds of success are low. However, tack an amendment onto an existing law and the road is easier. There is a simple amendment procedure to International Law. Higgins's proposal was to add ecocide as the fifth crime in what was originally called 'crimes against peace', which included war crimes, crimes against humanity, genocide, and recently, crimes of aggression. Polly Higgins sold her house, ditched her well-paid job as a barrister and went on a pioneering crusade. In 2010 she submitted a preliminary definition of ecocide to the United Nations Law Commission: 'Serious loss or damage or destruction of ecosystems such that peaceful enjoyment by the inhabitants has been or will be severely diminished.' Inhabitants. I like that word. It's inclusive and egalitarian with just the slightest sleight of hand. The stipulations were: Is it serious? Is it widespread? Is it long-term? Is it severe? In 2011 she ran a mock trial to see if it could work, and it could.

Any head of state can propose an amendment to the International Criminal Court (ICC) statute, and once tabled there is no veto: leaders can abstain, or add their signature. To allow an amendment, two-thirds of member states must sign. That law is then enforceable in the countries that have ratified it, and they would be obliged to include it in their domestic legislation. Because there are many island states affected by climate change whose interests a law against ecocide would serve, the numbers are achievable. In 2017 Higgins and Mehta co-founded the Stop Ecocide campaign.

So what about those enterprises that pollute the planet, deforest the Amazon, dump dredging substrate on coral reefs? Well, here is the beauty of it: insurance. When island states get enough signatures to table the amendment, that is when insurance companies get nervous and go: *Maybe we shouldn't underwrite some of these polluting activities any more?* And that is when the banks go: *Uh oh, if this is on the horizon, maybe we shouldn't be backing those industries?* And that's when those industries might have to start complying with new regulations before they get insurance and money from the bank. And *that* is beautiful.

Next problem. The US, Brazil, China don't sign up, or countries pull out, so the laws don't apply to them. Until a corporate executive or government minister goes somewhere where the laws do apply. Remember Pinochet. Arrested on British soil in 1998 under the jurisdiction of our law on an arrest warrant of a Spanish judge for crimes against humanity. Okay, he was given a get-out-of-jail-free card by British Home Secretary Jack Straw, but that might not have happened. International criminal law begins to marginalise those not signed up. It restricts movement. Which, at the very least, could be inconvenient for business as usual. More pertinently, when something is criminalised it shifts our perception of what is acceptable. This is what Jojo Mehta calls 'flipping the normative'. Baffling that polluting a river delta or dredging the ocean floor was ever acceptable, but that is the power of lawlessness. There is a strong relationship between the law and our cultural morals. Slave trading was legal once upon a time. Next beautiful thing is the breadth of what the crime of ecocide can address, so there is no need to specify every possible threat: toxic run-off, life-diversity loss, air pollution, etc. A careful definition will address them all.

January 2021. Human rights barrister Philippe Sands QC co-chairs an international expert panel to agree a legal workable definition of ecocide. Habitué of international courts and intimate with the difficulties arising from definition in the crime of genocide, if anyone has the experience and know-how, he does. It turns out he has a unique motivation too.

I first met Philippe in 2017. We are both accustomed to the way that forensic threads from the past can pull uncannily into the present, and now it is the journey Philippe made into *East West Street* — a book he wrote unearthing the origins of 'genocide' and 'crimes against humanity' first brought into the courtrooms of Nuremberg 75 years ago — that is sounding a loud bell in his own ear. In response to Nazi atrocities the door had opened for powerful leaders to be put on trial before an international court, which had never happened before. This was the moment our modern system of international justice came into being. In 1945, however, there was no conception of *other* atrocities we should be safeguarding against (even though we had detonated atom bombs and seen their consequences). Fast forward to Rachel Carson's *Silent Spring* of 1962, exposing the catastrophic damage caused by the industrial, carefree deployment of DDT poisoning the biosphere. Only then did the environment become a serious issue. Carson explained how the blanket use of deadly chemicals reached far beyond their target. If DDT could make its way up the food chain to cause the disappearance of American eagles — by thinning their eggshells — it's likely to be bad for us too. Humans are not separate from ecosystems. Sixty years on we are still fouling the nest, but now we know far more. With the continuation of unchecked anthropogenic ecological harm there is a dire urgency to criminalise atrocities against the environment. It is as if *East West Street* primed Philippe's mind.

I ask Philippe what made him accept the invitation to chair this panel.

'I just felt for me it was like Lemkin* in 1944/45. Something is going on that appears to be terrible; criminal law at the international level can be harnessed to address it. And I feel, you know, that I have an emotional connection and want to contribute to it.'

Spurred on by his children, Philippe wants to act. He tells me there are three issues to be solved. First is the nature of the harm: is it temporary or permanent, total or partial, localised or across borders? Second is the awareness or intent of the perpetrator — the *mens rea*, in legalese. Was the harm carried out with malign intent, or reckless, or negligent? These days it's possible to be found guilty without a *mens rea* requirement. Ignorance is not a defence. The third thing is if multiple acts or joint liability contributed to the problem, such as climate change, or the algal blooms of nutrient pollution from the run-off of excess chemical fertilisers. So. All the panel has to do is scoop this up into their definition.

---

* Raphael Lemkin was the Polish lawyer who coined the term *genocide*.

We have seen how fine-tuned, complex and interconnected ecosystems can be. How a market in otter pelts can wipe out a kelp forest
and all its inhabitants, including the gentle, toothless, 10-ton Steller's sea
cow. How hydroelectric dams cut off Chinook salmon spawning grounds
and force orca populations to switch prey. Thus, one collapsed ecosystem
can collapse another. We have seen how great plumes of whale faeces
bring iron and nitrogen to the surface of the ocean from the feeding
depths, providing the key nutrients that drive the whole marine food
web by fertilising the phytoplankton which sequester billion tons of
carbon dioxide a year, and provide 50% of the world's oxygen.[218] Blue
joinery. More whales, more phytoplankton, more carbon removed, more
food for fish, more life-diversity, more whales, more plumes . . . the
whale pump. (I know, repetition.) Take one blue whale. The largest
animal *ever* in the history of life – and we just happen to live alongside
them. They have a heart the size of a small car. Their calls, at 188 decibels, are louder than a jet engine (140 decibels). The heaviest recorded
blue whale weighed a colossal 190 tons, which is heavier than a Boeing
757 full of fuel and people. More than 30 elephants. A 30-metre pregnant leviathan could weigh even more. What is *her* ecological impact?
Or the impact of her loss? Over 90 years? Producing, say, 10 young,
who produce 10 young, who produce 10 young . . . Save the whale to
save the seas to save the planet. Whales might come to *our* rescue. But
the point of this detail is to ask: should killing whales be sentenced
proportionately? Japan, Norway, Iceland still commercially hunt whales.
And fishery by-catch and discarded fishing gear are responsible for the
death of more than 300,000 cetaceans a year. Don't forget the krill
harvest for the glucosamine health food trade. The international panel
of experts have their work cut out.

I'd like to charge the person who grubs out the hedge, baits the stink
pit and sloshes the Roundup, but then we might all be in the dock.
Criminalising ecocide is aimed at high-level responsibility for big-time
atrocities. The con of corporations has been to shift responsibility onto
the consumer: both in disposal of their polluting product *and* the blame
for consuming it, while they spend millions persuading us to do exactly
that. But the aim of this law is not about individual consumer responsibility, or to put executives behind bars. The aim, simply, is to get them
to STOP. And when we talk of punishment, we need to remember
reward. Green standards are an opportunity to provide countless green

jobs. The work and imagination waiting to be unleashed will be our new economy. Necessity *is* the mother of invention. That's why we *need* the law. So much ingenuity has been suppressed by corporate interests. We have to turn the boat around.

In June 2021, Philippe and his colleagues finalised the legal definition of ecocide as: 'unlawful or wanton acts committed with knowledge that there is a substantial likelihood of severe and widespread or long-term damage to the environment being caused by those acts'. Jojo Mehta said it was a historic moment, and that the definition was well pitched between what needs to be done to protect ecosystems and what will be acceptable to states.[219] For Philippe its real power will be in changing human consciousness, recognising that we are in a relationship with our environment, and that our well-being is dependent on it.

TYGERS

And a mouse is miracle enough to stagger sextillions of infidels.

Walt Whitman, 'Song of Myself'

'Good Heavens what insect can suck it?' Charles Darwin asked in a letter he sent in 1862 to his friend Joseph Hooker, then assistant director of the Botanical Gardens at Kew. Darwin was describing the flower of the spectacular star orchid, *Angraecum sesquipedale*, which he had been sent from Madagascar, with a preposterously narrow throat nearly a foot long. White, waxy, star-shaped flowers with this long green spur. To reach the nectar at the bottom, something out there must have a very long tongue. Accordingly, Darwin predicted the existence of an unknown species of pollinator that – it turned out – would take 41 years to be discovered, and 130 years to be filmed feeding from the orchid (in 1992). Enter the Madagascan moth who made Darwin's prediction come true: Morgan's sphinx moth, a 6–7-centimetre moth with a 16-centimetre wingspan and a 25-centimetre coiled proboscis, described in 1903 as *Xanthopan morganii praedicta*, tipping the hat to the science of prediction.*

Imagine yourself into that nocturnal moth world, approaching the seductive flower, uncoiling your 11-inch tongue (mid-flight remember, and that's more than four times longer than your body) and lassoing the tip precisely into the tiny opening and slipping it down the flower's throat to the well of nectar at the bottom. What aim, what vision, what dexterity, what a piece of work is moth. *And* orchid. The entomologist Ian Kitching suggests we try to aim a horsehair into a single buddleia flower. Impossible. Yet a hummingbird hawk moth does this accurately every half-second without repeating the same flower twice. That's *moth* intelligence.

The partnership of the Morgan's sphinx moth and Madagascar's star orchid is a beautiful illustration of natural selection and co-evolution, and

---

* Discovered and named by none other than our old friend Walter Rothschild (with colleague Karl Jordan, and collected by Charles Oberthür and Paul Mabille).

an explanation for the wonder of beauty in the colours, shapes and smells of the natural world, not so very long ago passed off as divine design. The moth that randomly varied with ever longer tongue was favoured by the orchid that randomly varied with the longer spur, the pollen from flowers with long nectaries was transmitted to others with long nectaries, and on it went, until they were basically made for each other. This lock-and-key specialisation creates a win–win partnership – the moth gets exclusive nectar rights, having the only tongue long enough, for which the orchid receives pollen delivery services exclusively to its own species.

It transpires that the fabulous moth proboscis has structural properties – a combination of regions that are liquid-attracting and liquid-repelling – that allows the organ to take up liquid as viscous as honey, while keeping perfectly clean. Properties which humans would like to harness to create fine bendable forensic probes or reusable self-cleaning vaccination needles . . . So there we go, two species co-evolving over millions of years. Lose one, and you lose the other, and a lot more besides.[220]

★

This morning a pair of red kites coil the wind in ever smaller circles over our house. The male (I assume) drops the prize he is carrying, then dives to catch it, retrieving and dropping again and again. What a show-off. What a flyer. I can hardly believe British skies can gift such a creature. In 1935 there were only two breeding pairs in Wales. And now we are here. A eulogy to human and bird tenacity, at 1,800 breeding pairs. Even in Britain, one of the most nature-depleted countries in the world,★ banished creatures are returning. The kites, sea eagles, pine martens, wild boar (some appear to come back by themselves). There are even common cranes, 4 feet tall, walking elegantly across Norfolk's wetlands with their ruffala tail feathers. Their trumpeting cry can be heard 3 miles away.★★ Beavers are damming Devon rivers, and white storks are nesting again in southern English oaks. Bald eagles, once on the brink, have barrelled back into American skies; wolf howls shiver over brows of Spanish hills; and blue whales have returned to the southern Atlantic seas around South

---

★ The UK ranks 189th out of 218 countries. According to a WWF Living Planet report, one in seven native species face extinction and more than half are in decline.

★★ In Homer's *Iliad* the calling of migrating flocks was compared to the sound of armies approaching in battle.

Georgia, including mothers and calves, 60 years after whaling so very nearly disappeared the largest animal that ever breathed on Earth.★

Life is what makes the world function. Diversity of life gives nature resilience. Biology is the mechanism, and physics the result. Ecosystems are dynamos of mutual interdependency. That's why we, above all else, must mend our relationship with the creaturely world, from bug to bear. The climate crisis, pollution, pandemics, floods, famines, habitat destruction – it all comes back to our broken relationship with and misunderstanding of the other inhabitants. If we think about *them* as the vital, individual beings they are, and as the miraculous eco-engineers they are, and safeguard their homes and their prospects, the seas, rivers, grasslands, wetlands and forests, far better fortunes and futures for all the young in the communities of life could follow. What are wild places destitute of fauna? We don't want to be punished for being too smart and expelled from the Garden. We don't want to live with the curse of the knowledge of Good and Evil – of what *could* have been. Animals are the key. Variety and abundance are the strengths. Our saviours are all around us. Let there be a golden joinery of tygers.

Burning bright

Many humans devote their lives to saving Earth's creatures. At the St Petersburg Tiger Summit in 2010 the goal was to double the number of tigers across five range countries – Bhutan, China, India, Nepal and Russia – by 2022, the Chinese Year of the Tiger.★★ India's tigers have more than doubled, to nearly 3,000. Nepal is on target; their 121 tigers are now 235. In Russia the Amur tiger has increased on the census to 540, 100 with cubs. In 2010 there were just 20 tigers recorded in the whole of China, now there are 55. Bhutan's tigers are up to 103; their well-protected forests have also helped the numbers of wild Asian elephants there increase from 513 to 678.[221]

There are many ways to live more peaceably with our 'non-human' neighbours. We might try to stretch our perception into their *Umwelt*, to

★ By 1970 there was just 1% of their pre-whaling population left. Then, after just two sightings in 50 years, 55 whales were spotted off South Georgia in 2020. The Antarctic blue whale has increased from 1,000 individuals in the 1960s to around 3,000 (in 1926 there were an estimated 125,000). The total blue whale population is thought to be between 10,000 and 25,000.

★★ Known as XT2. There are not 4,000 tigers in the wild, but a century ago there were 100,000.

put ourselves in their octopus skin or wolf fur, or wield a moth's proboscis into a flower spur. *Oikos*, the root of ecology and economy, means family, and home. Complex beings are fairly simple on one level, needing space, water, food, shelter and mates, like we do. We don't want to see the bucking fall of a flying walrus, a creature who should never be airborne. We don't want to eat our way through factories of animal mutants. If I wake in the middle of the night I think of that beaver dam, hidden away in Wood Buffalo National Park's 17,275 square miles, home to 5,000 wood bison, to the whooping crane, and one of the largest inland freshwater delta ecosystems in the world. In 1941 there were just 21 whooping cranes, and now there are 800. As the smartest animals ourselves, we do have the tools. Technology can play its part in the joinery workshop. Animals can inspire us – and do.

★

Jonathan and I are sitting in our campervan above a pier on the Isle of Mull off the west coast of Scotland, in a percussive downpour with a cup of tea, a piece of cake and the RSPB's Dave Sexton. For 20 years Dave has been monitoring the island's sea eagle population and protecting their nests; a reintroduction success story that brings the island economy £8 million a year in tourism. Sea eagles, or white-tailed eagles, are back to mesmerise us with their 2.5-metre barndoor fingery wingspan, their mustard-yellow feet and bone-cracking beaks, their swooping grabs and their cartwheeling courtship displays of locking claws mid-flight to then spiral earthwards and release each other just feet above the ground and soar upwards. Here we are, over a century since the last lone pale bird was shot sitting on a Shetland coastal crag waiting in vain for another sea eagle, in 1918. Their old name, erne, embedded in place names right across the land, reveals this great bird's historic range. Ern, Earn, Eas, Arn, Ar, Ayl are the lingering memories of white-tailed eagles: Arne, Arnwood; Arnecliff; Ayleford. Earley or Arley – eagle clearing; Earnshaw, Hernshaw – eagles wood; Arnold – eagle nook; Eglemont, Easdon – eagle hill; Yarnacombe – eagle valley; Yarnbury Castle; and of course, Lough Erne, Crewkerne . . .

In 1985 Dave was living in a tent, guarding the first wild chick to hatch in Scotland for 70 years around the clock in wind and rain until it fledged 12 weeks later. It was top secret. The cover story for Dave's presence was a red-throated divers survey. It was 'wet and 'orrible', he says.

I interrupt. 'Did anyone try and get them?'

'No!'

We laugh, in all that time, no excitement. Now there are organised nest tours with a special hide you can watch from. Mull Eagle Watch began as a protection scheme and developed into a tourism project that makes £25,000 a year which goes into the 'Eagle Fund' which local groups can apply for, youth clubs, girl guides – straight back into the community. *That* helps make people think eagles are worth having. The wind changes and people begin to see wildlife in a different way.

Not everyone. There is still the fixed conviction that sea eagles predate healthy lambs, supported by the NFU and promoted by the farming press. So here it is: they might, they can, they possibly do, but on a very small scale. Sea eagles are scavengers. So far there is no video evidence of a sea eagle killing a healthy lamb that Dave knows of. There *are* photographs of sea eagles carrying dead lambs, and of lambs in the nest larder – which are read as proof; but look closer to see little blood on the lamb, and eyes that have been pecked out by first-comers. One photograph peddled by the NFU, of a dead lamb and plenty of gore, turned out to have been staged with a known tame eagle. There is still little firm evidence of eagle-initiated attacks, which is inexplicable if they *were* the rampant lamb predators they are made out to be. Dave Sexton has been trying to witness the problem for 20 years. Around sea eagles day in and day out, he has hidden for 12 hours at a time near sheep flocks in lambing season. He has seen sea eagles swoop at greylag geese grazing among sheep. He has seen them chase a gannet to make them cough up a fish, ditto otters, but has never seen a sea eagle kill a healthy lamb or found an eyewitness attack.★ Yet nothing will persuade the convinced otherwise. It's easy to understand why; a massive flying scavenger is going to dominate the crime scene and will take both the lamb and the blame. It is easy to read a scene wrongly, and it is natural to look for a culprit, who now is *always* the sea eagle. Which is how and why we lost them 100 years ago. The myths are impossible to dislodge and no one wants to upset a shepherd who has lost lambs and seen an eagle sitting on one. In the Isle of Wight the NFU did everything to delay the reintroduction programme,★★ where the sea eagles' bad reputation came before them and farmers were up in arms. A meeting was set up for public

---

★   In the Netherlands researchers monitoring an expanding population have recorded no cases of eagles taking lambs. In Ireland there have been no cases.
★★ The project is a partnership between Forestry England and the Roy Dennis Wildlife Foundation. The plan is to release 60 eagles gradually over 5 years.

consultation with an open bar for angry questions to be fuelled by pints of beer; standing at the back, Dave was told to 'Get off our island'. But the clear majority of islanders could see the magic and wanted them back in their home range where they last bred in 1780.*

Sea eagles in Britain can teach us how to live with wild things once more. From the first nesting female they called Blondie,** Dave has watched her lineage grow and her great-great-grandchildren fly over the hills and lochs of Mull. Which is a life-affirming and hope-filled thing. In Mull, most farmers with nests on their land have come to accept them. Dave tells us about a farmer calling him to say one of the chicks was in trouble. Dave arrived to find the eaglet with a broken wing, being fed by his parents, calling loudly by the eyrie. He didn't think he had a chance but took him to his wildlife 'super-vet'. Two operations and six months of mending (in isolation so as not to become used to people) later, the young eagle's wing was fixed. Dave returned him, nervously, to his home ground on the spit looking out to the loch, where the most astonishing thing happened. The mother flew straight in, Dave fearing she might attack, but she landed and sat next to her grown chick. The two big birds side by side, calling-talking-eagling, a parent–child reunion.

In 2020, after a big storm ripped through, a concerned farmer alerted Dave to a nest hanging vertically with no sign of the chick. Dave assumed it had been predated. But a few weeks later he noticed a pair of adults behaving as if they were caring for young, confirmed by the food-begging call of a chick. He scanned the cliff with his telescope and there in front of his disbelieving eyes was the chick, who had not only survived the fall, but rowed with his unformed wings up the ridge to the safety of higher ground and thanks to his parents' ministrations was still alive. Dave did not see the chick again until late summer, when he heard a call and saw the young eagle in the sky, followed closely by his parents.

'For a couple of minutes all three circled lazily in the late summer sun, the perfect family picture, and then slowly drifted out of sight. It was one of the most satisfying experiences of my life and testament to their incredible ability to survive in this harsh and unforgiving landscape.'[222]

*

---

* Historically, sea eagles filled the vulture niche. Their presence naturally kept fish predators like cormorants in check, and so protected the salmon.
** One of the original juveniles released from Norway.

In a second bout of unseasonal freezing from the Beast from the East, Jonathan rang me from our nature reserve almost in tears. He had found a dead barn owl outside the tractor shed. We didn't know if this was a juvenile or one of the adult breeding pair who occupy the Barn Owl Hotel (5 star, fully insulated with balcony and clear access to excellent vole pasture). Jonathan could feel no weight in the feathered cradle of his hand. The barn owl's specialisation of silent flight is paid for by their soft feathers, which are not waterproofed, so they are particularly vulnerable to prolonged bouts of bad weather. In the old days, they could spend such nights in open hay barns helping the farmer control his mice. But such refuges are now few and far between. The following spring we were relieved and grateful to hear the familiar raspy-snot-noises of young in the hotel. Nevertheless our next mission was to build a dedicated open barn with perches, hay bales, chicken mash in mouse hutches, and a dark day-box for the adults when the owlets had taken over the main residence. Immediate success: both adults on camera, owl pellets under the perches, and on rainy nights if we sit quietly in the gloaming we can watch them swoop from their balcony straight into the barn.

★

Animals *are* good to think with. We use them as proxy for everything, metaphors, similes, ciphers, fables, morals, entertainment, company, refuge and even, weirdly, to define ourselves. Though what *they* think about *themselves*, we haven't a clue. In the twenty-first century, so our Promethean fire does not continue to backfire, we might consider and be considerate of them for their own narratives and intrinsic selves. For we are tygers too. With nostrils that can still smell faraway thunder and drenching rain, and eyes that can wonder how long those plum-heavy clouds can hold up. Run for shelter! It comes bold and godly. A fusillade of rain spears, the white noise of gallons per second, darkening the crooks of the branches above, exploding off leaves, until the aroma of soil pours into our noses . . . we snuff it up like the distant badgers we once were, it hisses through dry cracks, sleeks the skins of earthworms, spins its scent in explosions we remember from yesterday. We are alive. Here in the entangled bank, among birds hiding, insects sheltering, creatures looking out. Three billion years in the making, of 'endless forms most beautiful'.[223] I think of the octopus playing with the silver fish; I think of the calf lassoing a snowflake. Somewhere a wren opens his beak to let out his air-bubbling song.

LOVE, ACTUALLY

I linger at Dziedzinka. I know it is unlikely I will ever return. I think about all the loved creatures who have rolled in this grass. The wooden verandah where Simona and Lech sat, where Korasek perched, where Zabka slept, where Lech played with the badger cubs, somersaulting them into the sky, where the deer came to rest their noses, where a fox walked on Simona's shoulders, where the lynx, Agatha, rolled sun-dazed, claws stretching to the sun. Lech's garden pond is surrounded by wild strawberries. I make my way to the two birch trees in the clearing, the two trees that Simona and Lech planted when they first arrived at Dziedzinka, that had grown tall together, touching each other, that Lech saw as metaphor for Simona and himself. I lean into their bark and peer into the forest beyond.

Simona plants a kiss on a faun's nose, her strong jaw in profile, lips budding, eyelids down. The hidden loving-eye, yet utterly transparent is the camera's eye, and behind that, Lech Wilczek's eye. And floating somewhere above, like a space-probe, is my own. I think of the photo of Korasek upside down in the palm of Simona's hand, glint in his naughty eye, feet akimbo.

A select bibliography of resources used in the writing of this book can be found online at keggiecarew.co.uk and at canongate.co.uk/beastly

NOTES

1   Flannery, Tim. *Here on Earth: A Natural History of the Planet*. Allen Lane, 2010, pp. 96–7.

2   Hagenbeck, Carl. *Beasts And Men: Being Carl Hagenbeck's Experiences For Half A Century Among Wild Animals*, trans. Hugh S.R. Elliot & A.G. Thacker. Longmans, Green & Co., 1910, p. 64.

3   Hagenbeck, 1910, p. 291.

4   Flannery, 2010, pp. 59–60.

5   Hamilton, W.D. 'My intended burial and why'. Reprinted in *Ethology, Ecology & Evolution* 12, 2000, pp. 111–12. (Originally published in The Insectarium, 1991.).

6   New King James Bible, Isaiah 34:4–17.

7   Kemmerer, Lisa. 'The Great Unity: Daoism, Nonhuman Animals, and Human Ethics'. *Journal for Critical Animal Studies*, Vol. VII, Issue II, 2009, p. 74.

8   '. . . that plants are created for the sake of animals, and animals for the sake of Men; the tame for our use and provision; the wild, at least the greater part, for our provision also, or for some other advantageous purpose, as furnishing us with clothes, and the like. As nature therefore makes nothing either imperfect or in vain, it necessarily follows that she has made all these things for Men . . .' Aristotle. *A Treatise on Government*. Book I, Chapter 8.

9   Lopez, Barry. *Of Wolves and Men*. Scribner, 1978, pp. 236–9.

10 Lopez, 1978, p. 240.

11 While Suibhne's name first appears in the ninth century, the text of *Buile Suibhne* in its current form is dated around the end of the twelfth century.

12 Heaney, Seamus. *Sweeney Astray*. Faber & Faber, 1984.

13 Pope Francis addressing the International Association of Penal Law in the Vatican, November 2019.

14 Descartes, René. *Discourse on Method*. 1637.

15 Harrison, Peter. 'Descartes on Animals'. *The Philosophical Quarterly*, Vol. 42, No. 167 (April, 1992), pp. 219–27.

16 'Humpback Whale Shows AMAZING Appreciation After Being Freed From Nets'. Great Whale Conservancy, 2011, YouTube.

17 White, Gilbert. *The Natural History of Selborne*. Little Toller Books, 2014, p. 149. Barn owls were often called white owls.

18 Chuang Tzŭ. *Mystic, Moralist and Social Reformer*, trans. Herbert Allen Giles, Bernard Quaritch, 1889. The Project Gutenberg eBook of Chuang Tzŭ, by Chuang Tzŭ, Chapter XVIII.

19 Humboldt, Alexander von & Bonpland, Aimé. *Personal Narrative of Travels to the Equinoctial Regions of the New Continent, During the Years 1799–1804*, Vol. 4, pp. 505–6. Darwin Online <darwin-online.org.uk>.

20 Humboldt & Bonpland, *Personal Narrative*, Vol. 4, p. 217.

21 Bingley, Rev. William. *Animal Biography*. Vol. 1, 1829. Cited in Mabey, Richard. *The Perfumier and the Stinkhorn*. Profile Books, 2011, p. 78 and Thomas, Keith. *Man and the Natural World*. Penguin, 1984, p. 58.

22 Darwin, Francis (ed.). 'Letter to J.D. Hooker (11 January 1844)'. *The Life and Letters of Charles Darwin*. Vol. II. John Murray, 1887, p. 23.

23 Disraeli, Benjamin. *Tancred*. Vol. I. 1847, pp. 225–6. Cited in Barber, Lynn. *The Heyday of Natural History 1920–1870*. Doubleday, 1980, p. 215.

24 El-Zaher, Sumaya. 'The Father of the Theory of Evolution: Al-Jahiz and His Book of Animals'. Mvslim, <mvslim.com>, 9 October 2018.

25 Darwin, Charles. *On the Origin of Species by Means of Natural Selection*. Murray, 1859, p. 214. On Darwin Online as the 'Online Variorum of Darwin's *Origin of Species*', <darwin-online.org.uk>.

26 Darwin, 1859, p. 61.

27 Huxley in a letter to his friend Dyster, cited in Barber, 1980, pp. 274–5. The common mistake was the assumption we had *descended* from apes, rather than sharing a common ancestor. The divergence of ape and hominid is thought to have happened 6–10 million years ago.

28 Kingsley, Charles. *The Water Babies*. MacMillan & Co, 1891, p. 154.

29 Powys, Llewelyn. 'A Pond'. *Earth Memories*. Redcliffe Press, 1983, pp. 37–40; cited in Gray, John. *The Silence of Animals: On Progress and Other Modern Myths*. Allen Lane, 2013, pp. 175–6.

30 Chinese philosopher Chuang Tzŭ to the thinker Hui Tzŭ in the fourth century BCE. Chuang Tzŭ, Chapter XVII.

31 Foster, Charles. *Being a Beast*. Profile Books, 2016, p. 20.

32 Li, Gege. 'Weird caterpillar uses its old heads to make an elaborate hat'. *New Scientist*, 24 June 2020. Most caterpillars are genetically male or female, but do not develop sex organs until they are pupae.

33 Lorenz, Konrad Z. *King Solomon's Ring*. Methuen & Co. Ltd, 1952, p. 2.

34 Matt Somerville <beekindhives.uk>

35 Burkhardt, Richard, W. *Patterns of Behaviour: Konrad Lorenz, Niko Tinbergen, and the Founding of Ethology*. University of Chicago Press, 2005, p. 11.

36 'Robot spy gorilla infiltrates a wild gorilla troop'. *Spy in the Wild*. BBC, 23 January 2020.

37 Remes, Olivia. 'In Defence of: Harlow's Monkeys, Olivia Remes defends Harlow's experiments on the nature of love'. *Varsity*, 1 November 2013. Remes informs us that the longest period a monkey remained in the Pit of Despair was 15 years.

See also Bennett, Allyson J. 'Harlow Dead, Bioethicists Outraged'. Speaking of Research, <speakingofresearch.com>, 3 August 2014.

38 Pinker, Steven. *The Better Angels of Our Nature*. Penguin, 2012, p. 669.

39 Goodall, Jane. *Reason For Hope: A Spiritual Journey*. Grand Central Publishing, 2003, p. 66. (Although Thomas Savage, missionary to Liberia, had noticed chimpanzees cracking nuts with stones in 1844.)

40 Wilson, Michael L. et al. 'Lethal aggression in *Pan* is better explained by adaptive strategies than human impacts'. *Nature* 513, 18 September 2014, pp. 414–17.

41 Montgomery, Sy. *Walking With the Great Apes: Jane Goodall, Dian Fossey, Biruté Galdikas*. Chelsea Green Publishing, 2009, p. 123.

42 Montgomery, 2009, p. 185.

43 Montgomery, 2009, p. 190.

44 Montgomery, 2009, p. 54.

45 Fossey, Dian. 'Making Friends with Mountain Gorillas'. *National Geographic*, January 1970.

46 Brooks, Michael. *At The Edge of Uncertainty: 11 Discoveries Taking Science by Surprise*. Profile Books, 2015, p. 36.

47 de Waal, Frans. *Are We Smart Enough to Know How Smart Animals Are?* Granta Books, 2016, p. 15.

48 de Waal, 2016, p. 15.

49 Rutherford, Adam. *The Book of Humans*. Weidenfeld & Nicholson, 2018, p. 63.

50 De Waal, 2016, p. 228. The experiment was conducted by the American primatologist Sarah Boysen.

51 de Waal, 2016, pp. 145, 229.

52 Caruso, Catherine. 'Chimps May Be Capable of Comprehending the Minds of Others'. *Scientific American*, 6 October 2016.

53 'Primates'. *Protecting Primates*. Series 1. BBC, 1 May 2020.

54 Safina, Carl. *Beyond Words: What Animals Think and Feel*. Souvenir Press, 2016, p. 192.

55 Ted Turner, vice president of Dolphin Cay (2018), is 'an animal behaviourist' who believes in 'operant conditioning' – that if Kelly does something it is only because her behaviour has been reinforced with a reward. He does not subscribe to an animal having a personality. Eveleth, Rose. 'Kelly, the Sassy Dolphin'. Hakai magazine, <hakaimagazine.com/>, 2 October 2018.

56 Panksepp, Jaak & Burgdorf, Jeffrey. '"Laughing" rats and the evolutionary antecedents of human joy?'. *Physiology & Behavior* 79, 2003, pp. 533–47. In their 2010 paper, 'Laughing Rats?', they wrote, 'Our provisional conclusion is: Rats do laugh, and they certainly enjoy the frolicking that induces them to do so.'

57 de Waal, 2016, p. 10.

58 Stokstad, Erik. 'Q&A: Why fishery managers need to overhaul recreational rules'. Science, <Science.org>, 20 March 2019; 'Numbers of fish caught from the wild each year'. Fishcount, <Fishcount.org.uk>, 2019.

59   Marks, Paul. 'Green Machine: Wind farms make like a fish'. *New Scientist*, 23 September 2010; Harrington, Kent. 'Vertical Axis Wind Turbines Ready To Go Mano-A-Mano Against Industry Heavyweights'. ChEnected, <aiche.org>, 14 October 2015.

60   Scales, Helen. *Eye of the Shoal: A Fishwatcher's Guide to Life, the Ocean and Everything*. Bloomsbury Sigma, 2018, p. 153.

61   Scales, 2018, pp. 290–3.

62   'Joe Acheson on the Wren'. *Tweet of the Day*, BBC Radio 4, 19 February 2018.

63   *Springwatch 2020*, Episode 2. BBC 2, 28 May 2020.

64   'Beluga Whale Sounds Like a Human'. *World News Now*, ABC, 23 October 2012, YouTube.

65   Swift, Jonathan. *Gulliver's Travels*. Wordsworth Editions Ltd, 1992.

66   Tonti, Alexi. 'Herbert Terrace Studies Evolution of Language'. *Columbia College Today*, Winter 2012–13.

67   Singer, Peter. 'The Troubled Life of Nim Chimpsky'. *The New York Review*, 18 April 2011.

68   Marsh, James (dir.). *Project Nim*. Roadside Attractions, 2011.

69   Rabinowitz, Alan. 'Man and Beast'. In: Catherine Burns (ed.), *The Moth, 50 Extraordinary True Stories*. Serpent's Tail, 2015, pp. 43–50.

70   As *Time* magazine described him. Walsh, Bryan. 'The Indiana Jones of Wildlife Protection'. *Time*, 10 January 2008.

71   *Chambers Dictionary*, 1993.

72   Birkhead, Tim. *The Most Perfect Thing: Inside (and Outside) a Bird's Egg*. Bloomsbury, 2016, p. 230 (Note 17).

73   Durrell, Gerald. *My Family and Other Animals*. Penguin, 2016, p. 36.

74   From Apple dictionary for Mac online; the last line is a selection of phrases from various dictionaries.

75   Hughes, Ted. 'Swifts'. *Season Songs*. Faber, 1976; Bishop, Elizabeth. 'Moose'. *The Complete Poems, 1927–1979*. Farrar, Straus and Giroux, 1980.

76   Kimmerer, Robin Wall. *Braiding Sweetgrass: Indigenous Wisdom, Scientific Knowledge and the Teachings of Plants*. Penguin, 2020, p. 55. The end of the paragraph that follows is paraphrased from the same book, p. 211.

77   Mabey, Richard, 'The Nightingale and the Sonogram', *The Perfumier and the Stinkhorn*. Profile Books, 2011, p. 91.

78   Roberts, Callum. *Ocean of Life*. Allen Lane, 2012, p. 41.

79   Scales, Helen. *Spirals in Time: The Secret Life and Curious Afterlife of Seashells*. Bloomsbury Sigma, 2015, pp. 91–8.

80   Backhouse, Frances. *Once They Were Hats: In Search of the Mighty Beaver*. ECW Press, 2015, p. 85.

81   Frank Gehry quote cited in Crumley, Jim. *Nature's Architect: The Beaver's Return to Our Wild Landscapes*. Saraband, 2015, p. 1.

82   'Exploring Beaver Habitat and Distribution in Canada'. *EcoInformatics International Inc*, on geostrategis.com; 'World's Largest Beaver Dam', Wood Buffalo National Park, Parks Canada, <pc.gc.ca>.

83   Backhouse, 2015, p. 143.

84   Wulf, Andrea. *The Invention of Nature: The Adventures of Alexander von Humboldt, The Lost Hero of Science*. John Murray, 2015, p. 285.

85   Collier, Eric. *Three Against the Wilderness*. TouchWood Editions, 2007, pp. 15–16.

86   Heter, Elmo W. 'Transplanting Beavers by Airplane and Parachute'. *The Journal of Wildlife Management*, Vol. 14, No. 2. April 1950, pp. 143–7; Steve Liebenthal interviewed by Samantha Wright. 'Parachuting Beavers Into Idaho's Wilderness? Yes, It Really Happened'. Boise State Public Radio, 14 January 2015; *Parachuting Beavers*, Idaho Department of Fish and Game, c. 1950, YouTube.

87   Eyewitness account of Godfrey Morgan, Lord Tredegar. 'The Charge of the Light Brigade'. *Flintshire Observer, Mining Journal and General Advertiser, for the Counties of Flint and Denbigh*, 4 November 1897, p. 6.

88   *Today*, BBC Radio 4, 27 August 2019.

89   Sudbury, Dave. 'The King of Rome', recorded by June Tabor.

90   'The Secret Life of Pigeons'. *Open Country*. BBC Radio 4, 7 December 2019.

91   Because Pete had taught his pigeons to home back to the loft he'd built to fit on the back of his scooter, he and Nathanial were able to travel to events and give displays of their flying orchestra. 'The Pigeon Whistles', BBC Radio 4, 25 July 2017.

92   Letter from Wallace to Bates, 11 October, 1847. Wallace expresses his desire to study the mechanism for the theory of transmutation as it was known from Robert Chambers' *Vestiges* (then still anonymously published). J.M.S. Pearce, 'Alfred Russel Wallace', *Hektoen International*, <hekint.org>, 2019. See also John van Wyhe, 'A Delicate Adjustment: Wallace and Bates on the Amazon and "The Problem of the Origin of Species"', *Journal of the History of Biology* 47: 627–59, 2014 on Darwin Online, <darwin-online.org.uk>.

93   Berry, Andrew (ed.). *Infinite Tropics: An Alfred Russel Wallace Anthology*. Verso, 2002, pp. 141–5. Quotes from Wallace in this section appear on these pages of Berry's anthology.

94   Berry (ed.), 2002, pp. 136–8.

95   Harman, Kristyn. 'Explainer: the evidence for the Tasmanian genocide'. *The Conversation*, 17 January 2018.

96   Stockton, Richard. 'Australia's Centuries-Long Genocide Against Aboriginal People'. all that's interesting, <allthatsinteresting.com>, 16 November 2016.

97   First reported by the Dutch physician, Jacobus Bontius, working in Java in the seventeenth century. The name orangutan was picked up by his colleague, Nicolaes Tulp, who applied it to a chimpanzee from Angola. Following Tulp's essay, 'Homo sylvestris; Orang-Outang' in 1641, later writers used the name 'orangutan' for any human-like ape. European understanding of apes was a mixture of observation, anatomical study, hearsay, allegory, expeditionary tales, and mythology. Gorillas were not classified as a separate species until 1840, and bonobos, not until 1930. Wyhe & Kjaergaard, June 2015.

98   Darwin, Charles. Lines 79 and 196–7, Notebook C., 1838. Darwin's impressions of the hunter-gatherer people of Tierra del Fuego, the Yahgans, who he saw as

animal-like savages, narrowed the gap for him to see the great apes as the cousin of man. Darwin Online, < darwin-online.org.uk>.

99 Lyell to Darwin, 15 March 1863. The Darwin Correspondence Project, University of Cambridge, <darwinproject.ac.uk>.

100 Quoted by George W. Eveleth in his letter to Edgar Allan Poe, 19 January 1847. Edgar Allan Poe Society of Baltimore, <eapoe.org>.

101 The guillemot is a long-living bird who begins breeding at around six years old, laying one egg a season, with both parents feeding the chick. The oldest recorded guillemot was 43. Nicolson, Adam. *The Seabird's Cry: The Lives and Loves of Puffins, Gannets and Other Ocean Voyagers*. William Collins, 2017, pp. 156–7, 159.

102 Birkhead, 2016, p. 71.

103 Birkhead, 2016, pp. 209–10.

104 Community RSPB website, The Climmers of Bempton.

105 'The 1869 Sea Birds Preservation Act'. Iberia Nature, <iberianature.com>, 20 August 2009.

106 'A guillemot chick's heart-stopping leap from 400 feet'. *Highlands – Scotland's Wild Heart*. BBC Scotland, 18 May 2016.

107 Foster, John Bellamy & Clark, Brett. 'The Robbery of Nature'. *Monthly Review*, 1 July 2018.

108 'A Success Story, Johnston Atoll Chemical Agent Disposal System'. US Army's Chemical Materials Agency press release, 21 September 2005.

109 Haraway, Donna J. 'Apes in Eden, Apes in Space'. In: Donna J. Haraway, *Primate Visions: Gender, Race, and Nature in the World of Modern Science*. Routledge, 1989, p. 138, cites the Time Life nature series *The Primates* as the source of the quote.

110 'Miriam Rothschild'. *Desert Island Discs*. BBC Radio 4, 28 April 1989.

111 Rothschild, Miriam, Schlein, Y., Parker, K., Nevil, C. & Sternberg, S. 'The Flying Leap of the Flea'. *Scientific American*, Vol. 229, No. 5, 1 November 1973. Rothschild believed the knee was the flea's launchpad; recently Cambridge researchers have concluded that fleas launch from the toe.

112 Tucker, Anthony and Gryn, Naomi. 'Dame Miriam Rothschild', *Guardian*, 22 January 2005.

113 Cartmill, Matt. *A View to a Death in the Morning: Hunting and Nature Through History*. Harvard University Press, 1993, p. 165–6. See also pp. 160–88.

114 Kossak, Simona. *The Białowieża Forest Saga*, trans. Elżbieta Kowalewska. Muza SA, 2001, pp. 90–3.

115 *Mościcki I Goering na polowaniu – Białowieża 1935–37* ('Mościcki and Goering go hunting – Białowieża 1935–37'). Pathe39, 1939; *Goering at His Hunting Lodge*. British Pathé, 1938; *General Goering Gives A Shoot*. British Movietone News, 1936.

116 Dale-Harris, Luke. 'How the brown bear became public enemy number one in rural Romania'. *Guardian*, 22 November 2017.

117 Wilson, E.O. *On Human Nature*. Harvard University Press, 2004, p. 104.

118 Cartmill, 1993, p. 9, citing Robert Ardrey, *The Hunting Hypothesis*. Atheneum Books, 1976.

119 Cartmill, 1993, p. 225.

120 Bourjaily, Vance. *The Unnatural Enemy: Essays on Hunting*. Dial Press, 1963, p. 169.

121 Kingsley, Charles. *Prose Idylls, new and old*. 1873. Cited in Adams, Tim. 'Fantastic Mr Fox'. *Granta 90*, Summer 2005.

122 Pietersen, Kevin. *Beast of Man*. BBC podcast. Trophy Hunted. 7 June 2019.

123 'Trophy: The Big Game Hunting Controversy'. *Storyville*. BBC 4, 29 January 2018. An interview with unashamed trophy hunter Philip Glass, by Radhika Sanghani.

124 The Revive Coalition is a group of charities and organisations – Raptor Persecution UK, Friends of the Earth Scotland, OneKind, Common Weal, and League Against Cruel Sports – lobbying for grouse moor reform.

125 Figures from government and the industry's data. Wightman, Andy and Tingay, Ruth 'The Intensification of Grouse Moor Management in Scotland'. 2015.

126 Mbatha, Sicelo, with Bridget Pitt. 'Letting Go'. Channel, <channelmag.org>, 12 July 2021.

127 King, Angela, Ottaway, John & Potter, Angela. *The Declining Otter: A Guide to Its Conservation*. Friends of the Earth Otter Campaign, 1976, p. 30.

128 Samples of game meat from supermarkets reveal levels of lead up to 76 times higher than the legal limit set for beef, pork or chicken. 'Lead-contaminated game meat found for sale on Sainsbury's shelves', Wild Justice, <wildjustice.org. uk>, 9 December 2021; 'High lead levels in Waitrose and Harrods game meat', Wild Justice, <wildjustice.org.uk>, 18 December, 2021; Gill, Victoria. 'Most pheasants sold for food "contain lead shot"', BBC News, 24 February 2021.

129 Harvey, Denis. 'Plaiting the Magic', 1991. Unpublished manuscript, courtesy Dan Harvey.

130 Harvey, 1991.

131 Weihe, P., Grandjean, P., Debes, F. & White, R. 'Health implications for Faroe Islanders of heavy metals and PCBs from pilot whales'. *Science of the Total Environment* 186 (1–2), 16 July 1996, pp. 141–8. Weihe, P., Joensen, H. D. 'Dietary requirements regading pilot whale meat and blubber in the Faroe Islands'. *International Journal of Circumpolar Health* 71 (1), 10 July 2012.

132 'The Whale Hunters'. *Stacey Dooley Investigates*. BBC 3, 3 May 2020.

133 *Save This Shark*. THIS. Film Studio & National Geographic, July 2020. Trailer at <savethisshark.com>.

134 Weisman, Alan. *The World Without Us*. Virgin Books, 2007, p. 264. In Hong Kong shark fin soup can cost $100 a bowl. One hundred million sharks killed a year seemed so outlandish, but many sources including National Geographic, the Smithsonian, Al Jazeera, also quote this figure.

135 Leopold, Aldo. *A Sand County Almanac and Sketches Here and There*. Oxford University Press, 1968, pp. 130–2.

136 Leopold, 1968, p. 204.

137 'Ben Fogle on the Reality of Big Game Hunting.' *Good Morning Britain*, 20 April 2016.

138 The ivory burn was controversial – Kenya is a poor country, but President Kenyatta said he would rather wait for the judgement of the future generations. 'To lose our elephants would be to lose a key part of our heritage, and we quite simply will not allow it . . . We will not be the Africans who stood by as that happened.' Nuwer, 30 April 2016. The 1989 international ivory trade ban was working well, until it was undermined by allowing 'one-off' sales in 1999 and 2008, which stimulated fresh demand and provided cover to launder black market ivory, resulting in a catastrophic spike in poaching. Kenya's burnt ivory represented 5% of what was currently held in Africa's stockpiles. See Kathleen Garrigan, 'Behind the Legal Domestic Ivory Trade, a Black Market Flourishes', African Wildlife Foundation, <awf.org>. See also Morgan Kelly, 'After legal-ivory experiment, black markets thrive from greater demand, less risk'. Princeton University, <princeton.edu>, 14 June 2016.

139 Singh, Arjan. *Prince of Cats*. Jonathan Cape, 1982. Also, 'Meet Billy Arjan Singh', *Sanctuary Asia* 20 (9), September 2000, available on Sanctuary Nature Foundation, <sanctuarynaturefoundation.org>.

140 Eichelberger, John, Freymueller, Jeff, Hill, Graham & Patrick, Matt. 'Nuclear Stewardship: Lessons from a Not-So-Remote Island'. *Geotimes*, March 2002.

141 Carroll, Sean B. 'The Ecologist Who Threw Starfish'. *Nautilus*, 7 March 2016. See also Carroll, Sean B. *The Serengeti Rules: The Quest to Discover How Life Works and Why It Matters*. Princeton University Press, 2016.

142 Steller, Georg Wilhelm. 'De Bestiis Marinis'. *Typia Academiae Scientiarum*, 1751, p. 43. English translation by Jennie Emerson Miller & Paul Royster available at DigitalCommons@University of Nebraska – Lincoln, <https://digitalcommons.unl.edu>.

143 Katz, Brigit. 'Voracious Purple Sea Urchins Are Ravaging Kelp Forests on the West Coast'. *Smithsonian Magazine,* 25 October 2019.

144 'Daniel Pauly on Shifting Baselines'. TED Talk, April 2010, on Sea Around Us, <seaaroundus.org>.

145 McCarthy, Michael. *The Moth Snowstorm: Nature and Joy*. John Murray, 2015, pp. 108–9.

146 Goulson, Dave. 'Revealed: the chemical blitz of pesticides in our fields'. *Ecologist*, 30 January 2014. According to DEFRA figures, on average every field in the UK received 16.4 pesticide applications in 2016 (Goulson, 2021, p. 88). Buijs, Jelmer, Ragas, Ad & Mantingh, Margriet. 'Presence of pesticides and biocides at Dutch cattle farms participating in bird protection programs and potential impacts on entomofauna'. *Science of the Total Environment* 838, Part 3, 10 September 2022.

147 Goulson, Dave. *Silent Earth*. Jonathan Cape, 2021, pp. 89–90. Contaminants include DDT, organochloride insecticides, and polychlorinated biphenyls (PCBs): 'based on studies from diverse locations including Australia, Mexico, Ukraine and the Canary Isles'.

148 BirdLife International Data Zone, at <http://datazone.birdlife.org/sowb/casestudy/vultures-are-under-threat-from-the-veterinary-drug-diclofenac>. Also Juniper, Tony. *What Has Nature Ever Done for Us?* Profile Books, 2013. p. 135.

149 Barth, Brian. 'Bats, Beetles, Butterflies . . . And Other Pollinators That Aren't Bees (and How to Attract Them)'. *Modern Farmer*, 20 June 2017.

150 'Be concerned'. *Nature*, Vol. 511, Issue 7508, 10 July 2014, p. 126.

151 Yamamuro, Masumi, Komuro, Takashi, Kamiya, Hiroshi, Kato, Toshikuni, Hasegawa, Hitomi, Kameda, Yutaka. 'Neonicotinoids disrupt aquatic food webs and decrease fishery yields'. *Science*, Vol. 366, Issue 6465, 1 November 2019, pp. 620–3.

152 Barry Commoner sets out Four Laws of Ecology in *The Closing Circle*, 1971. They are: Everything is connected; everything must go somewhere; nature knows best; there is no such thing as a Free Lunch. Cited by Wilson, E.O. *Half Earth: Our Planet's Fight for Life*. Liveright, 2016, p. 106.

153 Wilson, 2016, p. 57.

154 Quammen, David. *The Song of the Dodo*. Scribner, 1996, p. 11.

155 'The Secret & Endangered Lives of Freshwater Mussels', living on earth (<loe.org>), 12 July 2019.

156 Sideris, Lisa H. & Moore, Kathleen Dean (eds). *Rachel Carson: Legacy and Challenge*. State University of New York Press, 2008, p. 20.

157 Carson, Rachel. *Silent Spring*. Penguin Modern Classics, 2000, p. 142.

158 Sánchez-Bayo, Francisco & Wyckhuys, Kris A.G. 'Worldwide decline of the entomofauna: a review of its drivers'. *Biological Conservation* 232, April 2019, pp. 8–27.

159 'Insectageddon'. *Apocalypse How*. BBC Radio 4, 2 December 2020.

160 Goulson, 2021, p. 91. Here's a trick: to fall outside current regulations change two components of the chemistry formula of your product and give it a new name.

161 Armitage, Simon. 'So the Peloton Passed'. *Guardian*, 21 November 2019.

162 Around 37 billion tons a year, which is 40% of the CO_2 produced and the equivalent sequestering ability of 1.7 trillion trees. Ozin, Geoffrey. 'A Whale of a Solution to Climate Change'. *Advanced Science News*, 9 October 2019. See also Lavenia Ratnarajah, Andrew Bowie & Indi Hodgson-Johnson. 'Bottoms up: how whale poop helps feed the ocean'. *Science Alert*, 11 August 2014. Shanahan, Mike. 'How Whale Poop Could Counter Calls to Resume Commercial Hunting'. *Scientific American*, 28 August 2018.

163 Groves, Danny. 'Financial worth of whales revealed'. Whale and Dolphin Conversation (WDC), <uk.whales.org>, 27 September 2019.

164 'Animals are the main victims of history, and the treatment of domesticated animals in industrial farms is perhaps the worst crime in history.' Harari, Yuval Noah. 'Industrial farming is one of the worst crimes in history'. *The Guardian*, 25 September 2015.

165 McGlone, John J. *The Crate (stall, case, cage, box, etc) – Its History and Efficacy*. Pork Industry Institute, Texas Tech University, 2002.

166 Kaufman, Marc. 'The Use of Narrow Crates Puts Hog Farming Under Scrutiny'. *Washington Post*, 8 July 2001, Laboratory of Animal Behavior, Physiology and Welfare at <depts.ttu.edu>.

167 Prescott, Matthew. 'Pork Producers Prohibit Painful Pig Pens'. *Live Science*, 20 June 2014.

168 McGlone, *The Crate*, 2002, p. 11. Quote from a newspaper photo caption.

169 Pig farmer interviewed on *Farming Today*. BBC Radio 4, 11 February 2021. For current DEFRA guidelines see, 'Code of practice for the welfare of PIGS', <gov.uk>.

170 Lockley, Ronald M. *Early Morning Island*. George G. Harrap & Co. Ltd, 1939, p. 96. In the original, the lobster is called Crayfish (possibly local usage) but because he is not a crayfish I have taken the liberty to rename him.

171 Goedde, Lutz, Horii, Maya & Sanghivi, Sunil. 'Pursuing the global opportunity in food and agribusiness'. McKinsey & Company, <mckinsey.com>, July 2015.

172 Safran Foer, Joanathan. *Eating Animals*. Penguin Books, 2010. Inside the cover, J.M. Coetzee writes: 'The everyday horrors of factory farming are evoked so vividly, and the case against the people who run the system presented so convincingly, that anyone who, after reading Foer's book, continues to consume the industry's products must be without a heart, or impervious to reason, or both.' *Eating Animals* should be compulsory reading. It informs this section throughout.

173 Pacelle, Wayne. 'Banned in 160 Nations, Why is Ractopamine in U.S. Pork? (Op-ed)'. LiveScience, <LiveScience.com>, 26 July 2014.

174 Meriwether, Lewis. *Journals of the Lewis and Clark Expedition*. See 17 September 1804. Available at lewisandclarkjournals, < lewisandclarkjournals.unl.edu/>.

175 Brown, Gabe. *Dirt to Soil: One Family's Journey into Regenerative Agriculture*. Chelsea Green, 2018, p. 64.

176 Young, Rosamund. *The Secret Life of Cows*. Faber & Faber Ltd, 2017.

177 Timothy Cummings speaking to poultry producers in Starkville, Mississippi, at a Turkey Meeting sponsored by the Arkansas Poultry Federation, 21 September 2007. 'Debeaking Birds Has Got to Stop'. United Poultry Concerns (UPC), *Poultry Press*, Winter 2007.

178 'The Welfare of Laying Hens'. RSPCA, August 2018, <www.rspca.org.uk>, p. 2.

179 Morelle, Rebecca. '400-year-old Greenland shark "longest-living vertebrate"'. BBC News, 12 August 2016; 'Clam-gate: The Epic Saga of Ming'. BBC News, 14 November 2013.

180 Flannery, 2010, p. 159.

181 Flannery, 2010, pp. 154–9.

182 Fawbert, Dave. '"Eco anxiety": how to spot it and what to do about it'. *Our Planet Matters*. BBC 3, 27 March 2019.

183 Clare, John. 'The Lament of Swordy Well', lines 81–4.

184 McAuliffe, Kathleen. 'If Modern Humans Are So Smart, Why Are Our Brains Shrinking?' 'Mind', *Discover Magazine*, 20 January 2011.

185 Adams, Douglas & Carwardine, Mark. *Last Chance To See*. William Heinemann, 1990.

186 Cited in Mirsky, Jonathan, *China's Assault on the Environment*. ChinaFile NYRB China Archive <chinafile.com>, 18 October 2001. See also Judith Shapiro, *Mao's War Against Nature: Politics and the Environment in Revolutionary China*. Cambridge University Press, 2001.

187 1949 recording made on a 78 rpm disc, a copy held in the Ngá Taonga Sound and Vision archives.

188 Szabo, Michael. 'Huia, the Sacred Bird'. *New Zealand Geographic*, No. 20, October–December 1993, p. 38.

189 Szabo, 1993, p. 37.

190 Kolbert, Elizabeth. *The Sixth Extinction: An Unnatural History*. Bloomsbury, 2014.

191 Little Barrier Island. Avibase – The World Bird Database, <avibase.bsc-eoc.org>.

192 Galapagos Conservancy. 'Escaping Extinction: The Long Road Home for the Española Tortoise Species'. 28 June 2020, YouTube. See also Galapagos Conservancy, 'Española Tortoises Return Home Following Closure of Successful Breeding Programme', <galapagos.org>, 15 June 2020.

193 Crumley, Jim. 'Nature needs less technology not more'. *The Courier*, 14 June 2016.

194 The One Health concept. World Organisation for Animal Health (OIE).

195 Rozenbaum, Mia. 'The Increase in Zoonotic Diseases: the WHO, the Why and the When?', 6 July 2020, and 'Why we must stop Covid-19 gaining a foothold in animals', 25 November 2020, both Understanding Animal Research, <understandinganimalresearch.org.uk>. See also van Tulleken, March 2021.

196 Attributed to George Osborne, Chancellor of the Exchequer, 2012. '"Environmental Taliban" is the latest in a series of insults aimed at the greens'. *The Guardian*, 19 October 2012.

197 Wilson, E.O. *The Meaning of Human Existence*. Liveright, 2015, p. 178: 'Selfish activity within the group provides competitive advantage but is commonly destructive to the group as a whole'.

198 Justin Rowlatt. 'Greta Thunberg: Climate change "as urgent" as coronavirus'. BBC News, 20 June 2020.

199 Wilson, E.O. *Nature Revealed: Selected Writings, 1949–2006*. Johns Hopkins University Press, 2006, p. 618. Quoted in Norman Myers (ed.). *The GAIA Atlas of Planet Management for Today's Caretakers of Tomorrow's World*. Pan Books, 1985.

200 Lynas, Mark. *The God Species*. Fourth Estate, 2011, p. 51. In 2010 the IPBES had a 10-year Strategic Plan called 'Living in harmony with nature' which directed governments 'to mainstream biodiversity concerns' in society, to take direct action to restore habitats and promote conservation and species recovery programmes. In 2022 we can't even ban plastic grass.

201 Howard, Len. *Living with Birds*. Collins, 1956, p. 119.

202 Howard, Len. *Birds as Individuals*. Collins, 1952, pp. 166–7.

203 British Trust for Ornithology (BTO) figures.

204 'Spider and window glass'. *30 Animals That Made Us Smarter*. With Patrick Aryee. BBC Radio 4, 29 July 2019.

205 'The Good, The Bad and the Cuddly. Attenborough, Eustice and Betty'. Betty Badger Blog, <bettybadger.blogspot.com>, 5 September 2020.

206 Langton, Thomas E.S., Jones, Mark W. & McGill, Iain. 'Analysis of the impact

of badger culling on bovine tuberculosis in cattle in the high-risk area of England, 2009–2020'. British Veterinary Association, *Vet Record*, 18 March 2022. See also Langton, 2019. Langton's answer to the problem is to roll out Actiphage testing with SICCT tests: Langton, T. 'Is DIVA bovine TB test a breakthrough?'. *Ecologist*, 9 September 2020.

207 Oliver, William J. (dir), *Beaver People* (9 mins), National Parks of Canada, 1928. (The date is disputed to be 1930, quite likely in consideration of the size of the grown beavers.)

208 Grey Owl. *Pilgrims of the Wild*. Dundurn Press, 2010, pp. 62–3.

209 Grey Owl, 2010, p. 185.

210 Grey Owl, 2010, p. 185.

211 Grey Owl, 2010, p. 226.

212 Leopold, Aldo. *Round River*. Oxford University Press, 1972, p. 165.

213 'Idi Amin's abandoned lodge'. *Earth's Great Rivers*, Series 1, 'Nile'. BBC 2, 2 January 2019.

214 Wendle, John. 'Chernobyl and Other Places Where Animals Thrive Without People'. *National Geographic*, 9 October 2015.

215 Mwenda, Mike. 'Yacouba Sawadogo, the African farmer who stopped the desert'. Lifegate, <Lifegate.com>, 19 July 2019.

216 Bob Newhart in 'Stop It'. *MADtv* 6 (24), 12 May 2001, YouTube.

217 'IMF sees cost of COVID pandemic rising beyond $12.5 trillion estimate'. Reuters.com, 20 January 2022.

218 Ozin, Geoffrey. 'A Whale of a Solution to Climate Change'. Advanced Science News, <advancedsciencenews.com>, 9 October 2019. See also Jessica Aldred, 'Priceless poo: the global cooling effect of whales'. China Dialogue Ocean, <chinadialogueocean.net>, 24 April 2020.

219 Siddique, Haroon. 'Legal experts worldwide draw up "historic" definition of ecocide'. *Guardian*, 22 June 2021.

220 Ardetti et al, July 2012; McAlister, Erica. 'Mighty Mouthparts'. *Metamorphosis – How Insects Transformed Our World*. BBC Radio 4, 2 March 2021.

221 'Endangered tigers have made a remarkable comeback in five countries'. *New Scientist*, 28 July 2020.

222 Amos, Ilona. '"Miracle" sea eagle chick flying high after nest destroyed'. *The Scotsman*, 28 June 2020.

223 The phrase occurs in the closing statement of Charles Darwin's 1859 book *On the Origin of Species*: '. . . endless forms most beautiful and most wonderful have been, and are being, evolved.'

ACKNOWLEDGEMENTS

My eternal gratitude to the joyous and generous spirit of Joanna Kossak; from the depths of Białowiez΄a, Joanna welcomed us into her forest home to tell me the stories of her aunt, Simona Kossak, and Lech Wilczek – without whose photographs this book would be so much the poorer. Thank you, Ida Matysek, for permission to share the fruits of Lech's extraordinary eye. To the wildlife photographer Jarosłow Chyra for putting me in touch with Joanna and trawling through the many photographs in Lech's archives. A big thank you to film-maker Hugo Smith. My thanks to writer and broadcaster Adam Wajrak, and campaigner Augustyn Mikos. To James and Rachel de Candole for taking such good care of us in Romania. Thanks also to Mihai Grigore, Katharina Kurmes, Nat Page and the extraordinary Bob Gibbons. Thank you, Paulo Martinho of Wild Côa, for showing us the Faia Brava Reserve in Portugal. Special thanks to Dave Sexton in Mull, Andy and Gay Christie at the Hessilhead Wildlife Rescue Centre, and to the peerless Polly Pullar for connecting us. I lament that lockdown stymied my visit to you, Polly, and the thrill of being clambered all over by red squirrels.

Thank you, Will Travers of the Born Free Foundation, for letting me pick your brains for longer than you bargained for. Thank you, Philippe Sands, for talking ecocide and explaining the imperatives of precise legal definition. Thank you Robbie Marsland for your knowledge and data, and Carolyn Robertson for your hospitality. Thanks to Heather Stewart for my London room, which was a head-clearing walk along the canal to the Zoological Society of London's Prince Philip Zoological Library. To the wonderful ZSL librarians, with particular thanks to James Godwin who pointed me towards Arjan Singh; and Tiger Haven Society in New Delhi, for the astonishing photograph of Arjan.

A month before lockdown, *BEASTLY* and I were cocooned in the impossibly romantic Hawthornden Castle, with my fellow Hawthornden Fellows: Felicia Yap, Lydia Syson, Chris Meredith, Carin Clevidence and

Nina von Staffeldt, with whom I was privileged to share the crackling fire, the swirling snow, the sight of the peregrine barrelling down the gorge, delicious dinners and so much laughter. To director Hamish Robinson, I bow. My gratitude to the late philanthropist Drue Heinz, who made it possible. Thank you all. In 2018 I was fortunate to be Writer in Residence at the beautiful Gladstone's Library, my thanks to Peter Francis, Louisa Yates and Amy Sumner, also to Alan and Robyn Cadwallader, and to Professor David Clough.

Thank you, Dan Harvey, for trusting your father's story in my hands. Thank you, Matt Somerville, for your wild bee expertise. Special thanks to the inspirational Sue Clifford and Angela King for otters and Common Ground. To Susie Alexander for your presence and your presents. To Michael Brooks for insights and bower birds. For many varieties of help and hospitality my thanks to Tom Langton, Anna Maria Tuckett, Sophie Ejsmond, Richard Kerridge, Richard Bowler, David and Sarah Burnett, Laurie Benenson, Pip Morgan, Jonathan and Lesley Cavendish, Gustavo Montes De Oca, Gareth Evans, Charlie Burrell, Isabella Tree, Catherine Symonds, Dave Goulson, Joe Hashman, Gareth Harris and Simon Smart. To good Twitter folk like the indefatigable John Oberg (@JohnOberg) and Science Girl (@gunsnrosesgirl3) for endlessly fascinating animal things. To Dr Siobhan O'Sullivan for her wonderful podcast *Knowing Animals*.

Navigating copyrights and permissions for quotes and photographs (that have been so carefully chosen to say what is unsayable in prose) for a book like this can be dispiriting, so a very special thank you to the generosity of David Harsent, Simon Armitage, John Burnside, Benjamin Zephaniah, John Bevis, Michael McCarthy, Robinson Jeffers' granddaughter Lindsay Jeffers and to the family of my late friend, the poet Grace Ingoldby: Lucy Hart, Patrick Abelson, Edward Abelson, thank you. For the priceless tea party photograph, thank you, Ann Mark, daughter of the great late Ronald Lockley. To Gurcharan Roopra for the white rhino and keeper that so warms my heart. My gratitude to the George A. Smathers Libraries, University of Florida; Te Papa Tongarewa, Museum of New Zealand; Lisa Moore; Carol Beckwith, Angela Fisher and Anne Sheridan.

Thank you to all things Canongate, this fine independent publisher. To Jamie Byng, Francis Bickmore, Anna Frame, Jenny Fry, Lorraine McCann. To Aisling Holling and Alice Shortland. To copy editor Helen Bleck, such a pleasure to work with. A special thanks to the exceptional

Claire Reiderman. And the excellent Leila Cruickshank, so many thank yous. And of course my editor, Simon Thorogood, the sanity clause in the contract, the steady beacon emanating calm in my book-building swirling sea. Thanks also to Jamison Stoltz at Abrams Press in the US for taking *BEASTLY* on, and over there.

At Pew Literary, I thank John Ash and Margaret Halton, but I have no idea how to thank my agent, the incomparable Patrick Walsh, who makes such a difference to my life. That he is my agent *and* dear friend is my great good fortune. So this book is dedicated to Patrick, without whom it might have remained just a bother in my head.

And Jonathan. To Jonathan for your unshakeable belief, your beautiful nature and your love, I thank you.

Long live the wild world and all its inhabitants.

PERMISSION CREDITS

IMAGES

Unattributed images are from the author's own collection. All images of Simona Kossak, Korasek and Zabka copyright Lech Wilczek, courtesy Ida Matysek.

I: WILD THING

Woolly mammoth found in Siberia, 1903. Copyright New York Public Library / supplied courtesy of Science Source.

III: THE ANIMAL WITHIN

Konrad Lorenz. Wilamette Biology on Flickr, flic.kr/p/7wNbPp/ (CC BY-SA 2.0).
Dian Fossey with Coco and Pucker. Copyright Robert I.M. Campbell / supplied courtesy of Nat Geo Image Collection, with thanks to Heather Campbell.

V: DAM NATION

Beaver drop images. Copyright Idaho Department of Fish and Game / Wildlife Restoration Program.
Pigeon. Bundesarchiv, Bild 183-R01996 / CC-BY-SA 3.0.
Four men with ropes and baskets of eggs, on the cliffs at Flamborough, East Riding of Yorkshire, July 1926. Source: Historic England Archive.
Chimpanzee Ham in Biopack Couch for MR-2 flight. National Aeronautics and Space Administration (NASA) MSFC 6100114 / Wikimedia Commons.
Walter Rothschild riding a tortoise at Tring. Copyright the Trustees of the Natural History Museum, London / supplied courtesy of NHM Images.

VI: IF YOU KILL IT, YOU HAVE TO EAT IT – 1

Goering. Bundesarchiv, Bild 146-1979-145-04A / CC-BY-SA 3.0.
August and Sylvia von Speiss, unknown photographer, circa 1911.
Billy Singh crosses the river. Photographed by Mike Price. From the collection of the Tiger Haven Societ, with thanks to Brinda Dubey.

VIII: IF YOU KILL IT, YOU HAVE TO EAT IT – 2

Dinka Child Climbing Among Horns, South Sudan. Copyright Carol Beckwith and Angela Fisher / supplied courtesy of Photokunst.

Young Ann Lockley with Lobster and Buzz. From *Early Morning Island* by Ronald Lockley / courtesy of Ann Marks (née Lockley) and with thanks to Little Toller Books.

A sow at a factory farm. Copyright Jo-Anne McArthur / supplied courtesy of Animal Equality and We Animals Media.

IX: DAMNED

Huia Onslow, 1891, New Zealand, by unknown maker. Te Papa (O.009440).

Rhino with caregiver. Copyright Gurcharan Roopra / supplied courtesy of the photographer.

Len Howard, Surrey, 1957 by David Moore. Copyright Lisa, Michael, Matthew and Joshua Moore / supplied courtesy of the Moore estate.

X: GOLDEN JOINERY

Beaver in canoe with Grey Owl, 1931. Library and Archives Canada, Accession no. 1989-455, e008300732.

TEXT

'So the Peloton Passed' by Simon Armitage, copyright © Simon Armitage, 2019. Used with the permission of the poet and David Godwin Associates.

An A-Z of Birdsong by John Bevis, copyright © John Bevis, 1995. Used with permission from the author and Coracle Press.

'The Fair Chase' from *Black Cat Bone* by John Burnside, copyright © John Burnside, 2011. Used with the permission of the author, RCW and The Permissions Company LLC on behalf of Graywolf Press, graywolfpress.org.

'Sacrilege' by Nina Cassian, translated by Petre Solomon/William Jay Smith, from *Life Sentence: Selected Poems* by Nina Cassian, edited by William Jay Smith. Copyright © 1990 by Nina Cassian. Used by permission of W. W. Norton & Company, Inc and Carcanet.

My Family and Other Animals by Gerald Durrell, reproduced with permission of Curtis Brown Group Ltd, London on behalf of The Beneficiaries of the Estate of Gerald Durrell. copyright © Gerald Durrell, 1956.

'My Intended Burial and Why' by W.D. Hamilton in *Ethology, Ecology and Evolution*, copyright © Dipartimento di Biologia, Università di Firenze, Italia, reprinted by permission of Taylor & Francis Ltd, http://www.tandfonline.com on behalf of Dipartimento di Biologia, Università di Firenze, Italia.

INDEX

References to images are in *italics*.